MOLLUSCS:
PROSOBRANCH AND PYRAMIDELLID GASTROPODS

A NEW SERIES

Synopses of the British Fauna
Edited by Doris M. Kermack AND R. S. K. Barnes

The *Synopses of the British Fauna* are illustrated field and laboratory pocket-books designed to meet the needs of amateur and professional naturalists from sixth-form level upwards. Each volume presents a more detailed account of a group of animals than is found in most field-guides and bridges the gap between the popular guide and more specialist monographs and treatises. Technical terms are kept to a minimum and the books are therefore intelligible to readers with no previous knowledge of the group concerned.

Volumes 1–28 inclusive are available from the Linnean Society of London, Burlington House, Piccadilly, London W1V 0LQ.

All subsequent volumes, second editions and revisions may be obtained from E. J. Brill, Publishing Company, Leiden, The Netherlands. All volumes in print are available from Natural History Book Service, Totnes, Devon TQ9 5XN

Synopses of the British Fauna (New Series)
Edited by Doris M. Kermack and R. S. K. Barnes
No. 2
(Second Edition)

MOLLUSCS: PROSOBRANCH AND PYRAMIDELLID GASTROPODS

Keys and notes for the identification of the species

ALASTAIR GRAHAM, F.R.S.

Department of Pure and Applied Zoology
University of Reading, Reading, RG6 2AJ, England

1988
Published for
The Linnean Society of London
and
The Estuarine and Brackish-water Sciences Association
by
E. J. Brill/Dr W. Backhuys
Leiden · New York · København · Köln

Library of Congress Cataloging-in-Publication Data
Graham, Alastair, F.R.S.
    Molluscs: prosobranch and pyramidellid gastropods: keys and
notes for the identification of the species/Alastair Graham.—
2nd ed.
    p.   cm.—(Synopses of the British fauna, ISSN 0082–1101:
new ser. no. 2)
    Rev. ed. of: British prosobranch and other operculate gastropod
molluscs. 1971.
    Bibliography: p.
    Includes index.
    ISBN 9004087710 (Pbk.)
    1. Prosobranchia—Great Britain—Identification.   2. Mollusks—
Identification.   3. Mollusks—Great Britain—Identification.
I. Graham, Alistair, F.R.S. British prosobranch and other
operculate gastropod molluscs.   II. Linnean Society of London.
III. Estuarine and Brackish-Water Sciences Association.   IV. Title.
V. Series.
QL255.S95 n.s., no. 2 1988
[QL430.4]
591.941 s—dc19
[594′.32′0941]

ISSN 0082-1101
ISBN 90 04 08771 0

Printed in Great Britain at The Bath Press, Avon

A Synopsis of the
Molluscs: Prosobranch and Pyramidellid Gastropods
ALASTAIR GRAHAM, F.R.S.
Department of Pure and Applied Zoology
University of Reading, Reading, RG6 2AJ, England

Contents

Foreword

Molluscs: Prosobranch and Pyramidellid Gastropods is the title of the second edition of *Synopsis No. 2, British Prosobranchs*, which has been out of print for some years. Those readers familiar with the latter book will not need to be told that this new edition is much more comprehensive and detailed than the earlier one.

This new edition covers the same geographical limits as those used by the Conchological Society of Great Britain and Ireland in the *Sea Area Atlas of the Marine Molluscs of Britain and Ireland* (ed. D. R. Seaward, 1982) except that sea-area 48, which includes the Faeroes, has been excluded. By the kind permission of the Malacological Society of London, this *Synopsis* contains the beautiful pen and ink drawings of shells by the Danish artist Poul Winther. The *Biology* section has been considerably enlarged and this in turn has been illustrated with drawings by the author which appeared in his Ray Society Monograph with Dr V. Fretter entitled *Prosobranch Molluscs*. The editors thank both these Societies for generously giving permission for all these figures to be reproduced.

Molluscs: Prosobranch and Pyramidellid Gastropoda 'encapsulates' Professor Graham's life-time's work involving many, many hours spent in the field along with many in the laboratory and at the desk. Generations of undergraduate students have made their contribution to the accuracy and 'workability' of the keys and descriptions. For those past undergraduates it is hoped that this new edition will revive memories of field-trips with their inspired and enthusiastic teacher. All readers will appreciate that this *Synopsis*, although written in pocket field- and laboratory-guide format, owes much to the dedication and scholarship of the author.

The editors express their sincere thanks to the author for the work and care he has taken in the preparation of this new edition; they know that they will be joined by all those who share his love and interest in this diverse group of molluscs.

R. S. K. Barnes
Estuarine & Brackish-Water
Sciences Association

Doris M. Kermack
The Linnean Society
of London.

Introduction

The Plan of this Book, and how to use it

Pages 4–18 give a general description of the shell and body of a prosobranch gastropod mollusc and introduce the technical terms which are used later. Definitions of these terms may also be found in a glossary (p. 637). After the introductory pages there are given (p. 28) some hints on the collection and preservation of the animals and their shells.

For identification a series of keys is provided as the inclusion in a single key of nearly three hundred species is not practical. The keys are based primarily on the characters of fully grown shells, which allows identification of empty shells, but some of the animals' external features are included wherever they offer important aids to identification. The first key (Key A, p. 44) leads to the identification of terrestrial and fresh and brackish water *species*. If the animal to be identified is marine a start should be made with Key B (p. 48) which in most cases directs the user to others (Keys C, D, or E). All these, however, lead only to the *family* to which the snail belongs, and for identification to genus and species reference must be made to further keys located at appropriate places in the text and indicated in the main keys. Since variation within a family is often considerable some families (e.g. Rissoidae, Buccinidae) appear several times in the main key to families; and since there may sometimes be doubt as to whether an animal should be regarded as brackish water or marine, these species are included in both keys. A complete classified list of all the species described appears on p. 32.

At the head of the description of each species there is given the currently accepted name with its author and date. There follow, first (if different), the name by which the species was originally described, then some superseded names, some in general use if different from that used here, and the names under which descriptions will be found in the following books: Forbes & Hanley (1849–53), *A History of British Mollusca and their Shells*; Jeffreys (1862–9), *British Conchology*; Ellis (1926), *British Snails*; Winckworth's (1932) list of *British Marine Mollusca*; McMillan (1968), *British Shells*; Seaward (ed.) (1982), *Sea Area Atlas of the Marine Molluscs of Britain and Ireland*; and Kerney (ed.) (1976), *Atlas of the Non-Marine Mollusca of the British Isles*. Except for some of the original names all those used here and in these books appear in the systematic index (p. 656) so that anyone familiar with, for example, Jeffreys' names, but not with those used here, may easily find the description.

Each account of a species starts with its most useful diagnostic characters: these should allow confirmation of the identification reached via the keys. Thereafter there is given a fuller description, first of the shell, then of the animal, and finally such information about habitat, food, and breeding as is available, the general distribution of the species and some relevant references. For more detailed information on distribution within the British Isles than is given here the reader is referred to the maps in the works of Seaward and Kerney mentioned above.

In the keys and in the descriptions shell proportions are frequently given. It is important, in using these for identification, that they should be based on actual *measurements* rather than on *estimates* made by eye as the latter can often be misleading. Measurements and proportions given in the text refer to grown shells and proportions may well be different in juveniles. The same is true of some other features, such as teeth developed within the outer lip, which do not appear until growth is completed. It is useful to note, when examining shells, that *ornament* is most easily seen in *dry* shells with the lighting angled to emphasize relief, whereas *colour patterns* are most obvious in *wet* ones.

Abbreviations used

H.W.N.T., high water of neap tides.
H.W.S.T., high water of spring tides.
L.W.N.T., low water of neap tides.
L.W.S.T., low water of spring tides.
M.H.W.N. or M.H.W.N.T., mean high water of neap tides.
M.H.W.S. or M.H.W.S.T., mean high water of spring tides.
M.L.W.S., mean low water of spring tides.
M.T.L., mean tide level.
m, metre.
mm, millimetre.
μm, micrometre $= 1 \text{ mm}^{-3} = 1 \text{ m}^{-6}$.

Prosobranchs

The phylum Mollusca, to which prosobranchs are assigned by zoologists, includes the animals popularly known as snails, slugs, mussels, squids and cuttlefish, together with others less familiar. In particular, prosobranchs are classified with other snails and slugs in a group called Gastropoda, the largest and in many ways the most successful of the main divisions into which the phylum is split. Though similar in many respects to the well-known snails of land and fresh water, prosobranchs differ in their organization in such ways as to convince the zoologist that they are a more primitive group from which the other types of gastropod, the opisthobranchs (described by T. E. Thompson and G. H. Brown in *Synopsis No. 8*) and pulmonates (described by R. A. D. Cameron and M. Redfern in *Synopsis No. 6*) have been derived. Their more primitive nature is also reflected in their habitat as most prosobranchs are marine, the ocean being accepted as the ancestral home of all molluscs. A few kinds have invaded brackish and freshwater habitats, and a still smaller number have become terrestrial, though because of their sensitivity to dehydration their choice of habitat is restricted.

The following account refers primarily to a prosobranch such as the edible winkle, *Littorina littorea*, found abundantly on almost every rocky shore around the British Isles. Continual reference, however, is made to other prosobranchs so as to reflect the diversity of their organization and to introduce the terms used in the keys and the descriptions of the various species.

General organization

The shell

The prosobranch body is normally enclosed within a **shell**, secreted by the surface of the animal, in the form of a hollow tube known as the helicocone, which is nearly always coiled in a three-dimensional or helicoid spiral (see Fig. 1). The shell is closed at its **apex**, this being the first part secreted by the young animal. At the other end is the most recently formed part, that around the **aperture**. In limpets the shell is a cap-shaped structure, but this can be easily shown to be derived from such a spiral. The shell is composed chemically of a matrix of a protein, conchiolin, impregnated with calcium salts, and a skin of the matrix, known as the **periostracum** is sometimes visible as the outermost layer, overlying the calcareous material, but always subject to erosion.

The spiral consists of a number of coils each known as a **whorl**, the visible line of contact between successive whorls forming a **suture**. The whorls are normally of steadily increasing dimensions, the smallest and oldest, produced early in life, lying at the apex, the youngest, and usually also the largest, forming the **last whorl** (sometimes called the **body whorl**) which terminates at the aperture. When a shell is examined closely it is possible to distinguish two groups of whorls, often differing in superficial appearance: the first group constitutes the **protoconch**, a short initial part of the shell secreted before hatching (embryonic shell or protoconch I) and during larval life if the animal has a free larval stage (larval shell or protoconch II). The second group, the **teleoconch**, makes up the greater part of the shell; its formation begins after the end of larval life and metamorphosis to the adult form, or directly after hatching if there is no free larval stage in the life history.

In most prosobranchs the direction of shell coiling is **dextral**, that is, the shell forms a clockwise-going spiral when it is viewed from above the apex; in some the direction of coiling (and with it the disposition of the parts of the body) is reversed, when the shell is said to be **sinistral**. This may occur as a regular specific character or as an occasional oddity. If a shell be held with the apex uppermost and the aperture towards the observer, the aperture lies right of the axis of coiling in dextral shells and left of it in sinistral ones. In the shells of some groups (for example, pyramidellids, see Fig. 233) the protoconch and teleoconch may appear to coil in opposite directions, the one sinistral, the other dextral; such a shell is said to be **heterostrophic**. The appearance is, in fact, deceptive, both parts coiling in a clockwise direction. In the complete shell, comprising protoconch and teleoconch, however, the

4

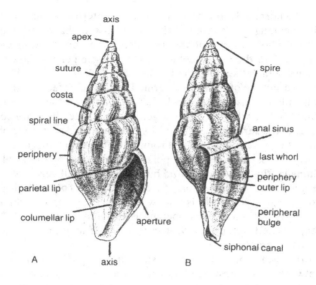

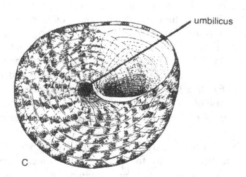

Fig. 1. A & B. A typical gastropod shell to show the terminology of the different parts. C. The shell of *Gibbula* to show the umbilicus, with umbilical groove leading to it alongside the columellar lip.

protoconch coils upwards from its beginning to the point which had been reached when the larva metamorphosed, whilst the teleoconch coils downwards from that point. It is this change in the polarity of the axis of coiling that makes the one appear sinistral and the other dextral. Throughout the life history of an animal with a heterostrophic shell the anatomy is unchanged and is that of a normal dextral animal. The term heterostrophic is also broadened in its meaning to describe shells in which the change in polarity of the axis is only 90°, instead of the 180° described above. Both types are most clearly seen in the pyramidellids in which some genera show the one, and other genera the other pattern of coiling.

The exposed surface of the shell is formed by the outer faces of the whorls, though it is only the last whorl which shows its outer surface in its entirety; in the other, older, whorls, which together constitute the **spire**, only the adapical part of the surface of each is visible, the abapical part being obscured by growth of the next younger whorl round it. The inner walls of the whorls are normally not seen, but form a pillar in the centre of the shell known as the **columella**; this may be exposed by cutting the shell vertically. In most shells the inner walls of the whorls touch so that the columella is solid; in others, however, the columella is hollow and its cavity opens to the exterior near the base of the shell. This opening is the **umbilicus** (Fig. 1C) and it is often approached by an **umbilical groove** which may persist even when the umbilicus is absent.

The whorls of the spire may be flat-sided in profile (Fig. 2A), when the sutures between them must be shallow, or they may be tumid or swollen (Fig. 2B), when the sutures become deep and the whorls dip to them. Whorls often show a rather flat area below the suture, lying more or less at right angles to the shell axis, the **subsutural shelf** or **ramp** (Fig. 2C); when present this gives the spire a profile described as **turreted**. The broadest part of the entire shell, or of each constituent whorl, is called its **periphery**. The profile of the spire (neglecting the various irregularities introduced by the convexity of the whorls) may be strictly rectilinear (Fig. 2A), or it may be concave, when it is described as **coeloconoid** (Fig. 2D), or convex, when it is said to be **cyrtoconoid** (Fig. 2G). In some species, when the shell becomes full-grown, the last whorl expands so as to grow over and conceal the whole spire (Fig. 2E,F): the shell is then called **convolute**. In other species the whorls all lie in one plane, when the shell is said to be discoidal or **planorboid**. Since the central part of the disk is occupied by the oldest and smallest whorls the disk is biconcave.

The prosobranch shell grows by the addition of strips of material to the edge of the aperture and by the plastering of calcareous matter on to its internal surface so as to thicken and strengthen the wall. The secretion of

---

Fig. 2. The profiles of different gastropod shells: A. *Jujubinus exasperatus* where the whorls of the spire are flat-sided; B. *Jujubinus montagui* where the whorls are tumid, ventricose or swollen; C. *Gibbula magus* where the whorls show a flat area below the suture, the *subsutural shelf*; D. *Calliostoma granulatum* where the profile of the spire is concave and described as *coeloconoid*; G. *Gibbula umbilicalis* where the profile of the spire is convex and described as *cyrtoconoid*; E & F. *Trivia monacha*, an example of a *convolute* shell, all older whorls (=spire) hidden by overgrowth of the last, the youngest.

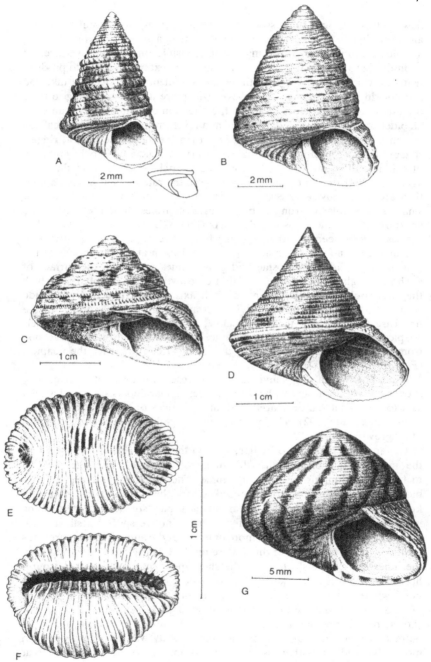

new shell is phasic, often associated with a tidal rhythm in marine forms, and sometimes with a pause between outbursts of activity. Each phase leaves a delicate **growth line**, often microscopic, visible on the shell surface. In addition there may be further elaboration of the external surface to produce features which collectively constitute the **ornament**. They usually take the form of ridges alternating with grooves which are named according to their orientation. If they are orientated in the direction of coiling and so run from an older to a younger part of the shell and result from a continuous localized secretion of material as the shell grows, then they are called **spiral ridges**, **striae** or **lines**, according to their size; if they run at right angles to the direction of coiling and so are confined to an area of shell which is all of the same age, and result from a temporary increase in the rate of shell secretion then they are called **costae** or **costellae**, according to whether they are large or small. A particularly strong or broad costa often lies along the edge of the aperture: this is known as a **labial varix** (Figs 82B, 86). Occasionally other varices may be seen on other whorls which represent previous positions of the aperture, and usually indicate pauses in growth at earlier periods of the animal's life (Fig. 4C). Costae and growth lines may cross the surface of a whorl at right angles to the direction of growth, when the plane in which they lie would, if projected on the shell axis, nearly or actually coincide with it, or they may lie at an angle to it and to the axis of coiling. In the first case they are said to be **orthocline**, in the second, **prosocline** if their adapical end is in advance of the abapical (i.e. further down the helicocone from the apex) and **opisthocline** if the abapical end precedes the adapical (Fig. 3).

When both spiral ridges and costae are present on a shell its surface may show a network of ridges with square or rectangular depressions, and is then said to be **reticulated** or **cancellated**, and tubercles or spines often develop at the nodes (Fig. 192). The periostracum may also show a similar pattern of outgrowths (Fig. 122).

The aperture is bounded by lips, that on the outer wall of the last whorl the **outer lip**, that on the inner the **inner lip**; in many species the two join to form a continuous rim, the **peristome**. The inner lip is divisible into a basal part lying alongside the exposed part of the columella and therefore known as the **columellar lip**, and an adapical part spread over part of the last whorl, the **parietal lip**. In a few, very primitive species a **slit** may lie at the periphery of the outer lip, more or less over the anus (Fig. 4A); previous positions of the blind end of this slit are marked by a **slit-band** running from the inner end of the slit towards the shell apex. In many others the outer lip shows a vestige of a slit in the form of a bay, usually at or near the point where it arises from the last whorl: this, too, is close to the position of the anus and it is therefore known as the **anal sinus**. The outer lip frequently shows a peripheral bulge below the anal sinus. In more primitive prosobranchs outer and inner lips run into one another basally without interruption; in more advanced forms there is found at this point a notch, or a semitubular

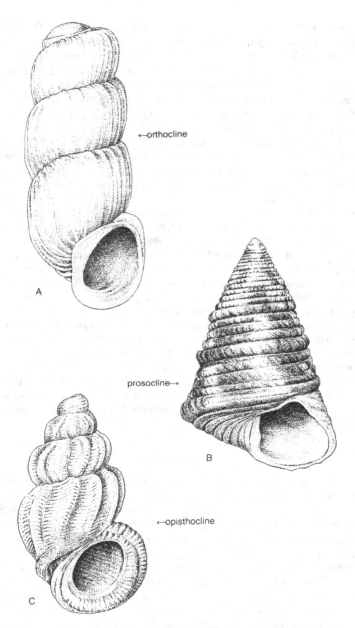

Fig. 3. Costae and growth lines in gastropod shells. A. orthocline, as in *Truncatella subcylindrica*; B. prosocline, as in *Jujubinus exasperatus*; C. opisthocline, as in *Manzonia crassa*.

canal, which may sometimes be considerably elongated. This is the **siphonal canal** (Fig. 4B) which supports the **siphon** leading the respiratory inflow into the mantle cavity. **Teeth**, in the form of tubercles or short ridges or folds, may occur within the outer lip of some species of prosobranch (Fig. 4B) and, more rarely, on the columella and inner lip. These are usually absent in juvenile shells, appearing only at maturity, but they are liable to be formed when growth is stopped for any reason (Bryan, 1969). The outer lip is thin in juveniles and may remain so in adults, but in some species it thickens markedly, and, as with teeth, this may happen whenever growth stops for any reason. In some shells with a siphonal canal a thick, spiral ridge marking previous positions of the tip, runs from the tip towards the columella, often lying alongside an umbilical groove: this is the **siphonal fasciole** (Fig. 4C).

In such prosobranchs as *Littorina littorea* the shell (Fig. 60) is a relatively simple structure, showing little ornament beyond growth lines when full grown, though juveniles have obvious spiral ridges alternating with grooves. The whorls are a little tumid, the spire sharply pointed and usually a trifle coeloconoid in profile because of the large size of the last whorl. The lips of the aperture are a little everted basally but otherwise show no complicating features – no anal sinus, no internal teeth, no siphonal canal – whilst the eversion of the columellar lip covers the umbilical groove and blocks the umbilicus. In some prosobranchs such as *Lunatia* (p. 334) and *Trivia* (p. 326) the shell is covered by extensions of the foot or mantle as the animal crawls, though these are withdrawn if it is disturbed; in the genus *Lamellaria* (p. 314), however, the mantle flaps are permanently fused together, the shell has become internal, and the animal looks like a dorid. In these animals the parts of the mantle which overlie the shell continue to secrete shelly material on its outer surface; as a result of this and of their protective action these shells often have a glossier appearance than do those of species in which this does not happen.

The body

The prosobranch **body** is divisible into two main parts: (1) the **head-foot**, (2) the **visceral mass** or **hump** (Fig. 5). The former, ventrally placed, is primarily sensory and locomotor in function, and comprises the **head**, which bears the mouth and major sense organs, and the **foot**, the muscular part of the body on which the snail crawls and which bears a posterodorsal horny plate (sometimes strengthened with calcareous matter), the **operculum** (Fig. 6). This exhibits a spiral coiling comparable to that of the shell but in the reverse sense, so that it coils anticlockwise when the shell is dextral and clockwise when it is sinistral. It may have many coils (**multispiral**) or few (**paucispiral**). Other variants of opercular structure occur, and the pattern may be a **concentric** series of rings rather than a spiral (Fig. 6).

The visceral mass or visceral hump is dorsally placed on the head-foot, is spirally coiled, and houses the main viscera. When a prosobranch snail

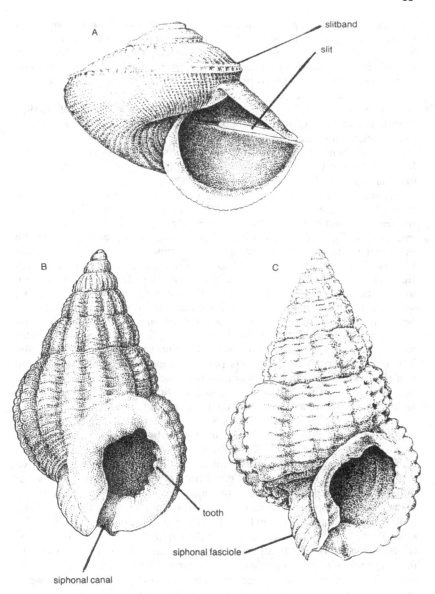

Fig. 4. The aperture of the shell. A. *Scissurella crispata*, with slit; B. *Hinia incrassata*, with siphonal canal and teeth within outer lip; C. *Hinia pygmaea*, with siphonal fasciole.

is creeping and feeding the head and foot are extended from the shell, but the visceral mass cannot be so extruded. If disturbed the whole animal may be retracted into the shelter of the shell by contraction of a **columellar muscle** running from the columella into the head-foot; the aperture is then blocked by the operculum. For this to be possible there must be a space which acts as a compensation sac and which is occupied by the head-foot when that is retracted but fills with water as it is extended. This is the **mantle cavity**, a pocket-shaped space which lies above and behind the head of a creeping snail, under the shelter of the shell. Its floor is formed by the dorsal surface of the head-foot, its roof by a fold of tissue, the **mantle skirt**, which arises from the face of the visceral hump, and from the edge of which new shelly material is secreted.

The mantle cavity, however, is more than a mere compensation sac, for its walls bear the gill or **ctenidium**, a chemosensory organ known as an **osphradium**, and the openings of the gut, the kidney and the reproductive organs (Fig. 7). In the most primitive prosobranchs, a group known as the Archaeogastropoda or Diotocardia, and represented by such animals as ormers (*Haliotis*), an element of bilateral organization is present, although the shell and body usually exhibit a spiral curvature. In these animals there are two ctenidia, right and left, each with an osphradium alongside, as well as two kidneys, though these differ in structure and function. In most proso-branchs, however, only a single gill, osphradium, and kidney occur, the com-pression of the animal's right side consequent upon its position on the inner side of the spiral having led to loss of those on the right, so that only the left members of the original pairs persist. These prosobranchs are placed in a group known as the Caenogastropoda or Monotocardia, which is usually accepted as showing two levels of organization, a less advanced meso-gastropod grade and a more advanced neogastropod grade.

The head of a prosobranch (Figs 8 and 9) bears the mouth at the end of a short snout. It also carries a pair of **cephalic tentacles**, at the base of each of which an eye lies laterally. In many diotocardians the eye is borne on an **eye stalk** separate from the tentacle but in most monotocardians eye stalk and tentacle have fused and the former is represented by only a small bulge. Cephalic tentacles are the only tentacles present in most monotocar-dians, but others, especially in diotocardians, may lie along the side of the foot. They may arise separately or from a continuous ridge known as the **epipodium**, for which reason they are called **epipodial tentacles**. The epipo-dium may expand on each side anteriorly at the entrance to the mantle cavity, forming **neck lobes** which guide the water current into and out of the mantle cavity; it may also give rise to smaller lobes, **cephalic lappets**, near or between the cephalic tentacles. Other tentacles may occur on the edge of the mantle skirt (**pallial tentacles**), or on the posterior part of the foot, especially under-neath the operculum (**metapodial tentacles**).

In male winkles a **penis**, carrying a **seminal groove** from the **genital opening** deep in the mantle cavity, lies on the head behind the right tentacle, though

13

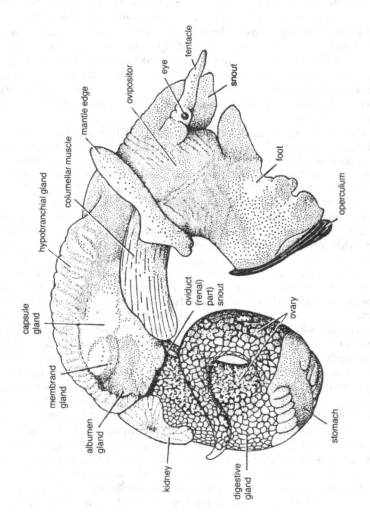

Fig. 5. *Littorina littorea.* Female removed from shell and seen from right side. (From Fretter & Graham, 1962.)

it is anatomically part of the foot (possibly a modified epipodial tentacle) (Figs 8 and 11). When not in use it is turned backwards into the mantle cavity, and it may bear glandular papillae which help to hold it in position during copulation. Female winkles have a glandular pad in much the same position as the penis of males. This is an **ovipositor** which guides the **egg capsules** out of the mantle cavity. In most monotocardians the seminal groove is closed and the male pore lies near the tip of the penis; females usually lack any special ovipositor, using the foot to do such manipulation of eggs as is necessary, but in many neogastropods and a few of the most advanced mesogastropods there is an ovipositor in the form of a glandular papilla lying in a pit on the sole of the foot which is used to fasten egg capsules to an appropriate substratum.

In *Littorina* the opening on the head is a true mouth and leads directly into the gut (Figs 8 and 10), where lies the ribbon of chitinous teeth known as the **radula**, which is extruded on a tongue-like structure, the **odontophore**, to rasp the substratum during feeding. Figure 10B shows the radula in the resting position (the base of the figure) when the teeth lie flat, folded into a groove on the odontophore. During feeding the radula is pulled outwards over the tip of the odontophore, and as they pass this point all the teeth erect and those more laterally placed also move sideways. The broadened radula is then applied to the substratum. On retraction over the odontophoral tip the reverse movement takes place, and, as indicated by the arrows, the lateral teeth swing back towards the mid-line, scraping the substratum as they do so, and all are folded back into the median groove. This is the essential feeding action of the radula in winkles and in most other prosobranchs. Figure 10C shows the marks made on the substratum by a feeding winkle. The whole grouping was produced by the application to the substratum of a series of rows of radular teeth, one after another. The straight marks in the middle were made by the central teeth as they moved from an erect to a prone position, the long curved ones by the most lateral teeth as they moved from an outstretched to a more median position. Intermediate marks are due to intermediate teeth. In some other gastropods (for example wentletraps and eulimids) it is the opening of an introvert, that is, an inturned part of the body wall with the true mouth at its inner end; it is, in fact, a greatly elongated snout which has been invaginated for protection and convenience. When the animal feeds the introvert is turned inside out so that the mouth lies at its tip and its original condition of elongated snout made obvious once more. This kind of structure is known as an **acrembolic proboscis**. In neogastropods, too, the opening seen on the head is not the true mouth but that

---

Fig. 6. *Littorina littorea* to show the relationship of operculum to shell, A, in a creeping winkle, B, in a retracted animal, and also the types of operculum. C, multispiral (*Gibbula cineraria*); D, paucispiral (*Littorina littorea*); E, concentric (*Viviparus*); F, with terminal nucleus (*Hinia reticulata*). (Fretter & Graham, 1962.)

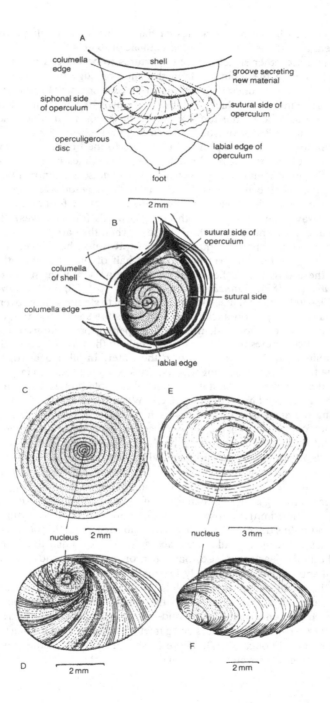

of a sac in which a proboscis lies partially retracted but only partially intro-verted: this type is known as a **pleurembolic proboscis**.

There are other structures and apertures to be noticed on the head-foot in addition to the mouth. Along the anterior edge of the foot there is a groove which contains the opening of an **anterior pedal gland** from which comes the mucus on which prosobranchs crawl. Its upper lip is derived from an anterior lobe of the foot known as the **propodium** which increases in prominence in some prosobranchs to form a shelf-like structure known as a **mentum**. The lower lip of the groove is formed by the anterior edge of the sole of the foot. Prosobranchs creep sometimes by means of cilia on the sole, more commonly by means of waves of muscular contraction travelling along the sole; the mechanism has been described by Jones & Trueman (1970). In some species the waves travel from the front of the foot to the rear (retro-grade waves), in others from the rear forwards (direct waves). The waves may affect the entire breadth of the sole, when they are described as mono-taxic, or at any one moment two sets of waves may be visible, out of phase with one another, one set on the right half of the sole, the other on the left: these waves are called ditaxic. In a few prosobranchs such as *Pomatias elegans* (p. 182), *Truncatella subcylindrica* (p. 200) and *Aporrhais* spp. (p. 298–300) different mechanisms, more akin to stepping, have been evolved. In many small prosobranchs such as rissoids the sole of the foot may show, towards its posterior end, the median opening of a **posterior pedal gland** which also secretes mucus, most commonly in the form of a thread on which the animal climbs up and down in the water. In all neogastropods placed in the family Muricidae (dog whelks, drills, sting winkles), so far as is known, a further opening on the anterior part of the sole houses a structure known as the **accessory boring organ** (ABO). When these animals feed, usually by boring the shell of a bivalve to reach the flesh, this is everted against the surface of the shell of the prey; its action is chemical and alters the shell so that it is more easily rasped by the radula. Boring consists of alternate periods of radular rasping and chemical attack (Carriker, 1981). A comparable organ is found in naticid mesogastropods, which also bore bivalves, but here it is located near the ventral lip of the mouth, not on the foot.

Respiratory gaseous exchange is effected at the surface of the **ctenidia** (Fig. 7). In primitive prosobranchs there are two of these, lying along the left and right sides of the mantle cavity, and each consisting of an axis bearing a series of lamellar leaflets on each side. The ctenidial axes are attached to the body near their base, but much of their distal ends lies free. Since the mantle cavity is effectively internal it must be ventilated: this is done by cilia on the sides of the leaflets. Water enters the mantle cavity on its right and left sides, washes over the osphradium and then the ctenidium on each side and leaves in the mid-line in close proximity to the anus and genital aperture so that faeces and gametes escape without fouling the respira-tory surface (Yonge, 1947). There is frequently a slit or hole here to allow easy escape of the water away from the neighbourhood of the head. In trochid

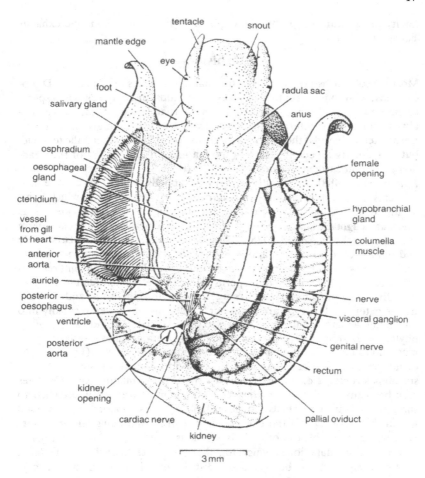

Fig. 7. *Littorina littorea*. Animal removed from shell and mantle cavity opened mid-dorsally to display its contents. Some structures are seen by transparency. (Fretter & Graham, 1962.)

diotocardians and in monotocardians the right ctenidium is typically lost, only the left persisting; in the monotocardians it, too, has been reduced by loss of its left series of lamellae and its axis is fused to the mantle skirt along its whole length. The respiratory stream in these animals now enters the mantle cavity on the left and leaves on the right, but still close to the anus and genital opening. Though the kidney opens deep within the mantle cavity its aperture lies right of the gill so that the urine joins the exhalant current (Fig. 7).

Life history

Most prosobranchs are gonochoristic, that is, the sexes are separate. Diotocardians, with some exceptions such as the neritids, discharge eggs and sperm to the external medium and fertilization is external. In monotocardians generally, fertilization is internal. In one or two more primitive groups (e.g. the freshwater viviparids) the penis is a modification of the right cephalic tentacle, but it is usually a pedal outgrowth though located on the side of the head. Some monotocardian species (*Turritella* (p. 292), *Cerithiopsis* (p. 466), *Epitonium* (p. 488), *Janthina* (p. 496)) have aphallic males but, despite this, it seems that fertilization is still internal, some of the cells which arise in the testis being converted into heavily ciliated structures known as **spermatozeugmata** (sing. spermatozeugma), to which the functional sperm become attached (Ankel, 1926). Though no true copulation occurs, male and female become closely approximated so that when spermatozeugmata are shed they pass into the mantle cavity of the female in the inhalant respiratory current and there liberate sperm to enter the female genital tract. In primitive prosobranchs eggs and sperm are shed into the water and fertilization is external. The eggs are numerous and small and develop rapidly into larvae known as **trochophores**, similar to those of annelids and various other phyla. Their free life is short and they soon metamorphose and settle. In most monotocardians the eggs are laid in protective capsules (Figs 76, 150) or masses of jelly (Fig. 12A) secreted by oviducal glands which form conspicuous structures on the extreme right part of the mantle skirt. Capsules may be planktonic (*Littorina littorea*) but are most frequently attached to some firm substratum, weeds, stones or shells. The trochophore stage is passed within the capsules and the eggs develop to free-swimming **veliger larvae**, each with a shell, head-foot and visceral mass, and swimming by means of two ciliated cephalic lobes which together constitute the **velum**. This larva may be long-lived and be a valuable distributive phase in the life history. In many prosobranchs, however, the veliger stage too is passed within the capsule and juvenile snails emerge (Figs 150, 155). As in other phyla this direct development is characteristic of species living where a free-swimming larval stage would be disadvantageous – freshwater, terrestrial or semi-terrestrial habitats, cold climates, and the like. A few prosobranchs, such as *Littorina saxatilis* (Fig. 12C–F), *Viviparus* spp. (p. 146–8) and *Potamopyrgus jenkinsi* (p. 194) are **ovoviviparous**, retaining the fertilized eggs within the oviduct until the juveniles emerge. These do not appear to have been provided with any food by their mother apart from the small quantity of yolk within the egg and some albumen. In many neogastropods, however, most of the eggs within a capsule fail to develop and these **food eggs** are eaten by a small number of the others which alone develop to hatching (Portmann, 1925; Fioroni, 1966).

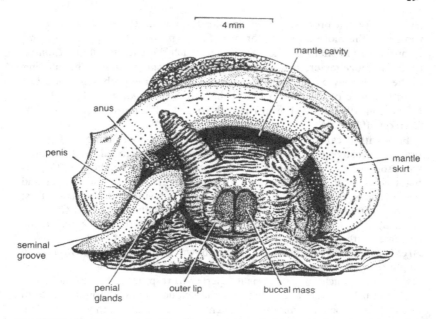

4 mm

mantle cavity

anus

penis

mantle skirt

seminal groove

penial glands

outer lip

buccal mass

Fig. 8. *Littorina littorea*. Anterior view of an anaesthetized animal removed from the shell. The mouth is partly open and the penis would normally be retracted into the mantle cavity. (Fretter & Graham, 1962.)

Mode of Life

Prosobranch gastropods show an extraordinary degree of variation in their way of life and have undergone an adaptive radiation which has enabled them to occupy many ecological niches with success. Many of these adaptations relate to their mode of feeding.

The most primitive members of the group are grazers and detritivores, scraping diatoms and small algae off the substratum with the radula, and sometimes rasping at the surface of larger weeds as well. This type of feeding is found in ormers, limpets, top-shells, winkles, and many kinds of small prosobranch, marine and freshwater. The same mechanism may be employed to take animal food, provided that the prey is sessile and incapable of escaping: such prosobranchs as keyhole limpets (*Diodora*, p. 70) and slit limpets (*Emarginula*, p. 65) feed in this way on sponges (Graham, 1939), whilst some top-shells (*Calliostoma*, p. 124) ingest hydroid and *Alcyonium* polyps (Perron, 1975; Perron & Turner, 1978), and cowries (*Trivia*, p. 326) eat tunicates (Fretter, 1951a).

Further adaptation, however, is necessary to allow prosobranchs to become more successful predators. One such modification is the elongation of the snout to form a proboscis. This enables the animal to reach into tunicate zooids for the more nutritious organs (*Erato* (p. 324), Fretter, 1951a), and to probe within shells or exoskeletons to reach the enclosed flesh (*Lunatia* (p. 334), Ziegelmeier, 1954; *Nucella* (p. 336); *Hinia* (p. 408)). The opening of the shells may be achieved by a simple thrust of the proboscis (*Nucella*), or of the edge of the shell between the valves (*Buccinum* (p. 400), Pearce & Thorson, 1967), or by boring (naticids (p. 334–346), muricids (p. 360–374), Carriker, 1981).

Prey with a calcareous skeleton is also attacked by a number of prosobranchs, mostly not found locally, by means of an acid saliva, sometimes with, sometimes without an included poison (Hughes & Hughes, 1981). This habit is found in some genera from warm waters such as the bonnet and helmet shells, the tuns and frog shells so much in favour with shell collectors; a few specimens of helmet shells of the genera *Cymatium* (p. 352), *Galeodea* (p. 350), and *Charonia* (p. 356) have been encountered in waters off the extreme south-west of the British Isles. Bonnet and helmet shells (*Galeodea*, *Argobuccinum*, *Charonia* have been shown to produce a saliva rich in sulphuric acid with which they attack echinoderms, making a hole in the test through which the proboscis is inserted to rasp flesh. The saliva is very acid (pH 0.13 to 1.1) and penetrates the echinoderm skeleton at a rate of about 0.1 mm per minute; this is much faster than the boring rate, 0.3 to 0.5 mm per day, achieved by the less acid secretion of the ABO (pH 3.8 to 4.0) of an oyster drill. There is no ABO in helmet shells and the role of the radula is limited to the removal of the calcium sulphate produced by the acid acting on the skeleton.

In the genus *Cymatium* the saliva also contains a poison (Houbrick & Fretter, 1969), and the animals attack other gastropods. These are sprayed with saliva, then the proboscis is rapidly thrust into the mantle cavity and the prey bitten close to the heart; all activity on the part of the prey stops and it is eaten at leisure. One local species of prosobranch, *Neptunea antiqua* (p. 396), the red whelk or buckie, so called because the flesh is often slightly reddish, also produces a poisonous saliva which aids it in overcoming its prey, mainly small polychaete worms or bivalves.

There are persistent, though largely unproven, suggestions that other prosobranchs kill their prey, or at least make it insensitive to their attack, by means of toxic secretions. *Janthina* (p. 496) and *Nucella* species have been said to use the secretion from the hypobranchial gland, and *Epitonium* (p. 488) species their saliva, for this purpose.

It is not possible for any of these predators to capture very active animals: within the British fauna this ability has been successfully acquired by only one family, the turrids (p. 420–465), which, though slow-moving themselves, can eject the proboscis at great speed. In these animals a poison gland opens to the buccal region of the gut; its secretion comes to fill the radular teeth, which have been modified to form barbed, tubular structures lying free in the buccal cavity. When prey, usually a polychaete worm, is attacked a tooth

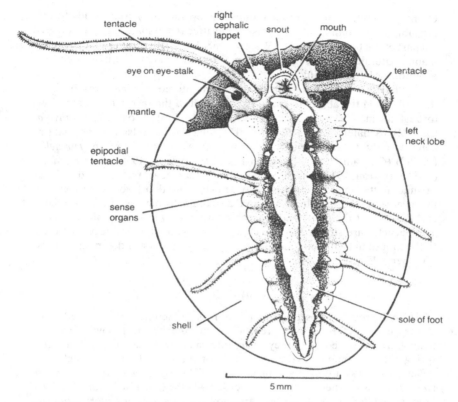

Fig. 9. Ventral view of the trochid *Jujubinus clelandi* showing various formations associated with the head and foot. (Fretter & Graham, 1962.)

filled with poison is manipulated so that it is held at the tip of the proboscis, projecting from the mouth, and is stabbed forcibly into the body of the worm, immobilising it and permitting its ingestion whole (Shimek & Kohn, 1981). This method can be successfully used against prey as active as fish. Though turrids are too small for their bite to have any effect on human beings that of their relatives, the cone shells of warmer regions, may do so and there are records of about sixteen persons being killed by the bite of a cone.

Eulimids are small prosobranchs which live on the surface of such large animals as echinoderms and use their proboscis to enter the host's body, sucking its fluids as food (Warén, 1983a). This trend towards parasitism is also seen in pyramidellids which, though not strictly prosobranchs, are dealt with below. They have no radula and have transformed the jaw into a pointed stylet which can be driven into the prey – usually a sedentary or sessile animal – allowing the attacker to suck its blood or other body fluid (Fretter &

Graham, 1949). In some other groups this evolutionary trend leads to the prosobranch becoming totally parasitic and attached to some external or internal part of the body of the host (Warén, 1983a). As with most similar parasites many anatomical features have become so transformed that its recognition as a mollusc becomes difficult.

Another type of feeding mechanism encountered in some prosobranchs depends upon the ability to strain particles out of the respiratory water flow through the mantle cavity. In some – *Viviparus* (p. 146) (Cook, 1949), *Bithynia* (p. 202) (Schäfer, 1952, 1953b) – this ciliary food-collecting mechanism appears only to supplement ordinary radular grazing, but in *Turritella* (p. 292) (Graham, 1938), *Calyptraea* (p. 308) (Yonge, 1938) and *Crepidula* (p. 310) (Orton, 1912) it is the sole source of food apart from what dissolved organic matter may be directly absorbed. Its adoption leads to an increase in the size of the mantle cavity and ctenidium to increase the water flow and the area of the filter, and it accompanies a very sedentary mode of life. The particles are strained by the ctenidium, agglutinated in mucus, and led to the mouth in a ciliated groove running along the floor of the mantle cavity (Werner, 1952, 1953, 1959).

Longevity

Most small prosobranchs are probably annuals or biennials. If produced early in the year they grow, mate, reproduce and die by the following winter; if they are produced later they overwinter and reproduce in the following spring. Larger animals live longer, at least for several years, though it is difficult to assess ultimate age in species which stop growing once a certain size has been reached; amongst these dog whelks (*Nucella lapillus*, p. 366) seem to live for 5–6 years. There are, however, records of *Littorina littorea* (p. 166) living for over twenty years in an aquarium. Limpets of the genus *Patella* (p. 76) are also, on the whole, long lived. In a population studied by Wright & Hartnoll (1981) in the Isle of Man nearly half of the animals were over 5–6 years old; the oldest which they collected was calculated to have lived for 17 years.

Hosts for platyhelminth parasites

Prosobranchs are well-known as the intermediate host of many parasitic platyhelminths, and their visceral mass is often found to be full of cercariae. Amongst local forms this is most frequent in *Littorina* species (p. 166–181), *Hydrobia ulvae* (p. 188) (from which more than twenty different kinds have been recorded), *Turritella communis* (p. 292) and *Buccinum undatum* (p. 400); the percentage infestation is rarely 10% and is usually much lower. The life history of the platyhelminth usually includes a fish as a second intermediate host and the definitive host is a bird. In other parts of the world the involvement of prosobranchs in the transmission of human disease is serious:

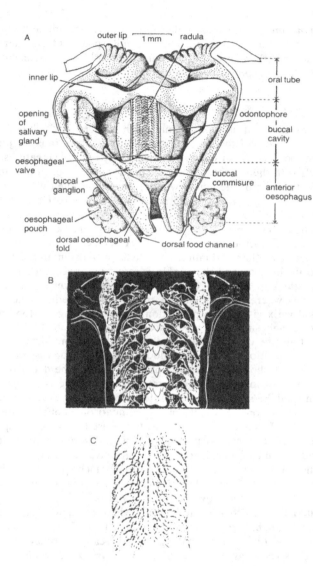

Fig. 10. *Littorina littorea*. A. Buccal cavity and anterior oesophagus opened mid-dorsally to expose the odontophore and radula. (Fretter & Graham, 1962.) B. Dorsal view of the anterior half of the odontophore partly protruded through the mouth as during feeding, showing the radula lying on the dorsal surface, the posterior teeth in the resting position, the anterior ones erected and spread. (After Ankel, 1938.) C. A tracing of the marks made on the substratum by the radula of a feeding winkle. (After Ankel, 1938.)

species of *Oncomelania* are intermediate hosts of the blood fluke *Schistosoma japonica*, a number of prosobranchs transmit species of the lung fluke *Paragonimus* and others carry the liver fluke *Clonorchis sinensis*, an important parasite of fish-eating populations in the far east.

Predators

Prosobranchs are used as food by many fish and sea birds, but the main predators, especially of the smaller sorts, are probably crabs, to judge by the frequent occurrence on shells of damage done by these creatures. Crabs attack shells with their pincers: if the shell is small enough (less than 9 mm in height in the case of the edible winkle), the whole shell will be crushed and its occupant eaten in about 2.5 minutes according to Hadlock (1980). If the shell is not sufficiently fragile for this type of attack to succeed (9–18 mm in height) the crab will proceed to crack pieces of shell away at the aperture, moving further and further up the spire until it can reach behind the operculum of the retreating mollusc and pull it out. Shells over 18 mm in height cannot be cracked in this way and the winkle is safe from crab predation. Snails avoid crab attack in a variety of ways: those with thicker shells are more likely to survive where crabs abound, so strengthening the edge of the aperture with a labial varix gives some protection, whilst foraging for food may be confined to daylight since crabs are most active at night.

Man must also be counted amongst the predators of molluscs and, in the light of the excavations of kitchen middens composed of vast numbers of shells, sometimes showing the effects of cooking, must be deemed to have been so since prehistoric times. Apart from the pulmonate escargots, mainly the Roman snail *Helix pomatia*, his preference has always been for marine species, freshwater ones being insipid by comparison, though considerable quantities of the freshwater clam *Corbicula* are used for food in Japan, the Philippines and Indonesia. At present his liking, differing in different parts of the world, is mainly for cephalopods and such bivalves as oysters, scallops, cockles, and mussels, and the number of prosobranch species used is much less, their catch in 1977 being less than 2% of the total marine molluscan fishery (Boyle, 1981). Only two kinds of prosobranch are at all frequently caught round the British Isles at present, winkles and whelks, and much of their use is for bait rather than for human food. Limpets, once eaten extensively, may still be occasionally gathered locally. In other parts of the world abalones (*Haliotis*, p. 62) and conchs (*Busycon*, *Strombus*) are the main prosobranchs used for food.

Prosobranchs are normally a safe and enjoyable sea food, which is not invariably true of bivalves, which have been convicted of causing a series of outbreaks of food poisoning (Pain, 1986), mainly due to their ingestion of bacteria and viruses from sewage which has been fed untreated into the sea, or to their uptake of such unicellular algae as *Gonyaulax*, which contain toxins (Clark, 1968), as they filter the water passing over their gills. Since

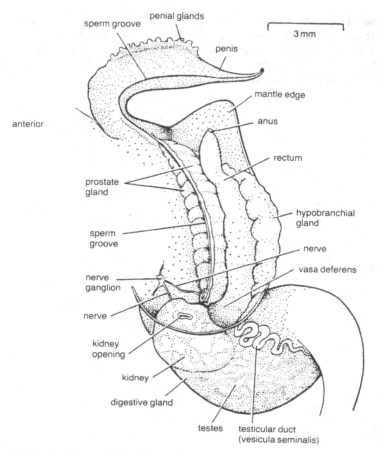

Fig. 11. *Littorina littorea*. Dissection to show the male reproductive system. (Fretter & Graham, 1962.)

neither winkles nor whelks feed in this way, and since, unlike oysters, they are always well cooked before being eaten, such dangers are almost unknown. Other factors enter, however, since the words 'winkle' and 'whelk' designate more than one species of animal. Winkles destined to be eaten should be the edible winkle, *Littorina littorea* (p. 166), rather than the rough winkle (*L. saxatilis*, p. 178), which is gritty because the females are ovoviviparous and contain large numbers of shelled young. A more serious reason underlies the necessity for choosing the correct whelk, which is *Buccinum undatum* (p. 400), rather than the superficially similar and, off northern British coasts, the equally common red whelk, *Neptunea antiqua* (p. 396). The salivary glands of the latter contain a poison, tetramethylammonium hydroxide (tetramine), which is used by the whelk to kill its prey. If these glands are eaten with the rest of the animal poisoning can result (Fleming, 1971) producing effects mimicking those of curare, and fortunately, in the cases so far recorded, transient – but not to be assumed to be invariably so!

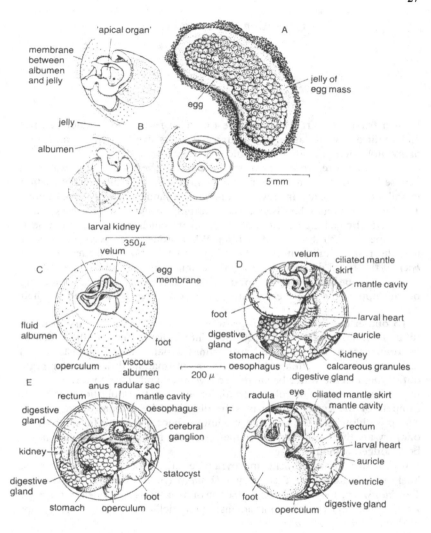

Fig. 12. *Littorina obtusata*. A, spawn on weed; B, developing embryos. *Littorina saxatilis*. Embryos from the brood pouch. C, veliger stage; D, late veliger; E, just before hatching; F, hatching by rasping through the egg membrane. (Fretter & Graham, 1962.)

Collection and Preservation

Most marine prosobranchs are to be found on rocky and boulder beaches and sublittorally. Relatively few intertidal kinds live in sand or mud, but many sublittoral ones are to be found on soft bottoms.

On rocky shores some, such as limpets, may be seen at low water on exposed surfaces but most spend this period sheltering against desiccation in and under seaweeds, in crevices, under ledges and stones, or in rock pools. The larger sorts may be discovered by turning over the stones, by shaking the weeds, by picking out of pools or crevices, which may on occasion need to be opened with the help of a chisel. When stones have been turned over it is the responsibility of every field biologist to *make certain to turn them back*: if this is not done not only will the animals on the underside, especially sessile ones, be destroyed by exposure but also the plants which were living on the upper side will die and the whole local area be polluted by their decay.

To obtain the smaller sort of prosobranch it may be necessary to pluck tufts of the finer weeds and wash them into a dish, either on the beach or on return from it. These, and such things as laminarian holdfasts, set in dishes of sea water, often produce numbers of small molluscs which migrate out of them as the water becomes deoxygenated and climb towards the surface, where they may be collected with forceps, pipette or plastic tea strainer. Limpets of the genus *Patella* may be dislodged by a knife or, better, by a surprise blow with the heel of a Wellington boot, but if this fails to remove one, try another – an alarmed limpet increases its grip and can then never be dislodged without damage.

Carnivorous prosobranchs and pyramidellids of rocky beaches are most likely to be found near their prey – *Diodora* (p. 70), *Emarginula* (p. 65), *Cerithiopsis* (p. 466), *Triphora* (p. 480) on or near sponges; *Lamellaria, Trivia, Erato* on or near compound ascidians; pyramidellids alongside large groups of *Pomatoceros, Sabellaria* and *Mytilus*.

Any prosobranchs to be taken home may be placed in plastic bags with a little water – just enough to keep them damp – and as much air as can be enclosed, or in tubes or jars with weed to stop them drying out: any intertidal mollusc can easily survive emersion provided it is kept damp – they do it at every low tide.

On sandy shores the most abundant molluscs are bivalves, but their predator *Lunatia* (p. 334) may be turned up whilst digging for them, or it may be detected under a small hump in the sand and scooped up with fingers. Small turrids may occasionally be found in the sand at extreme low water and they and other, normally sublittoral animals, may be washed up after a storm. The surface of mud, especially in brackish places, is often covered with great numbers of *Hydrobia ulvae* (p. 188) which may be gathered in a plastic tea strainer.

Whilst it may be possible to gather some of the smaller sublittoral prosobranchs by diving, many live at depths which make dredging the only certain method of obtaining them. This may involve lengthy periods of painstaking search, especially if the smaller infaunal species are being looked for.

Freshwater prosobranchs live most frequently in quieter waters which are usually hard (20 or more parts of calcium per million), and where vegetation is abundant. In summer they creep over the plants and may be collected by examining washings and nettings; in winter they tend to burrow into the substratum from which the plants are growing. *Viviparus* spp. (p. 146–148) also burrow into gravel beds and *Potamopyrgus* (p. 194) and *Theodoxus* (p. 144) are found on stones, usually in running water. The terrestrial species thrive only in calcareous districts and often burrow shallowly in soil or leaf litter.

Finally, remember that wherever and whatever you collect, do not take more animals than are necessary for your immediate purpose. In that way a population is conserved and you give others the chance to get the same pleasure from observing the animals as you do.

The soft anatomy of prosobranchs is best studied with the animals in a relaxed state. This also allows the preservative to penetrate the tissues. A 7% (0.36M) magnesium chloride solution in distilled water or a mixture of equal parts sea-water and 7% magnesium chloride solution give good results with most marine prosobranchs. Soda-water from a siphon may be used to anaesthetize small freshwater prosobranchs. They may also be killed in an extended state by putting them into a screw-topped jar filled to the brim with water, which has been previously boiled and cooled out of contact with air. The jar should then be tightly closed and left until the snails are fully extended and do not respond to stimulation. This may take some hours.

If so desired, shells are most easily removed by cracking them in a vice, when the body of the snail can be extracted entire by scraping the shell muscle off the columella. If only the shell is to be kept, the snail may be killed by brief boiling in water and the body of the animal picked out with the traditional pin. The animal may also be allowed to remain in the shell and dried, which has the advantage of letting the operculum remain visible. Unfortunately it is extremely difficult to obtain both an intact body and an intact shell; but repeated freezing and thawing sometimes succeed in loosening the columella muscle so that the animal's body may be extracted without damage to the shell.

Shells are best kept dry, in plastic bags, preferably the kind that can be closed; animals may be kept in formalin (about 10%) or alcohol (70–90% industrial, not methylated spirit).

Collecting Lists

Set out below is a series of lists giving the names of species of prosobranch gastropod which one might reasonably expect to collect in certain habitats around the British Isles. The first group of lists gives those that might be found within tidemarks on good shores, rocky and sandy. The main list names those species likely to occur throughout the British Isles, the subsidiary lists give those which might be found in addition if collecting is being done in either the north or south. Some species are common only during summer.

The second group lists the additional species that might occur if dredgings are available.

Littoral animals

1. Found throughout the British Isles.

Emarginula fissura	*Hydrobia ulvae*
Diodora graeca	*Hydrobia ventrosa*
Patella vulgata	*Cingula trifasciata*
Patella aspera	*Onoba semicostata*
Helcion pellucidum	*Alvania semistriata*
Acmaea virginea	*Alvania punctura*
Calliostoma zizyphinum	*Rissoa parva*
Gibbula cineraria	*Rissoa lilacina*
Gibbula umbilicalis	*Barleeia unifasciata*
Tricolia pullus	*Rissoella diaphana*
Lacuna vincta	*Rissoella opalina*
Lacuna parva	*Omalogyra atomus*
Lacuna pallidula	*Skeneopsis planorbis*
Littorina littorea	*Bittium reticulatum*
Littorina saxatilis	*Cerithiopsis tubercularis*
Littorina arcana	*Lunatia alderi*
Littorina neglecta	*Lamellaria perspicua*
Littorina nigrolineata	*Trivia monacha*
Littorina obtusata	*Nucella lapillus*
Littorina mariae	*Hinia reticulata*
Littorina neritoides	*Hinia incrassata*

2. Additional species likely to occur in northern areas only.

Collisella tessulata	*Rissoa rufilabrum*
Margarites helicinus	*Velutina velutina*
Onoba aculeus	*Lamellaria latens*

3. Additional species likely to occur in southern areas only.

Patella depressa	*Triphora adversa*
Monodonta lineata	*Calyptraea chinensis*
Cingulopsis fulgida	*Ocenebra erinacea*

Sublittoral animals

1. Found throughout the British Isles.

Gibbula magus	*Capulus ungaricus*
Gibbula tumida	*Aporrhais pespelecani*
Jujubinus montagui	*Lunatia catena*
Alvania beani	*Lunatia montagui*
Alvania punctura	*Trivia arctica*
Pusillina inconspicua	*Colus gracilis*
Turritella communis	*Colus jeffreysianus*
Epitonium clathrus	*Buccinum undatum*
Melanella alba	*Hinia pygmaea*
Eulima trifasciata	*Mangelia brachystoma*
Vitreolina philippii	*Comarmondia gracilis*

2. Additional species likely to occur in northern areas only.

Puncturella noachina	*Boreotrophon truncatus*
Iothia fulva	*Trophonopsis barvicensis*
Jujubinus clelandi	*Neptunea antiqua*
Trichotropis borealis	*Oenopota turricula*

3. Additional species likely to occur in southern areas only.

Emarginula conica	*Crepidula fornicata*
Propilidium ancyloide	*Trophonopsis muricatus*
Calliostoma granulatum	*Mangelia nebula*
Jujubinus striatus	*Raphitoma purpurea*
Epitonium turtonis	

Classification

Nomenclature and classification of rissoids follow Ponder (1985); of turrids, in most respects, Bouchet & Warén (1980); that of eulimids is based on Warén (1983a). For the rest the classification and most names are those generally accepted, but some changes have been made, most of which have been explained by Fretter & Graham (1982) and Fretter, Graham & Andrews (1986).

Phylum MOLLUSCA
  Class GASTROPODA Cuvier, 1798
    Subclass PROSOBRANCHIA Milne Edwards, 1848
      Order ARCHAEOGASTROPODA Thiele, 1929
        Superfamily Pleurotomariacea Swainson, 1840
          Family Scissurellidae Gray, 1847
            Genus *Scissurella* Orbigny, 1824
              Scissurella crispata Fleming, 1828
          Family Haliotidae Rafinesque, 1815
            Genus *Haliotis* Linné, 1758
              Haliotis tuberculata Linné, 1758
        Superfamily Fissurellacea Fleming, 1822
          Family Fissurellidae Fleming, 1822
            Genus *Emarginula* Lamarck, 1801
              Emarginula fissura (Linné, 1758)
              Emarginula conica Lamarck, 1801
              Emarginula crassa Sowerby, 1813
            Genus *Puncturella* Lowe, 1827
              Puncturella noachina (Linné, 1771)
            Genus *Diodora* Gray, 1821
              Diodora graeca (Linné, 1758)
        Superfamily Patellacea Rafinesque, 1815
          Family Acmaeidae Carpenter, 1857
            Genus *Acmaea* Eschscholtz, 1833
              Acmaea virginea (Müller, 1776)
            Genus *Collisella* Dall, 1871
              Collisella tessulata (Müller, 1776)
          Family Patellidae Rafinesque, 1815
            Genus *Patella* Linné, 1758
              Patella vulgata Linné, 1758
              Patella aspera Röding, 1798
              Patella depressa Pennant, 1777

Genus *Helcion* Montfort, 1810
  Helcion pellucidum (Linné, 1758)
Family Lepetidae Gray, 1850
  Genus *Lepeta* Gray, 1847
    Lepeta caeca (Müller, 1776)
  Genus *Iothia* Gray, 1850
    Iothia fulva (Müller, 1776)
  Genus *Propilidium* Forbes & Hanley, 1849
    Propilidium ancyloide (Forbes, 1840)
Superfamily Cocculinacea Dall, 1882
  Family Lepetellidae Dall, 1881
    Genus *Lepetella* Verrill, 1880
      Lepetella laterocompressa (Ponzi & Rayneval, 1854)
Superfamily Trochacea Rafinesque, 1815
  Family Trochidae Rafinesque, 1815
    Genus *Margarites* Gray, 1847
      Margarites helicinus (Fabricius, 1780)
      Margarites groenlandicus (Gmelin, 1791)
      Margarites argentatus (Gould, 1841)
    Genus *Solariella* S. Wood, 1842
      Solariella amabilis (Jeffreys, 1865)
    Genus *Danilia* Brusina, 1865
      Danilia tinei (Calcara, 1839)
    Genus *Gibbula* Risso, 1826
      Gibbula cineraria (Linné, 1758)
      Gibbula umbilicalis (da Costa, 1778)
      Gibbula pennanti (Philippi, 1851)
      Gibbula tumida (Montagu, 1803)
      Gibbula magus (Linné, 1758)
    Genus *Monodonta* Lamarck, 1799
      Monodonta lineata (da Costa, 1778)
    Genus *Jujubinus* Monterosato, 1884
      Jujubinus exasperatus (Pennant, 1777)
      Jujubinus striatus (Linné, 1758)
      Jujubinus montagui (W. Wood, 1828)
      Jujubinus clelandi (W. Wood, 1828)
    Genus *Calliostoma* Swainson, 1840
      Calliostoma zizyphinum (Linné, 1758)
      Calliostoma granulatum (Born, 1778)
      Calliostoma occidentale (Mighels & Adams, 1842)
  Family Skeneidae Clark, 1851
    Genus *Skenea* Fleming, 1825
      Skenea serpuloides (Montagu, 1808)
      Skenea nitens (Philippi, 1844)
      Skenea cutleriana (Clark, 1849)
      Skenea basistriata (Jeffreys, 1877)

Family Turbinidae Rafinesque, 1815
   Genus *Moelleria* Jeffreys, 1865
      Moelleria costulata (Möller, 1842)
Family Tricoliidae Woodring, 1928
   Genus *Tricolia* Risso, 1826
      Tricolia pullus (Linné, 1758)
Superfamily Neritacea Rafinesque, 1815
Family Neritidae Rafinesque, 1815
   Genus *Theodoxus* Montfort, 1810
      Theodoxus fluviatilis (Linné, 1758)
Order CAENOGASTROPODA Cox, 1959
Superfamily Viviparacea Gray, 1847
Family Viviparidae Gray, 1847
   Genus *Viviparus* Montfort, 1810
      Viviparus viviparus (Linné, 1758)
      Viviparus contectus (Millet, 1813)
Superfamily Valvatacea Thompson, 1840
Family Valvatidae Thompson, 1840
   Genus *Valvata* Müller, 1774
      Valvata piscinalis (Müller, 1774)
      Valvata cristata Müller, 1774
      Valvata macrostoma Steenbuch, 1847
Superfamily Littorinacea Gray, 1840
Family Lacunidae Gray, 1857
   Genus *Lacuna* Turton, 1827
      Lacuna parva (da Costa, 1778)
      Lacuna vincta (Montagu, 1803)
      Lacuna crassior (Montagu, 1803)
      Lacuna pallidula (da Costa, 1778)
Family Littorinidae Gray, 1840
   Genus *Littorina* Férussac, 1822
      Littorina littorea (Linné, 1758)
      Littorina neritoides (Linné, 1758)
      Littorina obtusata (Linné, 1758)
      Littorina mariae Sacchi & Rastelli, 1966
      Littorina nigrolineata Gray, 1839
      Littorina neglecta Bean, 1844
      Littorina saxatilis (Olivi, 1792)
      Littorina arcana Hannaford Ellis, 1978
      Littorina tenebrosa (Montagu, 1803)
Family Pomatiasidae Gray, 1852
   Genus *Pomatias* Studer, 1789
      Pomatias elegans (Müller, 1774)
Family Aciculidae Gray, 1850
   Genus *Acicula* Hartmann, 1821
      Acicula fusca (Montagu, 1803)

Superfamily Rissoacea Gray, 1847
 Family Hydrobiidae Stimpson, 1865
  Genus *Hydrobia* Hartmann, 1821
   Hydrobia ulvae (Pennant, 1777)
   Hydrobia ventrosa (Montagu, 1803)
   Hydrobia neglecta Muus, 1963
  Genus *Potamopyrgus* Stimpson, 1865
   Potamopyrgus jenkinsi (Smith, 1889)
  Genus *Pseudamnicola* Paulucci, 1878
   Pseudamnicola confusa (Frauenfeld, 1863)
  Genus *Marstoniopsis* Altena, 1936
   Marstoniopsis scholtzi (Schmidt, 1856)
 Family Truncatellidae Gray, 1840
  Genus *Truncatella* Risso, 1826
   Truncatella subcylindrica (Linné, 1767)
 Family Bithyniidae Troschel, 1857
  Genus *Bithynia* Leach, 1818
   Bithynia tentaculata (Linné, 1758)
   Bithynia leachi (Sheppard, 1823)
 Family Iravadiidae Thiele, 1928
  Genus *Hyala* H. & A. Adams, 1852
   Hyala vitrea (Montagu, 1803)
  Genus *Ceratia* H. & A. Adams, 1852
   Ceratia proxima (Forbes & Hanley, 1850)
 Family Rissoidae Gray, 1847
  Genus *Rissoa* Desmarest, 1814
   Rissoa parva (da Costa, 1778)
   Rissoa guerini Récluz, 1843
   Rissoa lilacina Récluz, 1843
   Rissoa rufilabrum Alder, 1844
   Rissoa porifera Lovén, 1846
  Genus *Rissostomia* G. O. Sars, 1878
   Rissostomia membranacea (J. Adams, 1800)
  Genus *Pusillina* Monterosato, 1884
   Pusillina inconspicua (Alder, 1844)
   Pusillina sarsi (Lovén, 1846)
  Genus *Setia* H. & A. Adams, 1852
   Setia pulcherrima (Jeffreys, 1848)
  Genus *Alvania* Risso, 1826
   Alvania carinata (da Costa, 1778)
   Alvania beani (Thorpe, 1844)
   Alvania cimicoides (Forbes, 1844)
   Alvania semistriata (Montagu, 1808)

Alvania abyssicola (Forbes, 1850)
Alvania subsoluta (Aradas, 1847)
Alvania lactea (Michaud, 1830)
Alvania punctura (Montagu, 1803)
Alvania jeffreysi (Waller, 1864)
Alvania cancellata (da Costa, 1778)
Genus *Manzonia* Brusina, 1870
    Manzonia crassa (Kanmacher, 1798)
    Manzonia zetlandica (Montagu, 1815)
Genus *Cingula* Fleming, 1828
    Cingula trifasciata (J. Adams, 1800)
Genus *Onoba* H. & A. Adams, 1852
    Onoba semicostata (Montagu, 1803)
    Onoba aculeus (Gould, 1841)
Genus *Obtusella* Cossmann, 1921
    Obtusella intersecta (Wood, 1856)
Genus *Setia* H. & A. Adams, 1852
    Setia inflata Monterosato, 1884
Family Barleeidae Gray, 1857
Genus *Barleeia* W. Clark, 1855
    Barleeia unifasciata (Montagu, 1803)
Family Assimineidae H. & A. Adams, 1856
Genus *Assiminea* Fleming, 1828
    Assiminea grayana Fleming, 1828
Genus *Paludinella* Pfeiffer, 1841
    Paludinella littorina (Chiaje, 1828)
Family Cingulopsidae Fretter & Patil, 1958
Genus *Cingulopsis* Fretter & Patil, 1958
    Cingulopsis fulgida (J. Adams, 1797)
Family Rissoellidae Gray, 1850
Genus *Rissoella* Gray, 1847
    Rissoella diaphana (Alder, 1848)
    Rissoella opalina (Jeffreys, 1848)
    Rissoella globularis (Forbes & Hanley, 1852)
Family Omalogyridae Sars, 1878
Genus *Omalogyra* Jeffreys, 1860
    Omalogyra atomus (Philippi, 1841)
Genus *Ammonicera* Vayssière, 1893
    Ammonicera rota (Forbes & Hanley, 1850)
Family Skeneopsidae Iredale, 1915
Genus *Skeneopsis* Iredale, 1915
    Skeneopsis planorbis (Fabricius, 1780)
Genus *Retrotortina* Chaster, 1896
    Retrotortina fuscata Chaster, 1896
Family Tornidae Sacco, 1896

Genus *Tornus* Turton & Kingston, 1830
    Tornus subcarinatus (Montagu, 1803)
    Tornus exquisitus (Jeffreys, 1883)
    Tornus unisulcatus (Chaster, 1897)
Genus *Circulus* Jeffreys, 1865
    Circulus striatus (Philippi, 1836)
Family Caecidae Gray, 1850
  Genus *Caecum* Fleming, 1813
    Caecum imperforatum (Kanmacher, 1798)
    Caecum glabrum (Montagu, 1803)
Superfamily Cerithiacea Fleming, 1822
  Family Turritellidae Woodward, 1851
    Genus *Turritella* Lamarck, 1799
      Turritella communis Risso, 1826
  Family Aporrhaidae Gray, 1852
    Genus *Aporrhais* da Costa, 1778
      Aporrhais pespelecani (Linné, 1758)
      Aporrhais serresianus (Michaud, 1828)
  Family Cerithiidae Fleming, 1822
    Genus *Bittium* Leach, 1847
      Bittium reticulatum (da Costa, 1778)
      Bittium simplex (Jeffreys, 1867)
Superfamily Hipponicacea Troschel, 1861
  Family Trichotropidae Gray, 1850
    Genus *Trichotropis* Broderip & Sowerby, 1829
      Trichotropis borealis Broderip & Sowerby, 1829
    Genus *Torellia* Lovén, 1867
      Torellia vestita Jeffreys, 1867
  Family Capulidae Fleming, 1822
    Genus *Capulus* Montfort, 1810
      Capulus ungaricus (Linné, 1758)
Superfamily Calyptraeacea Blainville, 1824
  Family Calyptraeidae Blainville, 1824
    Genus *Calyptraea* Lamarck, 1799
      Calyptraea chinensis (Linné, 1758)
    Genus *Crepidula* Lamarck, 1799
      Crepidula fornicata (Linné, 1758)
Superfamily Lamellariacea Orbigny, 1841
  Family Lamellariidae Orbigny, 1841
    Genus *Lamellaria* Montagu, 1815
      Lamellaria perspicua (Linné, 1758)
      Lamellaria latens (Müller, 1776)
    Genus *Velutina* Fleming, 1822
      Velutina plicatilis (Müller, 1776)
      Velutina velutina (Müller, 1776)
      Velutina undata Brown, 1839

Family Eratoidae Gill, 1871
   Genus *Erato* Risso, 1826
      Erato voluta (Montagu, 1803)
   Genus *Trivia* Gray, 1837
      Trivia monacha (da Costa, 1778)
      Trivia arctica (Pulteney, 1799)
Superfamily Cypraeacea Rafinesque, 1815
  Family Ovulidae Fleming, 1822
   Genus *Simnia* Risso, 1826
      Simnia patula (Pennant, 1777)
Superfamily Naticacea Gray, 1840
  Family Naticidae Gray, 1840
   Genus *Lunatia* Gray, 1847
      Lunatia alderi (Forbes, 1838)
      Lunatia catena (da Costa, 1778)
      Lunatia montagui (Forbes, 1838)
      Lunatia fusca (Blainville, 1825)
      Lunatia pallida (Broderip & Sowerby, 1829)
   Genus *Natica* Scopoli, 1777
      Natica clausa Broderip & Sowerby, 1829
   Genus *Amauropsis* Mörch, 1857
      Amauropsis islandica (Gmelin, 1791)
Superfamily Atlantacea Orbigny, 1835
  Family Carinariidae Reeve, 1841
   Genus *Carinaria* Lamarck, 1801
      Carinaria lamarcki Péron & Lesueur, 1810
Superfamily Tonnacea Suter, 1913
  Family Cassidae Latreille, 1825
   Genus *Galeodea* Link, 1807
      Galeodea rugosa (Linné, 1771)
  Family Cymatiidae Iredale, 1913
   Genus *Cymatium* Röding, 1798
      Cymatium cutaceum (Linné, 1767)
   Genus *Ranella* Lamarck, 1816
      Ranella olearia (Linné, 1758)
   Genus *Charonia* Gistel, 1848
      Charonia lampas (Linné, 1758)
Superfamily Muricacea Rafinesque, 1815
  Family Muricidae Rafinesque, 1815
   Genus *Boreotrophon* Fischer, 1884
      Boreotrophon truncatus (Ström, 1768)
      Boreotrophon clathratus (Linné, 1758)
   Genus *Trophonopsis* Bucquoy, Dautzenberg & Dollfus, 1882
      Trophonopsis muricatus (Montagu, 1803)
      Trophonopsis barvicensis (Johnston, 1825)

Genus *Nucella* Röding, 1798
    Nucella lapillus (Linné, 1758)
Genus *Urosalpinx* Stimpson, 1865
    Urosalpinx cinerea (Say, 1822)
Genus *Ocenebra* Gray, 1847
    Ocenebra erinacea (Linné, 1758)
Genus *Ocinebrina* Jousseaume, 1880
    Ocinebrina aciculata (Lamarck, 1822)
Superfamily Buccinacea Rafinesque, 1815
Family Columbellidae Swainson, 1840
Genus *Amphissa* H. & A. Adams, 1853
    Amphissa haliaeeti (Jeffreys, 1867)
Family Buccinidae Rafinesque, 1815
Genus *Liomesus* Stimpson, 1865
    Liomesus ovum (Turton, 1825)
Genus *Beringius* Dall, 1887
    Beringius turtoni (Bean, 1834)
Genus *Volutopsius* Mörch, 1857
    Volutopsius norwegicus (Gmelin, 1791)
Genus *Colus* Röding, 1798
    Colus gracilis (da Costa, 1778)
    Colus islandicus (Gmelin, 1791)
    Colus jeffreysianus (Fischer, 1868)
Genus *Turrisipho* Dautzenberg & Fischer, 1912
    Turrisipho fenestratus (Turton, 1834)
    Turrisipho moebii (Dunker & Metzger, 1874)
Genus *Neptunea* Röding, 1798
    Neptunea antiqua (Linné, 1758)
    Neptunea despecta (Linné, 1758)
Genus *Buccinum* Linné, 1758
    Buccinum undatum Linné, 1758
    Buccinum humphreysianum Bennett, 1824
    Buccinum hydrophanum Hancock, 1846
Genus *Chauvetia* Monterosato, 1884
    Chauvetia brunnea (Donovan, 1804)
Family Nassariidae Iredale, 1916
Genus *Hinia* Leach, 1852
    Hinia reticulata (Linné, 1758)
    Hinia incrassata (Ström, 1768)
    Hinia pygmaea (Lamarck, 1822)
Family Fasciolariidae Gray, 1853
Genus *Troschelia* Mörch, 1876
    Troschelia berniciensis (King, 1846)
Superfamily Cancellariacea Ponder, 1973

Family Cancellariidae Forbes & Hanley, 1853
Genus *Admete* Kröyer, 1842
  Admete viridula (Fabricius, 1780)
Superfamily Conacea Rafinesque, 1815
Family Turridae Swainson, 1840
Genus *Spirotropis* Sars, 1878
  Spirotropis monterosatoi (Locard, 1897)
Genus *Haedropleura* Bucquoy, Dautzenberg & Dollfus, 1883
  Haedropleura septangularis (Montagu, 1803)
Genus *Oenopota* Mörch, 1852
  Oenopota turricula (Montagu, 1803)
  Oenopota trevelliana (Turton, 1834)
  Oenopota rufa (Montagu, 1803)
  Oenopota violacea (Mighels & Adams, 1842)
Genus *Typhlomangelia* Sars, 1878
  Typhlomangelia nivalis (Lovén, 1846)
Genus *Thesbia* Jeffreys, 1867
  Thesbia nana (Lovén, 1846)
Genus *Mangelia* Risso, 1826
  Mangelia nebula (Montagu, 1803)
  Mangelia brachystoma (Philippi, 1844)
  Mangelia powisiana (Dautzenberg, 1887)
  Mangelia attenuata (Montagu, 1803)
Genus *Cytharella* Monterosato, 1875
  Cytharella smithi (Forbes, 1844)
  Cytharella coarctata (Forbes, 1840)
  Cytharella rugulosa (Philippi, 1844)
Genus *Comarmondia* Monterosato, 1884
  Comarmondia gracilis (Montagu, 1803)
Genus *Raphitoma* Ballardi, 1848
  Raphitoma linearis (Montagu, 1803)
  Raphitoma purpurea (Montagu, 1803)
  Raphitoma asperrima (Brown, 1827)
  Raphitoma leufroyi (Michaud, 1828)
Genus *Cenodagreutes* Smith, 1967
  Cenodagreutes aethus Smith, 1967
  Cenodagreutes coccyginus Smith, 1967
Genus *Teretia* Norman, 1888
  Teretia teres (Reeve, 1844)
Genus *Taranis* Jeffreys, 1870
  Taranis moerchi (Malm, 1861)
  Taranis borealis Bouchet & Warén, 1980

*Superfamily Cerithiopsacea
  Family Cerithiopsidae H. & A. Adams, 1853
    Genus *Cerithiopsis* Forbes & Hanley, 1849
        Cerithiopsis tubercularis (Montagu, 1803)
        Cerithiopsis barleei Jeffreys, 1867
        Cerithiopsis metaxa (delle Chiaje, 1828)
        Cerithiopsis pulchella Jeffreys, 1858
    Genus *Eumetula* Thiele, 1912
        Eumetula costulata (Möller, 1842)
    Genus *Cerithiella* Verrill, 1882
        Cerithiella metula (Lovén, 1846)
    Genus *Laeocochlis* Dunker & Metzger, 1874
        Laeocochlis granosa (S. Wood, 1848)
Superfamily Triphoracea
  Family Triphoridae Gray, 1847
    Genus *Triphora* Blainville, 1828
        Triphora adversa (Montagu, 1803)
        Triphora pallescens Jeffreys, 1867
        Triphora erythrosoma Bouchet & Guillemot, 1978
        Triphora similior Bouchet & Guillemot, 1978
Superfamily Epitoniacea Berry, 1910
  Family Epitoniidae Berry, 1910
    Genus *Cirsotrema* Mörch, 1852
        Cirsotrema commutatum (Monterosato, 1877)
    Genus *Epitonium* Röding, 1798
        Epitonium clathrus (Linné, 1758)
        Epitonium turtonis (Turton, 1819)
        Epitonium trevelyanum (Johnston, 1841)
        Epitonium clathratulum (Kanmacher, 1798)
  Family Janthinidae Leach, 1823
    Genus *Janthina* Röding, 1798
        Janthina janthina (Linné, 1758)
        Janthina pallida Thompson, 1840
        Janthina exigua Lamarck, 1816
Superfamily Eulimacea
  Family Aclididae G. O. Sars, 1878
    Genus *Aclis* Lovén, 1846
        Aclis minor (Brown, 1827)
        Aclis ascaris (Turton, 1819)
        Aclis walleri Jeffreys, 1867
    Genus *Hemiaclis* G. O. Sars, 1878
        Hemiaclis ventrosa (Friele, 1874)

---

* This and all following superfamilies are often separated from other prosobranchs in an order commonly called HETEROGASTROPODA, but its name and limits are still subject to discussion.

Genus *Cima* Chaster, 1898
    Cima minima (Jeffreys, 1858)
Genus *Graphis* Jeffreys, 1867
    Graphis albida (Kanmacher, 1798)
Genus *Pherusina* Norman, 1888
    Pherusina gulsonae (Clark, 1850)
Family Eulimidae H. & A. Adams, 1853
Genus *Eulima* Risso, 1826
    Eulima glabra (da Costa, 1778)
    Eulima bilineata Alder, 1848
Genus *Haliella* Monterosato, 1878
    Haliella stenostoma (Jeffreys, 1858)
Genus *Melanella* Bowdich, 1822
    Melanella alba (da Costa, 1778)
    Melanella lubrica (Monterosato, 1891)
    Melanella frielei (Jordan, 1895)
Genus *Eulitoma* Laseron, 1955
    Eulitoma compactilis (Sykes, 1903)
Genus *Polygireulima* Sacco, 1892
    Polygireulima sinuosa (Scacchi, 1836)
    Polygireulima monterosatoi (Monterosato, 1890)
Genus *Vitreolina* Monterosato, 1884
    Vitreolina philippii (Rayneval & Ponzi, 1854)
    Vitreolina collinsi (Sykes, 1903)
    Vitreolina petitiana (Brusina, 1869)
    Vitreolina curva (Monterosato, 1874)
Family Stiliferidae H. & A. Adams, 1853
Genus *Pelseneeria* Köhler & Vaney, 1908
    Pelseneeria stylifera (Turton, 1825)
*Family Pyramidellidae Gray, 1840
Genus *Chrysallida* Carpenter, 1857
    Chrysallida obtusa (Brown, 1827)
    Chrysallida suturalis (Philippi, 1844)
    Chrysallida terebellum (Philippi, 1844)
    Chrysallida indistincta (Montagu, 1808)
    Chrysallida clathrata (Jeffreys, 1848)
    Chrysallida decussata (Montagu, 1803)
    Chrysallida eximia (Jeffreys, 1849)
Genus *Ividella* Dall & Bartsch, 1909
    Ividella excavata (Philippi, 1836)
Genus *Partulida* Schaufuss, 1869
    Partulida spiralis (Montagu, 1803)

* The precise relationship of this family to the subclass Prosobranchia is uncertain.

Genus *Tragula* Monterosato, 1884
    Tragula fenestrata (Jeffreys, 1848)
Genus *Evalea* A. Adams, 1860
    Evalea divisa (J. Adams, 1797)
    Evalea diaphana (Jeffreys, 1848)
    Evalea obliqua (Alder, 1844)
    Evalea warreni (Thompson, 1845)
Genus *Liostomia* G. O. Sars, 1878
    Liostomia clavula (Lovén, 1846)
    Liostomia oblongula (Marshall, 1895)
Genus *Jordaniella* Chaster, 1898
    Jordaniella nivosa (Montagu, 1803)
    Jordaniella truncatula (Jeffreys, 1850)
Genus *Brachystomia* Monterosato, 1884
    Brachystomia rissoides (Hanley, 1844)
    Brachystomia eulimoides (Hanley, 1844)
    Brachystomia albella (Lovén, 1846)
    Brachystomia lukisi (Jeffreys, 1858)
Genus *Odostomia* Fleming, 1813
    Odostomia plicata (Montagu, 1803)
    Odostomia turrita Hanley, 1844
    Odostomia conoidea (Brocchi, 1814)
    Odostomia acuta Jeffreys, 1848
    Odostomia unidentata (Montagu, 1803)
    Odostomia conspicua Alder, 1850
    Odostomia umbilicaris (Malm, 1863)
Genus *Noemiamea* Hoyle, 1886
    Noemiamea dolioliformis (Jeffreys, 1848)
Genus *Eulimella* Gray, 1847
    Eulimella scillae (Scacchi, 1835)
    Eulimella laevis (Brown, 1827)
    Eulimella ventricosa (Forbes, 1843)
Genus *Ebala* Gray, 1847
    Ebala nitidissima (Montagu, 1803)
Genus *Turbonilla* Risso, 1826
    Turbonilla lactea (Linné, 1758)
    Turbonilla acuta (Donovan, 1804)
    Turbonilla pusilla (Philippi, 1844)
    Turbonilla crenata (Brown, 1827)
    Turbonilla fulvocincta (Thompson, 1840)
    Turbonilla jeffreysi Forbes & Hanley, 1850
    Turbonilla rufescens (Forbes, 1846)

Systematic Part

KEY A

Key to British species of brackish water, freshwater, and terrestrial prosobranchs

1. Animal terrestrial or apparently so (e.g. living out of water at top of beaches) **2**

   Animal aquatic in fresh or brackish water, though may be exposed at low tides **8**

2. Animal living on calcareous soils inland **3**

   Animal living in muddy places or in rock crevices at top of marine beaches **4**

3. Shell about 2 mm high, nearly columnar, apex blunt; whorls nearly flat, last = 50% of shell height or less; spiral ridge along each suture; narrow grooves across whorls; aperture oval; operculum thin .... *Acicula fusca* (p. 184)

   Shell over 2 mm high, conical; whorls tumid, last = about 75% shell height; many spiral ridges; aperture circular; operculum thick, calcified externally .... *Pomatias elegans* (p. 182)

4. Shell columnar, apex truncated; whorls flattened peripherally, usually with narrow costae and labial varix; operculum narrow; snout long, eyes behind tentacles at its base ....
   .... *Truncatella subcylindrica* (p. 200)

   Shell conical, apex pointed, sharp or blunt **5**

5. Whorls tumid with thin costae, except 2 topmost which are smooth; last whorl = 40% of shell height; aperture narrow oval, labial varix absent; snout long, eyes behind tentacles at base ....
   .... *Truncatella subcylindrica* juvenile (p. 200)

   Ornament of growth lines only; last whorl = about 70% of shell height; aperture broad **6**

6. Shell not over 2 mm high, apex blunt, whorls swollen; umbilicus distinct; southern England only .... *Paludinella littorina* (p. 266)

   Shell usually over 2 mm high, apex sharp; whorls little swollen; umbilicus absent **7**

7. Shell thick, spire straight-sided or coeloconoid; periostracal flap along outer lip; dark with paler parts, columella and throat dark; eyes at tentacle bases; in rock crevices very high on shore .... *Littorina neritoides* (p. 168)

   Shell rather thin, spire cyrtoconoid; no periostracal flap; light brown, without pattern, columella and throat pale; eyes at tentacle tips; groove on each side of head to sole of foot; on salt marshes in East Anglia .... *Assiminea grayana* (p. 264)

8. Spire low, apex blunt; last whorl = 95% of shell height; ornament of growth lines; aperture expanded, half blocked by broad columella; operculum calcified internally ... *Theodoxus fluviatilis* (p. 144)

   Shell not like this, columella not broad **9**

9. Shell disk-like or with very low spire; umbilicus very wide; ctenidium bipectinate; bases of tentacles linked by fold; pallial tentacle on right; operculum multispiral (*Valvata*) **10**

   Shell with elevated spire; operculum multispiral, pavcispiral or concentric **11**

10. Shell a biconcave disk, all whorls visible on both sides *V. cristata* (p. 152)

   Shell with low spire .... *V. macrostoma* (p. 152)

11. In fresh water **12**

   In or alongside brackish or sea water **18**

12. Shell with 4–5 swollen whorls; aperture nearly circular; operculum multispiral; ctenidium bipectinate; cephalic tentacles linked at base .... *Valvata piscinalis* (p. 150)

   Ctenidium not bipectinate; tentacles not linked; operculum paucispiral or concentric **13**

13. Shell large with brown spiral bands; operculum concentric; last whorl = 80% of shell height; right tentacle of male acts as penis (*Viviparus*) **14**

   Shell less than 10 mm high without spiral bands of colour; operculum paucispiral or concentric; last whorl not over 75% shell height; penis of male in mantle cavity **15**

14. Spire blunt; whorls moderately tumid; aperture elongated, narrow apically; umbilicus small or absent; brownish .... *Viviparus viviparus* (p. 146)

   Spire with sharp tip; whorls very tumid; aperture round, not narrowed apically; umbilicus clear; greenish .... *Viviparus contectus* (p. 148)

15. Operculum concentric with groove from centre along one radius,
    calcified internally; penis with double tip (*Bithynia*) **16**

    Operculum paucispiral **17**

16. Shell with 5–6 whorls, sutures shallow; aperture angulated apically;
    umbilicus minute or absent .... *Bithynia tentaculata* (p. 202)

    Shell with 4–5 whorls, sutures deep; aperture rounded apically;
    umbilicus rather large .... *Bithynia leachi* (p. 204)

17. Spire high, 6 moderately tumid whorls sometimes with spiral keel
    or tufts of bristles, otherwise smooth; umbilicus usually blocked;
    broadest part of aperture close to its base; snout evenly dark
    dorsally with pale tip .... *Potamopyrgus jenkinsi* (p. 194)

    Spire low, apex blunt, 4 tumid whorls never with keel or bristles;
    umbilicus marked; aperture rounded; only in canals near
    Manchester .... *Marstoniopsis scholtzi* (p. 198)

18. Shell a broad cone, its breadth 70% of its height, apex sharp;
    umbilicus absent; eyes at tip of tentacles; groove down each side
    of head to sole of foot; in salt marshes in East Anglia ....
    .... *Assiminea grayana* (p. 264)

    Shell breadth 60% of height; umbilicus present or absent; eyes
    at base of tentacles; without grooves on sides of head **19**

19. Shell broad, breadth about two thirds height; last whorl = 70%
    shell height; umbilicus distinct; orange patch behind eye; only
    in upper tidal reaches of rivers between Sussex and Suffolk and
    in south Ireland .... *Pseudamnicola confusa* (p. 196)

    Shell breadth not over half height; last whorl less than 70% shell
    height; umbilicus at most a chink; without orange patch behind
    each eye **20**

20. Shell solid, 6–7 whorls, nearly flat-sided, last slightly keeled;
    aperture angulated apically; snout typically with pale tip behind
    which is a dark transverse bar from which dark lateral bands
    run back (Fig. 68); tentacles with dark band near tip; penis with
    blunt tip, its right edge smoothly curved .... *Hydrobia ulvae* (p. 188)

    Shell more delicate, whorls tumid **21**

21. Shell of 5–6 markedly tumid whorls; apex pointed; aperture oval,
    little or not angulated apically, its greatest breadth about the
    middle of its height; umbilicus a chink; snout without dark
    transverse bar anteriorly (Fig. 68); tentacles often with dusky
    longitudinal line near tip but no transverse bar; penis with
    flagellar tip .... *Hydrobia ventrosa* (p. 190)

    Shell and animal not like this **22**

22. Shell with 6 distinctly tumid whorls, sometimes with spiral keel or row of bristles, sometimes smooth; spire nearly straight-sided; umbilicus absent or a chink; aperture with its greatest breadth near base; snout with pale tip and uniformly dark behind that (Fig. 68); tentacles dark basally with pale central line, pale distally; males usually absent .... *Potamopyrgus jenkinsi* (p. 194)

Shell with 4–6 slightly tumid whorls, always smooth; spire narrow; umbilicus a chink; snout with dark transverse bar behind pale tip as in *Hydrobia ulvae*; tentacles with black conical mark near tip; penis with blunt tip, its right edge with hump-like projection .... *Hydrobia neglecta* (p. 192)

KEY B

Key to British families of marine prosobranchs

1. Shell internal, covered permanently by mantle; animal resembles a dorid but has smooth tentacles, a ctenidium in the mantle cavity and males have an external penis .... Lamellariidae (p. 314)

   Shell exposed, though it may be temporarily covered by mantle lobes which are withdrawn when touched **2**

2. Shell small, a short curved tube, perhaps with a small spiral coil at one end .... Caecidae (p. 288)

   Shell not like this **3**

3. Shell ear-shaped with a row of holes on the last whorl ....
   .... Haliotidae (p. 62)

   Shell not like this **4**

4. Shell conical (spire absent), cap-shaped, or kidney-shaped; aperture width equal to greatest diameter of shell or nearly so **5**

   Shell with raised spire of several whorls **6**

5. Shell with apical hole, or with slit at margin .... Fissurellidae (p. 64)

   Shell with neither apical hole nor marginal slit .... see Key C (p. 49)

6. Outer lip with marginal slit at its periphery; slitband along whorls of spire .... Scissurellidae (p. 60)

   Outer lip without such a slit; without slitband **7**

7. Outer lip forms a winglike expansion with 5 points ....
   .... Aporrhaidae (p. 298)

   Outer lip not like this; aperture width less than greatest diameter of shell **8**

8. Aperture edge entire basally .... see Key D (p. 50)

   Aperture edge with basal notch or canal .... see Key E (p. 56)

KEY C

Key to limpet families of British marine prosobranchs
(see Key B p. 48)

1. Shell more or less conical; aperture equals greatest diameter of shell   **2**
   Shell cap-shaped, perhaps with backwardly projecting apex **8**

2. Shell with large tongue-shaped internal septum; apex central, sometimes with minute spiral coil; neck of animal with lateral extensions; ctenidium present; animals solitary .... Calyptraeidae (p. 308)
   Shell without large internal septum; neck without lateral expansions;
   with or without ctenidium **3**

3. Shell with strong radial ridges (unless eroded); ctenidium absent,
   but pallial gills on mantle edge (Fig. 20) .... Patellidae (p. 76)
   Shell without strong radial ridges; with or without ctenidium **4**

4. Shell smooth or nearly so, with bright blue lines sometimes reaching
   to margin; ctenidium absent, but pallial gills on mantle edge except
   anteriorly; on weeds .... Patellidae (p. 76)
   Shell not like this; without pallial gills **5**

5. Shell with brown or pink rays, nearly or quite smooth; ctenidium
   present .... Acmaeidae (p. 72)
   Shell white or tawny; ctenidium absent **6**

6. Shell tawny, sculpture very slight; apex near anterior end Lepetidae (p. 84)
   Shell grey-white, apex central or nearly so **7**

7. Shell profile nearly regularly conical; apertural edge flat or concave
   in profile .... Lepetidae (p. 84)
   Shell apex central on upraised papilla, profile convex; apertural edge
   convex in profile, raised anteriorly and posteriorly; rare ....
   .... Lepetellidae (p. 90)

8. Shell cap-shaped, extremely delicate, with keel along major axis;
   animal cylindrical, transparent, not enclosed by shell, pelagic;
   foot a fin opposite visceral mass .... Carinariidae (p. 348)
   Shell solid; animal not pelagic **9**

9. Shell cap-shaped with initial spiral coil; periostracum thick, fringed
   at edges; animal with grooved proboscis; shell without internal
   septum .... Capulidae (p. 306)
   Shell kidney-shaped, with large D-shaped internal septum; neck of
   animal with lateral expansions; animals usually living in chains
   or stacks .... Calyptraeidae (p. 308)

KEY D

Key to asiphonate families of British marine prosobranchs
(see Key B p. 48)

1. Shell heterostrophic, protoconch either lying across apex or tucked
   upside down into it (Fig. 233) .... Pyramidellidae (p. 544)
   Shell not heterostrophic **2**

2. Shell sinistral .... Skeneopsidae (p. 278)
   Shell dextral **3**

3. Shell nacreous (look within aperture); operculum with many turns
   and central nucleus (Fig. 6C); animal with epipodial tentacles,
   neck lobes and eye stalks fused to tentacles (Fig. 9) **4**
   Shell and animal not like this **5**

4. Shell ornament mainly or entirely spiral (any axial ornament subsutural only); no complete peristome, outer lip markedly prosocline; operculum horny; animal with cephalic lappets (Fig. 9);
   posterodorsal surface of foot transversely ridged ..... Trochidae (p. 91)
   Ornament mainly axial; complete peristome present; outer lip only
   slightly prosocline; no cephalic lappets .... Turbinidae (p. 140)

5. Shell not over 1.5 mm high, with distinct spire, white or colourless;
   ornament none or of some spiral ridges; last whorl occupies about
   90% of shell height; umbilicus prominent; operculum with many
   turns; epipodial tentacles and neck lobes present ... Skeneidae (p. 130)
   Shell and animal not like this **6**

6. Last whorl occupies less than 75% of shell height **7**
   Last whorl occupies 75% or more of shell height **30**

7. Shell without ornament (apart from growth lines) **8**
   Shell with obvious ornament **24**

8. Breadth of shell less than half its height **9**
   Breadth of shell greater than half its height **13**

9. Peristome flares out round large aperture; throat often constricted;
   labial varix present (Fig. 83) .... Rissoidae (p. 210)
   Shell not like this **10**

10. Spire tall, regularly conical, tapering to a fine point **11**
    Spire either clearly cyrtoconoid or with nearly parallel sides; apex
    blunt **12**

11. Whorls tumid, sutures deep .... Aclididae (p. 502)
    Whorls flat, sutures very shallow; shell may be bent .. Eulimidae (p. 516)

12. Last whorl occupies 60% of shell height, aperture half ....
.... Iravadiidae (p. 206)
    Last whorl not more than half shell height, breadth not over 40%
.... Aclididae (p. 502)

13. Shell breadth 75% or more of shell height **14**
    Shell breadth less than 75% of shell height **18**

14. Shell as broad as high; from top of beach or salt marsh ....
.... Assimineidae (p. 264)
    Shell breadth less than its height; from lower half of beach or sub-
littoral **15**

15. Shell milk white, iridescent; opercular nucleus at mid-columellar
edge; snout deeply bifid; dark mark showing through last whorl
if alive .... Rissoellidae (p. 270)
    Shell not like this; snout not deeply bifid **16**

16. Whorls flat; 2 brown spiral bands on last whorl; operculum red,
its growth lines nearly concentric; no metapodial tentacle on
foot .... Barleeidae (p. 262)
    Whorls tumid; operculum pale with spiral growth lines; metapodial
tentacles present **17**

17. Whorls moderately tumid; 3 whorls; 3 brown spiral bands on last
whorl; outer lip prosocline .... Cingulopsidae (p. 268)
    Whorls very tumid; 6 whorls; transverse brown lines on last whorl,
plus, sometimes, 1 basal spiral one; outer lip orthocline ....
.... Rissoidae (p. 210)

18. Breadth and aperture height both equal half shell height; apical
whorls form narrow style; on echinoids .... Stiliferidae (p. 542)
    Shell not like this **19**

19. Shell milk white, iridescent; opercular nucleus at mid-columellar
level; snout deeply bifid; dark mark showing through last whorl
of live shells .... Rissoellidae (p. 270)
    Shell and animal not like this **20**

20. Whorls tumid; umbilicus present or absent; no labial varix **21**
    Whorls flat or nearly so; umbilicus usually absent; with or without
labial varix **22**

21. Spire short; umbilicus present .... Rissoidae (p. 210)
    Spire long; umbilicus absent .... Iravadiidae (p. 206)

22. Shell with 2 brown spiral bands on last whorl; operculum red with
nearly concentric growth lines; no metapodial tentacles on foot
.... Barleeidae (p. 262)
    Shell and animal not like this **23**

23. Labial varix present; apex violet and usually brown streaks on basal whorls; eyes at base of tentacles; on lower parts of beach ....
.... Rissoidae (p. 210)

Varix absent; apex not violet and usually 1 reddish spiral band on last whorl; eyes at tip of tentacles; a groove from mantle cavity to sole on anterior part of foot; high on beach or in salt marshes .... Assimineidae (p. 264)

24. Ornament (not necessarily present on every whorl) of spiral ridges only **25**

Ornament (not necessarily present on every whorl) of costae, or of costae and spirals **27**

25. Last whorl occupies more than half of shell height; spiral ridges strap-shaped .... Rissoidae (p. 210)

Last whorl occupies less than half shell height; spiral ridges keeled **26**

26. Spiral ridges mainly restricted to peripheral regions of whorls; umbilicus present; operculum spiral without marginal bristles; mantle edge without tentacles .... Aclididae (p. 502)

Spiral ridges not so restricted; umbilicus absent; operculum concentric with marginal bristles; mantle edge with pinnate tentacles
.... Turritellidae (p. 292)

27. Ornament of costae only, or of costae with one spiral ridge on base of last whorl **28**

Ornament of costae and spiral ridges in approximately equal numbers but one set may be more prominent than the other or occur on only part of shell or whorl **29**

28. 10 or more swollen whorls with well spaced laminar costae; aperture circular with thick peristome and no labial varix ....
.... Epitoniidae (p. 484)

Up to 9 whorls, commonly fewer, flat or swollen; costae undulate in section; aperture oval with labial varix .... Rissoidae (p. 210)

29. Shell breadth equals one quarter to one fifth of height; apex bulbous; whorls with many delicate flexuous costae with fine spiral ridges in their interstices .... Aclididae (p. 502)

Shell breadth 40% of height or more; apex pointed .... Rissoidae (p. 210)

30. Shell without ornament visible to the naked eye apart from growth lines **31**

Shell with clear ornament in addition to growth lines **52**

31. Shell breadth equals or exceeds shell height **32**

Shell breadth less than shell height **42**

32. Shell internal, delicate; animal looks like a dorid but has smooth tentacles, ctenidium and external penis .... Lamellariidae (p. 314)

Shell external **33**

33. Shell of 2–5 weakly calcified whorls (flexible); shape obscured by thick periostracum; no umbilicus; no operculum .... Lamellariidae (p. 314)

Shell not like this **34**

34. Shell of 2–3 calcified but transparent whorls covered with thin periostracum; spire short; aperture very large, oval; northern .... Lamellariidae (p. 314)

Shell not like this **35**

35. Shell discoidal, no spire visible; minute .... Omalogyridae (p. 274)

Shell not like this **36**

36. Shell without umbilicus **37**

Umbilicus obvious **39**

37. Shell fragile, pale purple; aperture circular; animal with bifid tentacles; pelagic, but may be cast ashore .... Janthinidae (p. 496)

Shell robust, animal benthic **38**

38. Shell with thick periostracum, usually with prosocline ridges; animal with 2 metapodial tentacles .... Lacunidae (p. 154)

Shell without thick periostracum; animal without metapodial tentacles .... Littorinidae (p. 162)

39. Shell nearly discoidal; umbilicus very large .... Skeneopsidae (p. 278)

Shell globose **40**

40. Aperture occupies about half shell height; columella rather broad; foot without anterolateral grooves; high up beach on rocky shores .... Assimineidae (p. 264)

Aperture more than half shell height **41**

41. Aperture round; columella thin; operculum with nucleus at midcolumellar level and with internal struts; animals live at L.W.S.T. on rocky shores .... Rissoellidae (p. 270)

Aperture D-shaped; operculum with nucleus near base, without internal struts; foot of active animal with large propodium and metapodium covering most of shell; animals live at L.W.S.T. or lower on sandy-muddy shores .... Naticidae (p. 332)

42. Shell with obvious umbilicus **43**

Shell without umbilicus **47**

43. Aperture less than half shell height; 4 whorls; apex blunt; 2–4 rows of brown marks on last whorl .... Rissoidae (p. 210)

Aperture more than half shell height; colour pattern different **44**

44. Columella grooved, leading to umbilicus; animal with metapodial tentacles .... Lacunidae (p. 154)

Columella not grooved **45**

45. Aperture occupies 70–80% of shell height; animals live on soft bottoms .... Naticidae (p. 332)

Aperture occupies not more than 60% of shell height **46**

46. Shell robust with thick, transversely ridged periostracum; whorls with subsutural flattening and deep sutures; foot with 2 metapodial tentacles; animals usually sublittoral on soft bottoms .... Lacunidae (p. 154)

Shell delicate, nearly transparent, periostracum not obvious; foot without metapodial tentacles .... Rissoellidae (p. 270)

47. Shell flexible, covered with thick periostracum hiding its shape; operculum absent .... Lamellariidae (p. 314)

Shell solid, with or without obvious periostracum **48**

48. Apical whorls form a narrow style; shell breadth and aperture height both equal half shell height; lives on echinoids .... Stiliferidae (p. 542)

Shell not like this; animals free-living **49**

49. Shell purple, fragile; animal with bifid tentacles; animals pelagic with float, but may be cast ashore .... Janthinidae (p. 496)

Shell solid, not purple; animal benthic **50**

50. Shell with red-brown colour pattern; operculum calcareous, hemispherical .... Tricoliidae (p. 142)

Shell not like this **51**

51. Shell without obvious periostracum; operculum flat with outer white calcareous layer; without metapodial tentacles .... Naticidae (p. 332)

Shell with thick periostracum; operculum horny; 2 metapodial tentacles .... Lacunidae (p. 154)

52. Shell without raised spire, last whorl occupying the total height **53**

Shell with distinct raised spire, even if low **54**

53. Shell discoidal, minute, of 3–4 whorls crossed by costae; periostracum inconspicuous .... Omalogyridae (p. 274)

Shell flexible, globose, of 2–3 whorls with spiral ridges; periostracum thick .... Lamellariidae (p. 314)

54. Shell with thick bristly periostracum folded along growth lines; fine spiral ridges present; animal with dorsally grooved proboscis .... Trichotropidae (p. 302)

Shell without obvious periostracum; animal without such a proboscis **55**

55. Shell purple, fragile, ornament of V-shaped transverse ridges (apex pointing up spire); aperture with notched outer lip; animal pelagic but may be cast ashore .... Janthinidae (p. 496)

Shell not like this **56**

56. Ornament of spiral ridges and grooves (and growth lines) only **57**

Ornament of spiral ridges and costae **58**

57. Shell nearly discoidal, apex blunt; right pallial tentacle bifid ....

.... Tornidae (p. 280)

Shell globose-conical, apex sharp; pallial tentacle absent ....

.... Littorinidae (p. 162)

58. Shell nearly discoidal, apex blunt; right pallial tentacle bifid ....

.... Tornidae (p. 280)

Shell tall; right pallial tentacle (if present) single .... Rissoidae (p. 210)

KEY E

Key to siphonate families of British marine prosobranchs
(see Key B p. 48)

1. Spire hidden, shell a cowrie or similar **2**

   Spire visible **3**

2. Shell a cowrie, aperture not extended into long apical and basal
   canals .... Eratoidae (p. 324)

   Aperture extended into long apical and basal canals ....
   .... Ovulidae (p. 330)

3. Shell sinistral (apex upwards – aperture to left) **4**

   Shell dextral (apex upwards – aperture to right) **5**

4. Shell brown, with tuberculated surface; siphonal canal a notch;
   spire cyrtoconoid .... Triphoridae (p. 479)

   Shell white, with spiral ridges; siphonal canal moderately long,
   bent to right; spire straight-sided .... Cerithiopsidae (p. 466)

5. Outer lip forms a wing-like process with points ..... Aporrhaidae (p. 298)

   Outer lip not like this **6**

6. Shell a tall, slender cone, last whorl occupying not more than half
   shell height, aperture not more than a third **7**

   Shell with different proportions **9**

7. Shell without ornament other than growth lines; base of aperture
   spoon-shaped .... Aclididae (p. 502)

   Shell with spiral ridges and costae which may form close-set
   squarish tubercles **8**

8. Penult whorl with 4 rows of tubercles; base of last whorl with many
   spirals; a sinus at base of aperture .... Cerithiidae (p. 294)

   Penult whorl with 3–4 rows of tubercles or with costae; a canal
   at base of aperture .... Cerithiopsidae (p. 466)

9. Last whorl occupies more than 80% of shell height **10**

   Last whorl not so large **12**

10. Shell with distinct spiral ornament .... Muricidae (p. 358)

    Shell smooth or with only microscopic spiral lines **11**

11. Aperture long and narrow; spire very short .... Eratoidae (p. 324)

    Aperture oval; spire moderately long .... Buccinidae (p. 378)

12. Periostracum thick with spiral rows of bristles; 4 large spiral ridges
    on last whorl (smaller ones also present); aperture triangular
    to pear-shaped; animal with dorsally grooved proboscis ....
    .... Trichotropidae (p. 302)

    Shell and animal not like this **13**

13. Shell with growth lines but neither spiral ridges nor costae ....
.... Buccinidae (p. 378)

Shell with growth lines and other ornament **14**

14. Shell with spiral ornament (even if very fine) but without costae
or tubercles **15**

Shell with costae or tubercles, with or without spiral ornament **22**

15. Spiral ridges keel- or cord-like .... Turridae (p. 420)

Spiral rides thread-like or very fine, numerous **16**

16. Shell breadth greater than half its height **17**

Shell breadth equal to or less than half its height **18**

17. Columellar lip projects freely over umbilical groove as a deep
flange; umbilicus large; teeth present within outer lip ....
.... Cassidae (p. 350)

Large flange and teeth absent .... Buccinidae (p. 378)

18. Apical whorls nearly styliform, perhaps with slightly bulbous tip;
siphonal fasciole absent .... Buccinidae (p. 378)

Apical whorls not styliform; with or without a fasciole **19**

19. Shell with siphonal fasciole; whorls slightly angulated; lip a little
thickened .... Buccinidae (p. 378)

Shell without siphonal fasciole; periostracum pronounced **20**

20. 3–4 more prominent spiral ridges at periphery of each angulated
whorl; siphonal canal rather long, widely open ....
.... Fasciolariidae (p. 416)

All spiral ridges about equal; whorls not angulated; siphonal canal
of varying length but always open **21**

21. Shell thin, semitransparent; spiral ridges beaded; protoconch
white; siphonal canal short, animal without operculum ....
.... Turridae (p. 420)

Shell thick, with obvious periostracum; spiral ridges not beaded;
siphonal canal may be long or short; animal with operculum
.... Buccinidae (p. 378)

22. Spiral ornament absent, costae prominent .... Muricidae (p. 358)

Spiral ornament and costae present, but one or the other may
be fine **23**

23. Shell breadth less than half its height **24**

Shell breadth equal to or more than half its height **27**

24. Siphonal canal slender, about as long as the aperture **25**

Siphonal canal short **26**

58

36. Costae curved; last whorl very broad; end of canal not forming
a notch in abapertural view; foot without metapodial tentacles
.... Buccinidae (p. 378)

Costae straight; end of canal forms conspicuous notch in abapertural view; 2 metapodial tentacles at posterior end of foot ....
.... Nassariidae (p. 408)

Family SCISSURELLIDAE Gray, 1847

Genus SCISSURELLA Orbigny, 1824

Scissurella crispata Fleming, 1828
(Fig. 13)

Diagnostic characters
A spirally wound shell with a slit at the mid-point of the outer lip continued up the spire as a closed slit-band with raised edges.

Other characters
The shell has 4–5 rather swollen whorls, the spire moderately high, decorated with many low prosocline costellae crossing the whorls and with still finer spiral ridges running from costa to costa. There is a wide umbilicus. The aperture is oval, the lip a little out-turned basally. White. About 1 mm high, 2 mm broad.

The animal has a short, broad snout. The head bears two setose tentacles, each with an eye at the base. A bay in the mantle skirt underlies the slit in the shell, providing an exhalant opening from the mantle cavity; its edge carries five tentacles, some of which project through the slit. The foot is truncated anteriorly, rounded behind. On each side it bears seven epipodial tentacles, two behind each cephalic tentacle, three grouped at the middle of the foot, two posteriorly by the operculum. All, except the anterior pair, are setose. Animal white, eyes black.

S. crispata occurs at depths of 8–2000 m off the south and west coasts of England and Ireland, the west and north coasts of Scotland on gravelly or shelly bottoms with admixed mud. It is circumpolar in general distribution, occurring in deeper water in lower latitudes. Its feeding and breeding are not known, but it is probably a grazer or detritivore with external fertilization.

Reference. Fretter & Graham (1962).

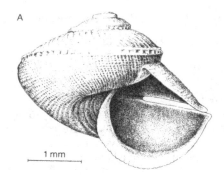

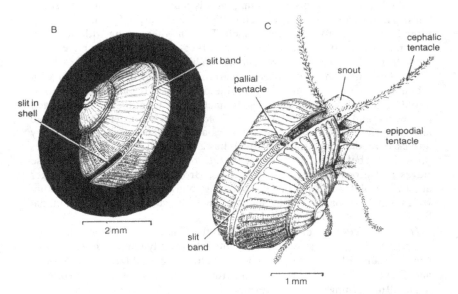

Fig. 13. *Scissurella crispata*. A & B, two views of the shell (in A the upper part of the outer lip is straighter than it commonly is); C, living animal.

Family HALIOTIDAE Rafinesque, 1815

Genus HALIOTIS Linné, 1758

Haliotis tuberculata Linné, 1758
(Fig. 14)

Diagnostic characters
Shell ear-shaped with very low spire; 5–7 small openings with out-turned lips lie in a spiral along the last whorl and are continued up the spire as a series of closed tubercles. Aperture very large; inside of shell with bright mother-of-pearl sheen. Foot without an operculum.

Other characters
The shell has 3–4 rapidly expanding whorls marked with many fine spiral lines. The aperture is oval, the outer lip with a sharp edge, the inner lip turned inwards to form a narrow shelf. Pinkish when young with light and dark mottling or bars; older shells are darker, reddish brown or greenish. Up to 90 mm long, 65 mm broad, and 20 mm high.

The head is small and flattened, carrying two long tentacles each with a bluish eye on a stalk on the outer side of its base. Cephalic lappets form a fold across the head dorsal to the tentacles. The mantle edge is thickened and warty and slit under the row of holes in the shell. The foot is large and powerful, with an oval sole; its sides are papillated ventrally and carry a thick epipodial ridge dorsally which has a scalloped edge. Many tentacles project from between the lobes. The posterior end of the foot bears transverse ridges and grooves on its dorsal surface. Brown or greenish, with darker and lighter blotches, sometimes alternating to give stripes on the epipodium which match the pattern of the shell and the substratum on which the animal lives.

H. tuberculata lives on rocky shores in the Channel Islands and Brittany, but not the British or Irish mainland, though recently imported to the latter commercially. Elsewhere ormers occur south to the Mediterranean, Azores, and Canary Islands. They are found from extreme low water to depths of about 40 m, feeding on encrusting algae.

The eggs escape through the holes on the shell, are fertilized externally, and give rise to trochophore larvae. After a very short free-swimming life these settle with a shell 2 mm long.

Vernacular name: the ormer.

References. Stephenson (1924); Crofts (1929, 1937); Forster (1962); Hayashi (1980a,b).

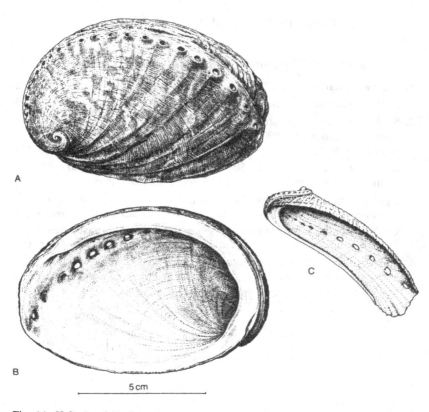

Fig. 14. *Haliotis tuberculata.* A, apical view; B, internal view; C, shell posed so as to show relation with typical shells.

Family FISSURELLIDAE Fleming, 1822

Key to British genera and species of Fissurellidae

1. Shell with marginal slit **2**

   Shell with apical slit or hole **4**

2. Shell with fine reticulated ornament; about 50 radiating ridges; breadth about two thirds length, height usually not more than half length .... *Emarginula crassa* (p. 66)

   Shell with coarse ornament; 30–40 radiating ridges; breadth about three quarters length, height usually more than half length **3**

3. Beak just posterior to shell centre, pointing away from base; shell grey-white .... *Emarginula fissura* (p. 65)

   Beak overhangs posterior shell margin, pointing towards base; shell often pinkish .... *Emarginula conica* (p. 66)

4. Hole slit-like, with underlying septum, anterior to apex, which is central with recurved beak .... *Puncturella noachina* (p. 68)

   Hole oval, without septum; apex in anterior half of shell ....
   .... *Diodora graeca* (p. 70)

Genus EMARGINULA Lamarck, 1801

Vernacular name: slit limpets

Emarginula fissura Linné, 1758
(Fig. 15A)

Patella fissura Linné, 1758

Emarginula reticulata Sowerby, 1813

Diagnostic characters
Shell conical with anterior margin of aperture marked by a slit continued to apex as groove. Apex curving posteriorly to end about three quarters of total length behind the anterior end, nearly level with highest point of shell. Surface reticulated by ridges. White. Animal with about twenty epipodial tentacles.

Other characters
A minute spiral coil of two whorls lies on the right posterior side of the apex. There are 25–35 radial ridges crossed by 20–30 others parallel to the edge of the aperture or to its previous positions. These are narrower than the radial ridges and are clearest between them. There are small knobs where the two sets cross. The aperture is oval, its edge made sinuous by the ends of the ridges. Up to 10 mm long, 8 mm broad, 6 mm high.

The animal has a broad snout and two rather thick tentacles, each with an eye on a short stalk on its outer side; the right cephalic tentacle has an epipodial tentacle immediately behind it. The mantle skirt is split under the slit in the shell and extends a little through it to form an exhalant siphon. The sole of the foot is oval; there is no operculum; the median epipodial tentacle posteriorly is longer than the others. Flesh white.

E. fissura is a sponge eater and is not uncommon on rocky coasts to about 150 m sublittorally and at L.W.S.T., usually on the underside of large stones supporting sponge growths and with a pitted surface in which the limpets rest. They avoid brackish water. They occur on all suitable shores in the British Isles except the east Channel and the southern North Sea. The species is found on all European Atlantic coasts.

Emarginula conica Lamarck, 1801
(Fig. 15B)

Emarginula rosea Bell, 1824

Diagnostic characters

The shell apex lies over or nearly over the posterior edge and below the highest point of the shell. Shell often pinkish. Animal with 24–28 epipodial tentacles.

Other characters

The shell is like that of the previous species but smaller and glossier. There are 35–40 radial ridges and about twenty parallel to the apertural edge. The open slit reaches about one fifth of the distance to the apex. Up to 6 mm long, 4 mm broad, 4 mm high.

   E. conica has been recorded from the Channel Islands, the Fleet (Dorset), south west England, Northumberland, and Arran. It also occurs on the west coast of Ireland and from the north coast of France south to the Mediterranean. It lives sublittorally, usually at depths down to 30 m in the same kind of habitat as *E. fissura* or in the interstitial spaces of gravel bottoms, though rare specimens may be found at L.W.S.T.

Emarginula crassa Sowerby, 1813
(Fig. 15C)

Diagnostic characters

Shell in general like that of other *Emarginula* spp. but the ornament is much finer so that it seems rather smooth. White. Animal with 30 epipodial tentacles.

Other characters

Most radial ridges are composed of triplets of smaller ridges with the openings of pits in the intervening grooves appearing as white dots. Up to 30 mm long, 20 mm broad, 15 mm high.

   The animal is like *E. fissura*. *E. crassa* is found under stones, rarely at L.W.S.T., more frequently sublittorally to 200 m. It has been recorded from south and west Ireland, the Clyde, and north to Shetland and Rockall. It is a northern species, commoner in Norway and Iceland.

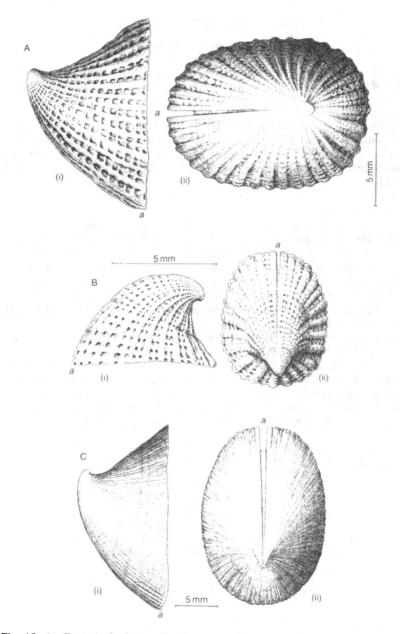

Fig. 15. A, *Emarginula fissura*; B, *E. conica*; C, *E. crassa*. (i) lateral views and (ii) viewed from above; *a* = anterior.

Genus PUNCTURELLA Lowe, 1827

Puncturella noachina (Linné, 1771)
(Fig. 16)

Patella noachina Linné, 1771

Diagnostic characters
Shell conical with small, backwardly curving apex nearly centrally placed
and with numerous radial ridges. A lanceolate opening lies on the anterior
slope near the apex, in empty shells seen to be partly blocked by a septum
which internally separates the apical part of the shell from the region of
the slit.

Other characters
The apex of the shell lies posteriorly on the right and ends in a spiral of
one whorl. There are 19–25 radial ridges; these are crossed by other ridges
lying irregularly parallel to the apertural plane and more prominent near
the oval aperture. White-cream; white spots (pits plugged with dirt) lie in
the furrows between the ridges. Up to 7 mm long, 5 mm broad, 4 mm high.

The snout is broad; at its base lie the two cephalic tentacles, each with
an eye basally on a lateral eye stalk, the tentacle on the right with an additional
bifid tentacle behind it. The foot is oval, has no operculum, and carries 20–23
epipodial tentacles, the more posterior ones longer than the others. The
body is white. The mantle skirt is slit under the apical opening.

P. noachina may be dredged, not uncommonly, on stones from hard
bottoms from 20 to about 150 m deep. It occurs from the north coast of
Ireland round Scotland to Rockall and the Northern Isles and south to Yorkshire in the North Sea. Abroad it is circumpolar in distribution, becoming
intertidal in high latitudes. It is probably a grazer or detritivore.

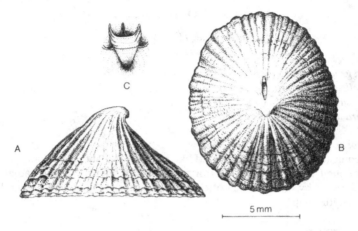

Fig. 16. *Puncturella noachina*, anterior end to left in A, above in B; C, internal view of apex, showing septum.

Genus DIODORA Gray, 1821

Diodora graeca (Linné, 1758)
(Fig. 17)

Patella graeca Linné, 1758
Fissurella graeca (Linné, 1758)
Fissurella reticulata (da Costa, 1778)
Diodora apertura (Montagu, 1803)

Diagnostic characters
Shell conical with apical hole shaped like a figure-of-eight, placed about one third of shell length behind anterior end. Surface with numerous radiating ridges crossed by others which often have an upturned edge and lie parallel to the apertural lip.

Other characters
The anteroposterior axis of the cone is longer than the transverse, and the posterior end broader than the anterior. 20–30 ridges radiate from the apex with smaller ones in furrows between the main ones. The aperture has a crenulated edge and its lip is marked internally by short ridges under the furrows of the outer surface. In side view the lip arches slightly towards the apex. White, with radiating bands of brown or green. Up to 25 mm long, 15 mm broad, 10 mm high.

The animal has a large head and snout with two tentacles at its base, each with a lateral eye stalk; that on the right has a tentacle behind the eye stalk. The mantle edge is thick and elongated to form a curtain over the sides of the foot; it bears warty papillae and others lie in the grooves at the edge of the shell. The mantle edge projects through the apical hole to form a short siphon. The foot is an elongated oval in outline, with 30–35 epipodial tentacles, alternately long and short. White, yellow, orange or red, often with darker speckling.

Keyhole limpets occur on rocky shores from L.W.S.T. to 250 m deep, usually under stones or ledges between tidemarks, on stones or shells sublittorally, feeding on sponges which they mimic in colour. They are not uncommon on most British and Irish coasts except those facing the North Sea. The species ranges from about 60° N to the Mediterranean.

The animals breed December to May (Plymouth). Eggs are passed from the genital opening to the anterior end of the mantle cavity then down the sides of the foot and are fastened to the substratum, usually the underside of a stone, partly by their own adhesive coat, partly by a secretion from glands in the foot. At some stage in this process they must be fertilized. There is no free larval stage and the young hatch as juveniles (Boutan, 1885; Lebour, 1937; Fretter & Graham, 1962).

Vernacular name: key-hole limpet.

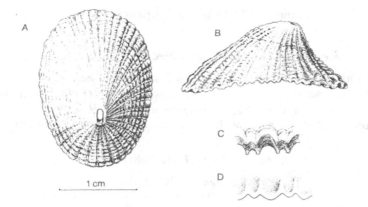

Fig. 17. *Diodora graeca*, anterior end below in A, to the right in B; C, external and D, internal views of apertural edge.

Family ACMAEIDAE Carpenter, 1857

Key to British genera and species of Acmaeidae

Shell with brown reticulate pattern; head scar brown, mantle edge
green; northern .... *Collisella tessulata* (p. 74)

Shell with pink rays; head scar unpigmented or with V-shaped red
mark, mantle edge banded with pink .... *Acmaea virginea* (p. 72)

Genus ACMAEA Eschscholtz, 1833

Acmaea virginea (Müller, 1776)
(Fig. 18)

Patella virginea Müller, 1776
Patelloida virginea (Müller, 1776)
Tectura virginea (Müller, 1776)

Diagnostic characters
Shell approximately conical without marginal slit, apical hole or internal
septum. Apex tilted forwards, about one third of shell length behind anterior
end. Surface smooth with pink lines on white-yellow background, the lines
often broken to give a chequered pattern; internally shell is white or pink
and smaller shells have a dark red V-shaped mark with the point near the
apex and the arms diverging posteriorly. Mantle edge with red bands under
those of the shell; no marginal tentacles and no pallial gills; a ctenidium
in the nuchal cavity. Foot without epipodial tentacles and operculum.

Other characters
The shell is smaller and smoother than that of *Collisella tessulata*, the beak
close to the anterior end, which is narrower than the posterior. In sublittoral
animals the apex is more central. Up to 10 mm long, 8 mm broad, 4 mm
high.

The head has a broad snout with two tentacles each with a small eye at
its base, but these are unpigmented in some animals and then do not show.
The lips round the mouth are drawn out into posterior lobes. Flesh white,
yellow or pale pink; mantle edge with glands which are, some chalk white,
others scarlet.

A. virginea may be collected at L.W.S.T. and to depths of 100 m. It is
common in and out of rock pools, attached to the underside of smooth,
fixed stones with a growth of red weeds, especially *Lithothamnion*, on which
it feeds. It occurs throughout the British Isles wherever the substratum is
suitable and is generally distributed on the Atlantic and Mediterranean coasts
of Europe. The animals breed in spring. The eggs are planktonic, giving
rise to free trochophore larvae which soon settle.

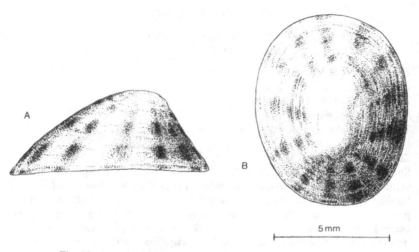

Fig. 18. *Acmaea virginea*, anterior end to right in A, below in B.

74

Genus COLLISELLA Dall, 1871

Collisella tessulata (Müller, 1776)
(Fig. 19)

Patella tessulata Müller, 1776
Acmaea tessulata (Müller, 1776)
Patelloida tessulata (Müller, 1776)
Acmaea testudinalis (Müller, 1776)
Tectura testudinalis (Müller, 1776)

Diagnostic characters
Shell approximately conical; without marginal slit, apical hole or internal septum. Apex tilted forwards, one third of shell length behind anterior end. Surface rather smooth. Irregular chocolate brown marks radiate from apex; pale sectors posterolaterally. Inner surface with brown and white margin, central area mainly chocolate. Mantle skirt green, edged with tentacles but no pallial gills; one ctenidium in nuchal cavity. Foot without operculum and epipodial tentacles.

Other characters
A minute spiral coil may be seen at the apex of shells about 1 mm long, but it is later lost. Many fine ridges radiate from the apex and are crossed by others parallel to the aperture edge. The aperture is oval, sometimes narrower anteriorly. The shell is basically cream or greenish, its radial bands of brown often bifurcating towards the edge. The central brown mark internally has usually a pale patch at the head end. Up to 20 mm long, 14 mm broad, 10 mm high; intertidal animals are commonly about half these dimensions.

The head has a broad snout with two tentacles each with a swollen base on the dorsal side of which is a small eye. The lips round the mouth are not drawn out into posterolateral processes as they are in *Acmaea virginea*. The mantle edge carries a double row of short tentacles. The foot has a large, oval sole. Cream with brown round the mouth; the red buccal mass shows by transparency.

C. tessulata is common on small stones that are firmly fixed on rocky shores and bear red weeds, on which it feeds. It occurs from L.W.S.T. (M.H.W.N.T. in pools) to depths of 50 m. In the British Isles it is found only north of a line from Dublin to Anglesey in the Irish Sea and north of Yorkshire in the North Sea; not on northern or western Irish coasts, but there is a recent report of its occurrence on the French coast south of the Channel Islands. The species is circumpolar in its distribution.

Eggs are laid in spring and early summer, fastened to a firm substratum in a mucous sheet in which they are embedded one layer deep. They hatch rapidly as trochophore larvae which metamorphose and settle after a short free life.

Vernacular name: tortoiseshell limpet.

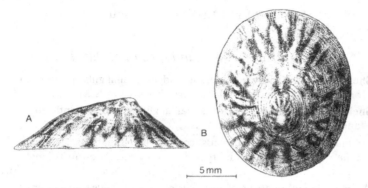

Fig. 19. *Collisella tessulata*, anterior end to right in A, below in B.

Family PATELLIDAE Rafinesque, 1815

Key to British genera and species of Patellidae

1. Shell smooth with only fine radiating ridges; pallial gills absent from anterior part of mantle edge; animals live on weeds **2**

   Shell with strong radiating ridges; pallial gills all round mantle edge; animals live on rocks **3**

2. Animals live on fronds of weeds, commonly laminarians; shell with blue rays from apex to margin; breadth about two-thirds length
   .... *Helcion pellucidum pellucidum* (p. 82)

   Animals live in holdfasts of *Laminaria*; shell has blue rays confined to apex; breadth greater than three quarters length ....
   .... *Helcion pellucidum laevis* (p. 82)

3. Posterior end of shell smoothly rounded; inner surface dull greenish brown; pallial tentacles without white pigment; sole of foot dusky; often high on shore .... *Patella vulgata* (p. 76)

   Posterior margin of shell often angulated; pallial tentacles with white pigment; never high on shore **4**

4. Sole of foot apricot colour; inner surface of shell pale with orange head scar; animals live low on beach in wet places *Patella aspera* (p. 80)

   Sole very dark; inner surface of shell with prominent coloured rays and yellow head scar; pallial tentacles opaque white; on exposed shores .... *Patella depressa* (p. 80)

Genus PATELLA Linné, 1758

Patella vulgata Linné, 1758
(Fig. 20)

Diagnostic characters
Shell conical, thick, its apex more or less central, its outer surface with radial ridges. Aperture oval, narrower anteriorly, rounded behind. Animal with pallial gills round the entire mantle edge; marginal pallial tentacles devoid of white pigment. Operculum absent.

Other characters
The radial ridges are usually single, sharply crested in young shells but often eroded in older ones. New ridges appear in the furrows towards the shell edge. There are also growth lines parallel to the edge and tubercles may occur where the two sets cross. The height of the shell varies with age and with position on the beach, those from higher levels being taller and usually with convex profiles, those from lower levels less high and straight-sided (Orton, 1929); estuarine animals also have higher shells (Nelson-Smith, 1967). The lip of the aperture is bevelled internally, crenulated by the ridges and grows to fit the particular place the animal uses as home. White or grey externally, grey-green or yellowish internally, white towards the centre; there are dark paired bands in the furrows, seen by holding it against a light and sometimes visible near the margin internally. Up to 50 mm long, 40 mm broad, 20 mm high.

The animal has a large head with two tentacles each with a small black eye dorsally on the swollen base. The marginal tentacles on the mantle skirt are of up to four lengths, representing successive families. The foot is large but has no glandular streak along its sides anteriorly as in *Helcion*. Grey-green, the cephalic tentacles and foot sole darker than the rest of the body.

This is a ubiquitous animal wherever it can find firm attachment between the upper part of the laminarian zone and M.H.W.N.T. or M.H.W.S.T., depending on exposure and shade (Bowman & Lewis, 1977). It tolerates salinities to about 25‰ (Arnold, 1957, 1972). It grazes the rocks on which it lives, making feeding excursions, often at night, and then returning to its home (Funke, 1964, 1968; Cook *et al.*, 1969; Hartnoll & Wright, 1977; Little & Stirling, 1985).

The animals are protandrous hermaphrodites (Orton, 1928; Orton, Southward & Dodd, 1956), a typical population containing 60% males, 20% females and 20% immature or spent (Choquet, 1967, 1970). Breeding occurs in autumn in the north, winter in the south (Bowman, 1981). Eggs are fertilized externally, are planktonic, and give rise to free-swimming trochophore larvae which settle with a shell length of about 200 μm, often in crevices or shallow pools from which they later migrate when able to cope with exposure (0.5–1 mm long low on the shore, 2–3 mm long at high levels). The youngest stages are identifiable by the presence of ribs running along the main anteroposterior axis and by the even pattern of their pigment stripes (Bowman, 1981).

The species has a range extending from the Mediterranean to northern Norway.

Vernacular name: common limpet.

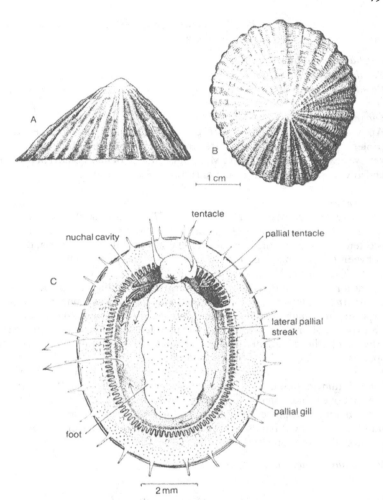

Fig. 20. *Patella vulgata*, anterior end to the right in A, below in B; C, living animal in ventral view; arrows show direction of water currents. (Fretter & Graham, 1962.)

Patella aspera Röding, 1798
(Fig. 21)

Patella aspera Lamarck, 1819
Patella athletica Bean, 1844
Patella depressa Jeffreys *non* Pennant

Diagnostic characters
Apex of shell clearly anterior to centre; ridges on outer surface often in triplets; posterior end of aperture straight; inside of shell bluish white with yellow-orange colour centrally. Marginal tentacles of mantle skirt in two series and always with some white pigment; sole of foot apricot yellow.

Other characters
Whilst resembling that of *P. vulgata* (p. 77) in general, this shell is always rather flat, the anterior end distinctly narrower than the posterior, which is often angulated by the ends of prominent posterolateral ridges and straight between their points. The tubercles on the radial ridges may be prominent. Up to 50 mm long, 40 mm broad, 20 mm high.

The animal is like *P. vulgata* but differs in its colouring – cream or yellowish. *P. aspera* lives lower on the beach than *vulgata*, prefers damp places, stands exposure to rough seas better and may be the common species of limpet on exposed shores. It is found on all suitable shores throughout the British Isles except those lying between the Thames and the Humber. It is a southern species extending to the Mediterranean.

P. aspera breeds throughout the year except in the depth of winter, with a maximum in summer. Their development is much as in *P. vulgata* with the spat overwintering in pools. They may be recognized by the lack of ridges in the main anteroposterior axis and by the presence of broad and prominent mid-lateral pigment bands which follow a straight course from centre to edge of the shell (Bowman, 1981).

Vernacular name: china limpet.

Patella depressa Pennant, 1777

Patella intermedia Jeffreys, 1865

Diagnostic characters
Shell like that of *P. aspera* (above) in shape and sculpture, usually smaller and narrower anteriorly; pigmented rays broad and conspicuous, especially internally and marginally, the central part yellow. Marginal tentacles on mantle skirt chalk white and sole of foot nearly black.

Other characters
Up to 30 mm long, 25 mm across, 12 mm high.

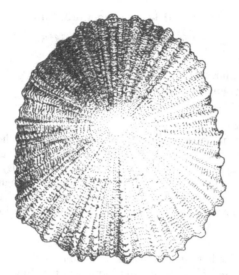

Fig. 21. *Patella aspera*, anterior end above.

This species is confined to exposed rocky shores and lives at about mid-tide level – higher than *aspera* but never as high as *vulgata* (p. 77) can survive. It is rather restricted in its distribution: the western Channel, the Irish Sea as far north as Anglesey. Outside this area it occurs along the Atlantic coasts of Europe to North Africa.

Like *P. aspera*, *P. depressa* is predominantly a summer breeder though it can breed throughout the year; unlike *P. aspera* it does not appear to have a winter break in its reproductive activity. The development is like that of *P. vulgata*, the young living in *Lithothamnion* in low pools and in wet areas after settlement until they have a shell length of 5 mm, when they migrate to their permanent home. Like *P. aspera* at this stage they have no median anteroposterior ridges on the shell, but are distinguishable from them because the lateral pigmented rays of the shell are skewed backwards on the right, forwards on the left (Bowman, 1981).

Vernacular name: black-footed limpet.

Genus HELCION Montford, 1810

Helcion pellucidum (Linné, 1758)
(Fig. 22)

Patella pellucida Linné, 1758
Patina pellucida (Linné, 1758)
Patina laevis (Linné, 1758)

Diagnostic characters

Shell cap-shaped or conical, without marginal slit, apical hole or internal septum. Outer surface smooth. Shell either translucent (var. *pellucidum*) when it has an anteriorly placed apex, or opaque (var. *laevis*) when the apex is nearly central. Apical area always with some radiating kingfisher blue lines or rays; some red ones may also occur, usually near the margin. Ctenidium absent, but the mantle edge bears pallial gills except anteriorly. Glandular streak on each side of foot anteriorly; no operculum.

Other characters

This limpet is typically found on plants of *Laminaria*. It exists in two forms: *H. p. pellucidum* and *H. p. laevis*, the former on the fronds and stipes, the latter in the holdfast under the stipe. Limpets may also occur on fucoids when adult and are regularly found on them when young, migrating to laminarians when their radula and buccal muscles have grown powerful enough to eat them (shell length about 4 mm) (Vahl, 1971). The shell of var. *pellucidum* (alive or fresh) is thin, horn-coloured, with a dark area parallel to the apertural edge and another at the apex. The lips of the aperture lie in one plane. It grows to a length of about 14 mm. In var. *laevis* the shell is nearly conical and thicker, with a rougher surface; the apertural lip does not lie in a plane; its apical area consists of a small shell with the characters of *pellucidum*. Blue rays are rare except in this area (Graham & Fretter, 1947). It measures up to 18 mm long.

The animal has a large head with two tentacles, each bearing an eye. The mantle skirt is edged with tentacles; the foot is large. Body white.

Blue-rayed limpets are found on all shores of the British Isles and on all Atlantic coasts between Portugal and northern Norway.

They breed maximally in winter and spring, the eggs being planktonic. The larvae settle in May with a shell 2 mm long, usually on fucoids (Vahl, 1971). Initially all spat are identical in appearance, like *Helcion pellucidum pellucidum*; later, when the shell is about 4 mm in length, the animals migrate to *Laminaria* plants, the tough tissues of which they have not so far been able to eat. Those juveniles which take up their position on the fronds retain the *pellucidum* characters of the shell whereas those that enter holdfasts acquire the features characteristic of *H. p. laevis*. The adult shell of the latter shows the features of *pellucidum* in its apical area developed before the migration to *Laminaria* occurred.

Vernacular name: blue-rayed limpet.

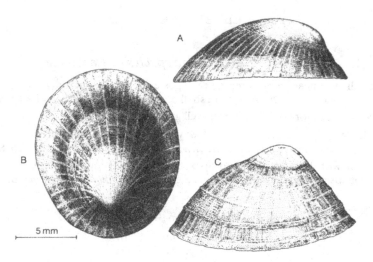

Fig. 22. *Helcion pellucidum.* A, var. *pellucidum*, anterior end to right; B, var. *pellucidum*, anterior end below; C, var. *laevis.*

Family LEPETIDAE Gray, 1850

Key to British genera and species of Lepetidae

1. Shell apex near anterior margin; the posterior profile markedly convex; without internal septum; shell orange yellow ... *Iothia fulva* (p. 86)

   Shell apex more or less central; shell greyish 2

2. Shell nearly conical, apex not tilted; yellowish grey; without internal
   septum .... *Lepeta caeca* (p. 84)

   Shell with backwardly tilted apex; grey; a small internal septum
   at apex .... *Propilidium ancyloide* (p. 88)

Genus LEPETA Gray, 1847

Lepeta caeca (Müller, 1776)
(Fig. 23)

Patella caeca Müller, 1776

Diagnostic characters
Shell conical, without marginal slit, apical hole or internal partition; apex
a little anterior to the mid-point; surface with many fine radial ridges crossing
concentric lines; yellowish grey. Animal without pallial gills or visible eyes.

Other characters
The shell is like a miniature *Patella* shell in shape but is much smoother.
The posterior sloping profile of the cone is usually convex, the lateral and
anterior profiles rather straight. The aperture is broadly oval, narrower anteriorly than posteriorly, bluntly rounded at each end, the sides nearly straight,
the margin not much affected by the ends of the ridges. The apex is usually
eroded. Up to 8–9 mm long, 6–7 mm broad, 3–4 mm high.

The animal is like a small patellid limpet. Though it lacks visible eyes
unpigmented ones are present. It is white.

L. caeca is a circumpolar species which extends southwards at increasing
depths. It is never intertidal and at the latitude of the British Isles is usually
found at depths of 6 m downwards on rocky and gravelly bottoms. It is a
detritivore. It is rare, but has been found off the west coast of Scotland,
the Northern Isles, and in the northern parts of the North Sea.

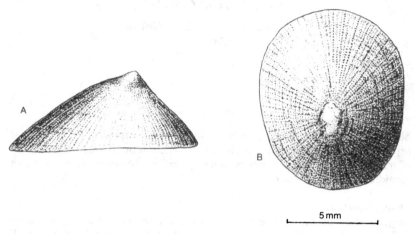

Fig. 23. *Lepeta caeca*, anterior end to right in A, below in B.

Genus IOTHIA Gray, 1850

Iothia fulva (Müller, 1776)
(Fig. 24)

Patella fulva Müller, 1776
Lepeta fulva (Müller, 1776)
Tectura fulva (Müller, 1776)

Diagnostic characters
Shell conical to cap-shaped, without marginal slit, apical hole or internal
septum; apex near anterior margin; surface finely reticulated by the crossing
of radial and concentric ridges. Orange-yellow. Animal white, without
ctenidium, pallial gills or operculum.

Other characters
In general shape the shell is like that of *Lepeta caeca* but it is less conical
and more cap-shaped, with the apex over the anterior margin, not eroded,
and often retaining a small larval coil. The posterior profile is markedly convex, the anterior a little concave. The aperture is oval, with rather straight
sides and rounded ends. Up to 7 mm long, 5 mm broad, 2.5 mm high.

   I. fulva is a northern species known from Iceland, the Faeroes, Norway
and the North Atlantic south to the Azores at increasing depths. In the British
Isles it occurs on stony grounds at depths from 5 m to about 200 m, never
intertidally. It has been found off Scilly, along the west coast of Ireland and
Scotland, off the Northern Isles and from Yorkshire northwards in the North
Sea. It is moderately common and is probably a detritivore.

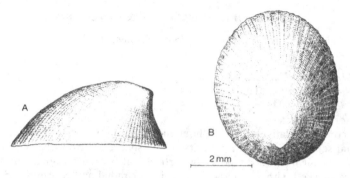

Fig. 24. *Iothia fulva*, anterior end to right in A, below in B.

Genus PROPILIDIUM Forbes & Hanley, 1849

Propilidium ancyloide (Forbes, 1840)
(Fig. 25)

Patella ancyloide Forbes, 1840
Propilidium exiguum (Thompson, 1844)

Diagnostic characters

Shell conical, relatively tall, without marginal slit or apical hole; a small internal septum separates the cavity of the shell apex from the rest. Apex forms a small backwardly-pointing vertical coil behind mid-point of shell. Surface marked with a shallow reticulation of radial and concentric ridges. White or grey-brown.

Other characters

The shell is a little transparent when fresh. Its anterior profile is convex, the lateral profiles straight, the posterior slightly S-shaped. The aperture is oval, the anterior end slightly narrower than the posterior. Shell white and shiny internally. Up to 4 mm long, 3 mm broad, 2.5 mm high.

The animal, cream in colour, has much the appearance of a small *Patella*, but lacks gills and visible eyes. The foot has an oval or shield-shaped sole, and has no operculum. The lateral margins of the dorsal lip are drawn out into short processes lying under the cephalic tentacles.

P. ancyloide is not common but has been found in most parts of the British Isles from Shetland and Rockall south to the Celtic Sea; it is absent from the Channel and from most of the North Sea. It does not occur between tidemarks but lives on hard, stony bottoms from 7 to 190 m deep. It is probably a grazing detritus feeder and has probably a brief trochophoral stage in its life history. Outside the British Isles *P. ancyloide* has been recorded from Scandinavian waters and, at greater depths, south to the Azores and Mediterranean.

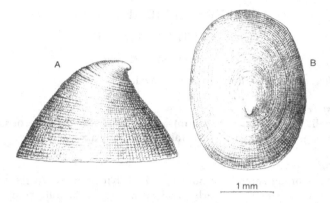

Fig. 25. *Propilidium ancyloide*, anterior end to left in A, above in B.

Family LEPETELLIDAE Dall, 1881

Genus LEPETELLA Verrill, 1880

Lepetella laterocompressa (Ponzi & de Reyneval, 1854)

Patella laterocompressa Ponzi & de Rayneval, 1854

Diagnostic characters
Shell small, patelliform, with central, raised apex; without trace of spire or spiral coiling. Apertural lips a little raised anteriorly and posteriorly.

Other characters
The shell is ornamented with numerous lines lying parallel to the edge of the aperture, which is oval, nearly circular, only slightly laterally compressed. White. Up to 1.5 mm long, 1.25 mm broad.

The animal has a short snout with terminal mouth. It lacks a ctenidium. The foot is oval. White. The species has a boreal range and the animals have been found only between Shetland and Norway, in deep water.

Family TROCHIDAE Rafinesque, 1815

Key to British genera and species of Trochidae

1. Whorls tumid; spiral and transverse ornament about equally developed; outer lip with varix; columella with sharp tooth at end of tall ridge; umbilicus absent .... *Danilia tinei* (p. 102)
   Shell not like this 2

2. Shell small, smooth to naked eye except for growth lines, iridescent, nacreous; animal with 6 pairs of epipodial tentacles ....
   .... *Margarites helicinus* (p. 94)
   Shell with ornament, mainly spiral, possibly fine 3

3. Shell conical with sharp apex and flat sides 4
   Shell approximately conical but without both sharp apex and flat sides 9

4. Breadth of shell about equal to height 5
   Breadth of shell about equal to half height 8

5. Shell with spiral and transverse ridges giving a reticulated surface; snout not papillated and without tubular mid-ventral extension; 3 pairs of epipodial tentacles .... *Jujubinus clelandi* (p. 122)
   Shell with spiral ridges which may be beaded, but no reticulate pattern; snout with papillae and mid-ventral extension; 3 to 5 pairs of epipodial tentacles 6

6. Shell of 7 whorls; 4–5 spiral ridges on apical half of last whorl
   .... *Calliostoma occidentale* (p. 128)
   Shell of 10–12 whorls; 7 or more spiral ridges on apical half of last whorl 7

7. Spiral ridges on last whorl smooth (except sometimes the subsutural ridge), those on topmost whorls beaded; columella with bulge; 4–5 pairs of epipodial tentacles .... *Calliostoma zizyphinum* (p. 124)
   Spiral ridges all finely beaded; columella without bulge; 3 pairs of epipodial tentacles; not intertidal .. *Calliostoma granulatum* (p. 126)

8. Last whorl with 4–5 slightly nodose spiral ridges ....
   .... *Jujubinus exasperatus* (p. 116)
   Last whorl with 8–9 smooth spiral ridges .... *Jujubinus striatus* (p. 118)

9. Whorl profile smoothly convex 10
   Whorl profile angulated, with subsutural shelf 17

10. Shell breadth less than its height **11**

Shell breadth and height about equal **12**

11. Last whorl equal to half shell height; 6–7 well spaced spiral ridges on penult whorl; flecked with brown .... *Jujubinus montagui* (p. 120)

Last whorl equal to three quarters of shell height; ornament slight, often eroded; spiral ridges strap-shaped, numerous; tooth on columella prominent .... *Monodonta lineata* (p. 114)

12. Ornament conspicuous; 5–7 prominent spiral ridges on penult whorl; 7 pairs of epipodial tentacles .. *Margarites groenlandicus* (p. 96)

Ornament slight **13**

13. Spiral ridges narrower than grooves; shell delicate, semitransparent; sutures deep; 5 pairs of epipodial tentacles ....

.... *Margarites argentatus* (p. 98)

Spiral ridges broader than grooves; shell solid; sutures shallow; 3 pairs of epipodial tentacles **14**

14. Shell with finely speckled dark pattern; prominent tooth on columella .... *Monodonta lineata* (p. 114)

Shell with coloured pattern of axial stripes **15**

15. Umbilicus closed; base of shell with reticulate pattern of purple; Channel Islands, not mainland Britain or Ireland ....

.... *Gibbula pennanti* (p. 108)

Umbilicus open; base of shell with stripes **16**

16. Colour pattern of many fine lines; 13–14 spiral ridges on base, the outermost not higher than the others; umbilicus small ....

.... *Gibbula cineraria* (p. 104)

Colour pattern of a few broad red lines; 8–11 basal spiral ridges, the outermost the biggest; umbilicus rather large ....

.... *Gibbula umbilicalis* (p. 106)

17. Upper part of each whorl without tubercles or costae; 16–18 spiral ridges between suture and periphery of last whorl; umbilicus usually small; shell straw-coloured with brown spots on ridges, iridescent .... *Gibbula tumida* (p. 110)

Upper part of each whorl tuberculated or with costae; umbilicus wide **18**

18. Shell very solid, with tubercles but without costae *Gibbula magus* (p. 112)

Shell pearly, with numerous costae on adapical half of each whorl; umbilicus with beaded margin; last whorl with 3 prominent peripheral spiral ridges; northern and rare ...... *Solariella amabilis* (p. 100)

Genus MARGARITES Gray, 1847

Margarites helicinus (Fabricius, 1780)
(Fig. 26)

Trochus helicinus Fabricius, 1780

Diagnostic characters
Shell with 4–5 swollen whorls, apparently smooth and glossy and with green
and purple refringence. Aperture prosocline and round; umbilicus large.
Head with ventrolateral ridges along snout; foot with six epipodial tentacles
on each side.

Other characters
Though appearing smooth to the naked eye the shell surface bears growth
lines and there are some spiral ridges and grooves on the base of the last
whorl. The aperture is slightly oval with the outer lip arising at right angles
to the surface of the last whorl; the umbilicus is sometimes partly covered
by an out-turning of the columellar lip. Horn-coloured; some shells show
a spiral brown band on the last whorl and perhaps on the penult as well:
all show refringence. Up to 4–5 mm high, 5–6 mm broad; last whorl occupies
about 90% of the shell height, the aperture about 45%.

The snout carries numerous papillae, and these also beset the sides of
the cephalic and epipodial tentacles. An eye stalk lies lateral to each cephalic
tentacle and a smooth-edged cephalic lappet lies on its medial side. A lobe,
with smooth edge, lies on each side of the neck. The foot is rather narrow
and has a longitudinal epipodial ridge on each side below which the tentacles
arise, each with a pigmented tubercle ventral to its base. The operculum
is multispiral and its outer surface is concave. Yellowish, with purple-brown
lines on the sides of the foot, the snout and the neck lobes.

M. helicinus is found on the lower parts of rocky shores and sublittorally
to depths of a few metres, from Yorkshire northwards on the east and from
North Wales, the Isle of Man and Dublin on the west; it also occurs on
north and west Irish coasts. The species has a circumboreal range. The animals
are to be found on *Fucus*, *Laminaria*, red weeds, in rock pools, and under
stones, eating detritus and weeds. Breeding occurs in spring, eggs being laid
on laminarian fronds. They are fertilized externally and about 100 are aggregated to form a mucous mass. There is no free larval stage and the young
emerge as juveniles (Fretter, 1955).

Fig. 26. *Margarites helicinus.* A, two shells; B, spawn mass; C, living animal showing
epipodium and neck lobes. (Fretter & Graham, 1962.)

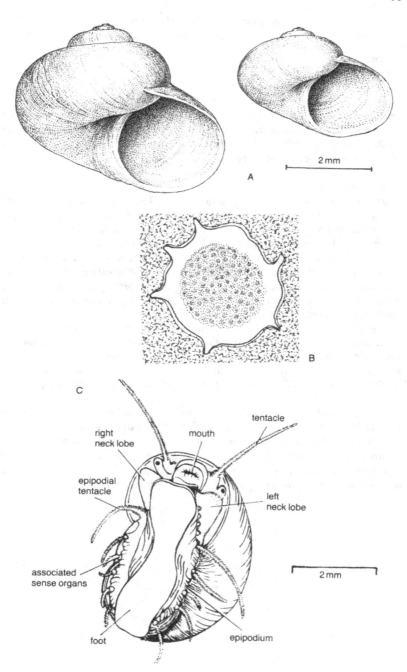

A

2 mm

B

C

tentacle

mouth

right
neck lobe

epipodial
tentacle

left
neck lobe

associated
sense organs

foot

epipodium

2 mm

Margarites groenlandicus (Gmelin, 1791)
(Fig. 27)

Trochus groenlandicus Gmelin, 1791
Trochus undulatus (Sowerby, 1838)

Diagnostic characters
Shell helicoid with large last whorl and moderately high spire; ornament variable but commonly there are well-developed spiral ridges, though in some shells these are limited to the periphery of the last whorl. Aperture prosocline; umbilicus large. Seven pairs of epipodial tentacles.

Other characters
There are 5–6 whorls which are rather tumid. The ornament is variable but usually there are 21–25 spiral ridges on the last whorl, 9–10 between the suture and the periphery and 12–15 between the periphery and the umbilicus; the furrows between ridges tend to be broader in the adapical half of the whorl and the most basal ridge forms a sharp boundary to the umbilicus. There are 5–7 ridges on the penult whorl, 4–5 on the others. The aperture is quadrangular. Cream, orange or pink, without refringence except in the throat. Up to 7 mm high, 5 mm broad; last whorl occupies about 85% of shell height, aperture about 55%.

The body of this animal has essentially the same structure as that of *M. helicinus* (p. 94) except in having an additional pair of epipodial tentacles; four pairs of the tentacles arise alongside the operculum. The flesh is cream and there are dark lines along the sides of the foot, on the tentacles and on the mantle skirt. The sense organs related to the epipodial tentacles are black.

M. groenlandicus is a northern species at its southernmost limits in the British Isles, where it occurs only off Scottish coasts. It is rare, living from L.W.S.T. to depths of 75 m, on weeds and under stones, feeding on the weeds or on detritus. The only observations on its breeding relate to East Greenland (Thorson, 1935) where spawn like that of *M. helicinus* was found fastened to weeds. The young emerged as miniature adults.

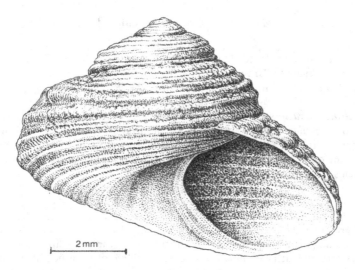

Fig. 27. *Margarites groenlandicus.*

Margarites argentatus (Gould, 1841)
(Fig. 28)

Margarita argentata Gould, 1841
Margarites olivaceus (Brown, 1827)
Trochus argentatus (Gould, 1841)

Diagnostic characters

Like *M. groenlandicus* but differs in having four swollen whorls with many fine spiral threads. Olive or grey with green or purple refringence. Animal with ventrolateral ridges on snout much reduced; five pairs of epipodial tentacles.

Other characters

The shell is delicate and semitransparent, with deep sutures. There are prosocline growth lines present in addition to the spiral threads. The umbilicus is large and partially blocked by an out-turning of the inner lip. Up to 3 mm high, 4 mm broad; last whorl occupies about 80% of shell height, aperture about half. The flesh is cream with some speckling of dark colour and there is a dark band along the edge of the mantle skirt.

    M. argentatus is an Arctic species reaching southwards on both sides of the Atlantic. It is sublittoral in its habitat and has been found at only a few places off Shetland and the west coast of Scotland.

Fig. 28. *Margarites argentatus*. Above, apical whorls in side view, ×100; below, shell in apical view, ×400.

Genus SOLARIELLA S. Wood, 1842

Solariella amabilis (Jeffreys, 1865)
(Fig. 29)

Trochus amabilis Jeffreys, 1865

Diagnostic characters

Shell glossy, slightly transparent, with a pearly lustre; a moderately tall cone of 5–6 whorls with a markedly stepped profile. Ornament of low prosocline costae and spiral ridges of which the most prominent (on the last whorl) are one near suture, three at periphery, and one round umbilicus. Aperture prosocline, round, its lip slightly angulated. Umbilicus large. Animal with tip of snout elongated into finger-shaped processes, with long cephalic tentacles, reduced neck lobes, but cephalic lappets are absent; the anterolateral processes of the foot are elongated and grooved. Three pairs of epipodial tentacles alongside operculum.

Other characters

Minor spiral ridges run between the main ones, most obviously on the base of the shell. Most spiral ridges are tuberculated by the numerous costae. These cross the whorls of the spire but on the last whorl usually end about level with the most adapical spiral; between the suture and that spiral the profile is flat and at right angles to the shell axis (producing the stepped profile) whilst between that spiral and the periphery the profile is flat or concave, below that, on the last whorl, convex. Up to 8 mm high, 8 mm broad; last whorl occupies 70–75% of shell height, aperture 40%.

The snout is depressed and broad, the cephalic tentacles not setose. The eyes are small, the left neck lobe forms a short tentacle, the right a small scroll fastened to the eye stalk. The foot is large and broad anteriorly, without epipodial sense organs (Fretter & Graham, 1977). Cream with brown on the sides of the foot.

S. amabilis has been recorded only a few times off the British Isles – off Shetland and in the Celtic Sea. It is widespread in the north east Atlantic, dredged from sandy or gravelly bottoms from 150 to over 1000 m deep. Its way of life is unknown: in view of its modified structure it deserves investigation. Females lay a few large yolky eggs surrounded by jelly; there is probably no free larval stage.

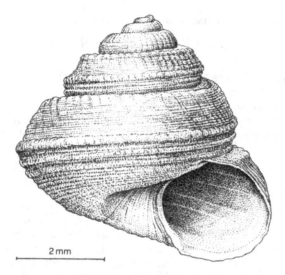

Fig. 29. *Solariella amabilis*.

Genus DANILIA Brusina, 1865

Danilia tinei (Calcara, 1839)
(Fig. 30)

Monodonta tinei Calcara, 1839

Diagnostic characters
Shell conical, sharply pointed, whorls with regular array of tubercles. Aperture nearly circular with varix along the outer lip; columella broad, with triangular depression and two teeth.

Other characters
The shell has 6–7 tumid whorls meeting at deep sutures. They bear spiral ridges and prosocline costae with tubercles where these cross. The costae are thin and undercut near the suture and the periphery and the tubercles are steeper on the side facing down the spiral. There are about 35 costae and eleven spiral ridges on the last whorl, 30 and seven on the penult. The aperture lies in a prosocline plane, the peristomial edge flares outwards a little, and the outer lip has a series of small longitudinal ridges on its inner surface. The umbilicus is absent. The shell is white with a mother-of-pearl sheen but bears a brownish periostracum. Up to 14 mm high, 10 mm broad; last whorl and aperture both occupy about 45% of shell height.

D. tinei is a southern species found from the Mediterranean northwards along the western European coast in rather deep water. The sole recent record from the British Isles is from off the west coast of Ireland.

Fig. 30. *Danilia tinei*, ×13.

Genus GIBBULA Risso, 1826

Gibbula cineraria (Linné, 1758)
(Fig. 31A,C)

Trochus cinerarius Linné, 1758

Diagnostic characters
Shell solid, smoothly dome-shaped with blunt apex, with numerous low spiral ridges and grooves crossed by fine prosocline growth lines. Last whorl distinctly keeled. Plane of aperture markedly prosocline; columella with slight bulge into aperture. Umbilicus egg-shaped. White-yellow, with many fine opisthocline lines of brownish purple. Snout with fringed ventrolateral ridges and median fold attached to ventral lip. Cephalic lappets and left neck lobe fringed, right neck lobe plain and fused to eye stalk. Foot with three pairs of epipodial tentacles.

Other characters
There are 5–6 slightly tumid whorls, the sutures not usually interrupting the smooth profile of the spire. The spiral ridges are slightly nodose. On the last whorl 8–10 commonly lie between suture and periphery and 13–14 on the base, with a group of low ones at the periphery; 8–9 lie on each of the penult and antepenult whorls. Shells erode easily, especially towards the apex. The narrow end of the umbilicus points to the origin of the outer lip. Up to 16 mm high, 15 mm broad; last whorl occupies 60–70% of shell height, aperture about 40%. Young shells are relatively lower and broader than the large ones to which these figures refer and some mature ones are unusually high.

Cephalic and epipodial tentacles are all long and slender; the eye stalks are separate. The foot sole has a papillated edge, a median longitudinal and many transverse grooves. Posteriorly its dorsal surface is grooved and ridged transversely. Yellow-grey with purple-brown marks, transverse on the snout, transverse and longitudinal on the tentacles, longitudinal on the sides of the foot. Animals may be sexed by the colour of the gonad: green in females, cream in males.

G. cineraria is one of the commonest intertidal molluscs on rocky shores with varied surface but without too great exposure, preferring flatter to steeper shores (Ebling *et al.*, 1948). It is found on and amongst weeds, under stones and ledges, and in pools, from about L.W.N.T. (sometimes higher in pools) to depths of 130 m; they live where the salinity may be as low as 25‰ (Nelson-Smith, 1967; Arnold, 1972). The species occupies a belt on the shore at a lower level than that occupied by *G. umbilicalis* (p. 106) where the two occur together. The animals are to be found on all suitable British and Irish coasts, often in very large numbers, especially in the North Sea and in the north, where the animals are both larger and darker in their coloration. They eat weeds and their epiphytes and are also detritivores. The species ranges from Gibraltar to Iceland.

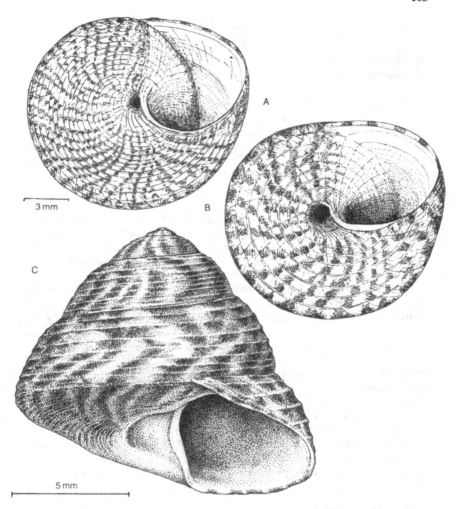

Fig. 31. Shells of *Gibbula*. A, *G. cineraria*; B, *G. umbilicalis* (Fretter & Graham, 1962); C, *G. cineraria* (in most shells the colour bands are both narrower and more numerous, and the spiral ridges are broader and less sharp than in the shell illustrated here).

Breeding occurs mainly in spring (Underwood, 1972). The eggs are fertilized externally and are planktonic, hatching into trochophore larvae which settle randomly on the shore after a short free life: only those which settle at low levels have any chance of survival.

Vernacular name: grey top shell.

Gibbula umbilicalis (da Costa, 1778)
(Figs 31B, 32)

Trochus umbilicalis da Costa, 1778
Trochus umbilicatus of authors

Diagnostic characters
Shell like that of *G. cineraria* (p. 104) in general aspect but with a relatively small number of wide red-purple bands on a background which is cream in adults, often greenish in small shells; with larger, nearly circular umbilicus; columella leans distinctly to left. Animal like *G. cineraria* but cephalic lappets and left neck lobe have smooth edges.

Other characters
The number of spiral ridges on the base of the shell is less than in *G. cineraria* (8–11), and the groove nearest to the periphery is often rather deep, especially in young shells, which are relatively lower and broader than adult ones. The spiral ridges are not nodose. Up to 16 mm high, 22 mm broad; last whorl occupies about 80% of the shell height, aperture about 50%.

G. *umbilicalis* is common, is found on the same kind of shore as *G. cineraria*, and has the same kind of diet. It tolerates exposure and resists emersion better, usually living at a higher level on the beach, M.L.W.S. to M.H.W.N., and still higher in pools. It is found under stones and ledges, amongst weeds, and in pools (Fischer-Piette & Gaillard, 1956; Moyse & Nelson-Smith, 1963; Nelson-Smith, 1967). It is a more southern species than *cineraria*, is absent from the North Sea and from the Channel east of the Isle of Wight and Calais (Crisp & Southward, 1958). It occurs on Irish coasts and west Scottish ones to Orkney but not Shetland.

The breeding season appears to vary from place to place: May–June in Cardigan Bay (Williams, 1965) but later, August–September, and more concentrated according to Garwood & Kendall (1985); summer and early autumn at Plymouth (Underwood, 1972). The eggs are laid singly and fertilized externally. From them hatch free-swimming trochophore larvae which soon settle low on the shore, moving upwards later to their normal level.

Vernacular name: flat top shell.

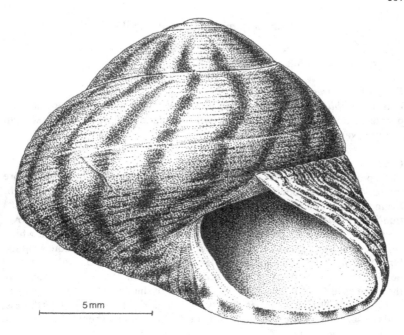

5 mm

Fig. 32. *Gibbula umbilicalis*.

Gibbula pennanti (Philippi, 1851)

(Fig. 33)

Trochus pennanti Philippi, 1851

Diagnostic characters

Like *G. umbilicalis* but without an umbilicus. Stripes of colour on the shell purple rather than red and there is a reticulate pattern on the base (but colour is variable).

Other characters

The apex is more pointed than in *G. umbilicalis* (p. 106), the whorls more tumid, the sutures deeper, so that the profile of the spire is not smooth. Young shells have an open umbilicus. The columella is more nearly vertical than in *umbilicalis*. The colour of the shell, though commonly as above, is variable and it may be reticulated all over; the background may be white, yellow, sometimes green, or nearly black, and some shells are without striping. Up to 16 mm high, 15 mm broad; last whorl occupies about 80% of shell height, aperture 50–60%. The external features of the animal's body are the same as in *G. umbilicalis*; the basic colour of the flesh, however, is greyish rather than yellow; neck lobes usually are a rather bright yellow nevertheless, and the same colour occurs on the sole of the foot. White speckles are present on the sides of the foot.

G. pennanti occurs on the same type of shore as *cineraria* and *umbilicalis*, lower than the latter. It seems to need damp surroundings when emersed and, if not in pools, lives close to them on wet rocks or amongst weeds (Fischer-Piette & Gaillard, 1956). It has not been found on mainland shores of Britain or Ireland but occurs in the Channel Islands, on the French coast west of Cherbourg and south to Spain.

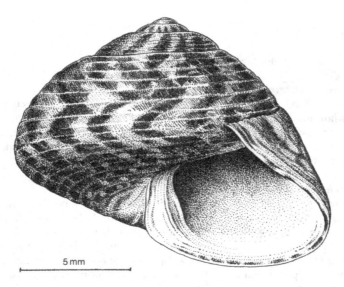

Fig. 33. *Gibbula pennanti*.

Gibbula tumida (Montagu, 1803)
(Fig. 34)

Trochus tumidus Montagu, 1803

Diagnostic characters
Shell a moderately high and sharply pointed cone with a stepped profile. Last whorl markedly keeled at periphery and flat-sided adapical of that; ornament of shallow spiral grooves and ridges. Columella bulging into aperture; inner lip everted to block wholly or partly the umbilicus, which, if open, is comma-shaped.

Other characters
The shell has 6–7 whorls. There are 14–18 spiral ridges on the base of the shell, 16–18 on the adapical part of the last whorl, 14–16 on the penult, the number reducing up the spire. Young animals show an open umbilicus but this is gradually closed by growth of the inner lip. Brownish with blue-green iridescence and scattered brown spots, most marked below the sutures; base paler. Up to 9 mm high, 10 mm broad; last whorl occupies about 75% of shell height, aperture about 50%.

The body of the animal agrees in most particulars with that of other *Gibbula* species. The snout is less papillated but has prominent ventrolateral ridges connected to the corresponding eye stalk which is also linked over the tentacle base to a cephalic lappet. The right eye stalk connects with a neck lobe with simple edge and then with the epipodial fold; the left eye stalk and neck lobe are separate, this lobe with a fringed margin. There are three pairs of epipodial tentacles each with a sense organ behind its base. The body is cream with brown and white lines and blotches.

G. tumida is a northern species found between Iceland and northern Norway and Spain. It has been recorded from off most British and Irish coasts; it is not intertidal but may be dredged to 1200 m both amongst weeds and on stones. Its way of life is like that of other *Gibbula* species. Breeding occurs (Helgoland: Gersch, 1936; Ankel, 1936) in spring, when eggs are laid, each in a jelly coat and aggregated into spawn masses.

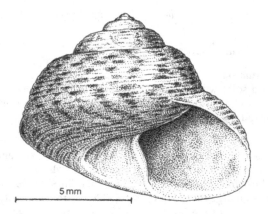

5 mm

Fig. 34. *Gibbula tumida.*

Gibbula magus (Linné, 1758)
(Fig. 35)

Trochus magus Linné, 1758

Diagnostic characters

Shell large and solid with markedly stepped profile and a row of tubercles on the step below each suture; last whorl keeled; umbilicus obvious; marked with irregular red streaks and spots.

Other characters

There are 7–8 whorls, the last one rather sharply keeled, and the sutures are usually deep. In addition to the tubercles, of which there are about 20 on the last whorl and 14–18 on the two previous, the shell bears spiral ridges, one of which is enlarged to form the keel. There are commonly 17–18 ridges on the adapical half of the last whorl, often alternately large and small, and 22–27, more equal in size, on its base. The penult whorl has 12–14, the others 6–8 each. The number is variable and ridges on the oldest whorls and on the shell base are often eroded; the ridges may be spiky in shells from deep water. The umbilicus is large, round and deep, approached by a comma-shaped groove. There is a distinct bulge on the columella. Up to 30 mm high, 35 mm broad; last whorl occupies 70% of shell height, aperture 40%.

The snout is densely papillated distally and has a ventrolateral fringe on each side. The cephalic lappets are large, each with a lobed margin, and connect over the tentacle base to the eye stalk; the right eye stalk is joined to a neck lobe with a smooth edge; the left neck lobe is not so joined and has a scalloped edge. Each neck lobe is linked to an epipodial ridge which runs along the foot; ventral to this on each side arise three epipodial tentacles, their bases in sheaths with papillated edges and with a spiky flap-like sense organ. The foot is blunt anteriorly, pointed posteriorly where its dorsal surface carries many transverse grooves. The flesh is yellow-pink or orange with black-purple blotches, similar lines over the snout and front of the foot, a central line with branches on each tentacle; epipodial tentacles are sulphur yellow, their sheaths and sense organs white; neck lobes and cephalic lappets yellow.

G. magus extends north from the Mediterranean and the Azores to the British Isles, where it has been found in the Channel as far east as Swanage, the Cotentin peninsula and the Channel Islands. It occurs all round Ireland and north to Shetland on the west coast of Britain; not in the North Sea except perhaps very occasionally. The animals like the same kind of shore as other top shells but usually live below tidemarks, though rare specimens may be found at L.W.S.T. Their mode of life is like that of other *Gibbula* spp.

Eggs are laid singly in spring and early summer, each in a jelly coat. They hatch to give trochophore larvae which settle after a brief free life.

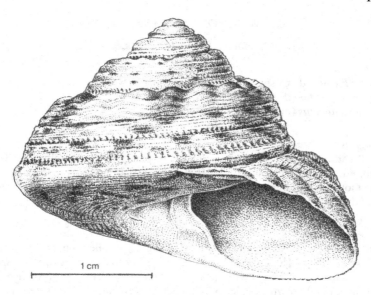

Fig. 35. *Gibbula magus*.

Genus MONODONTA Lamarck, 1799

Monodonta lineata (da Costa, 1778)
(Fig. 36)

Turbo lineatus da Costa, 1778
Trochus lineatus (da Costa, 1778)
Osilinus lineatus (da Costa, 1778)

Diagnostic characters

Shell rather large with pointed apex, the profile not (or only very little) stepped, often cyrtoconoid with the last whorl bluntly keeled. Dark with fine mottled pattern and a large white patch in the umbilical area. Columella with distinct tooth; umbilicus minute or closed.

Other characters

The shell is somewhat globose with 5–6 slightly swollen whorls meeting at shallow sutures. The general appearance is smooth but there are fine spiral ridges and prosocline growth lines, sometimes exaggerated and then representing annual growth checks (Williamson & Kendall, 1981). This ornament is clear on young shells but is less, and often eroded, on older ones. The columella is thick and the inner lip folds out over the umbilicus leaving a depression and sometimes a minute hole. Whilst the general colour of the shell is dark the ground colour is buff and this is marked extensively by streaks of brown, green or red running in a prosocline direction. Up to 30 mm high, 25 mm broad; last whorl occupies 75% of the shell height, aperture 35–40%.

The tip of the snout is slightly papillated, the cephalic tentacles gently lobed laterally, each with a broad basal eye stalk connected over the base of the tentacle to a smooth-edged cephalic lappet and, posteriorly, to a neck lobe. The left lobe has 10–12 marginal processes, the right is smooth but has one process on its underside. The foot has an epipodial fold along each side with three tentacles under it, each with two sense organs at its base; a grooved area lies under the operculum. The body is greyish green with many black or purple lines; each tentacle has usually a central dark line with divergent lateral branches. The foot sole and the epipodial sense organs are pale.

M. lineata is found on rocky shores, preferring areas with rock pools, avoiding those with loose boulders, sometimes with weed, sometimes where the rock seems bare, between L.W.N.T. and H.W.N.T. approximately. The animals browse on algae, mainly microphytes and detritus, rarely on macrophytes. Their distribution in the British Isles is, for unknown reasons, patchy, and they are confined to south west England, the Channel as far east as St Alban's Head (Hawthorne, 1965), the Irish Sea and the south and west coasts of Ireland (McMillan, 1944; Williams, 1965; Nelson-Smith, 1967).

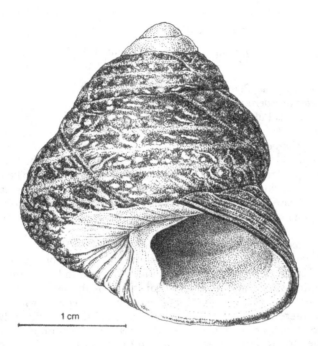

1 cm

Fig. 36. *Monodonta lineata.*

Abroad the animals are found on the western Channel coast of France and south to Portugal.

Spawning occurs in summer (May to August) when green eggs are laid, each surrounded by a jelly coat. The larva hatches rapidly and settles after 4–5 days, low on the beach, migrating later to a higher level (Williams, 1965; Desai, 1966; Underwood, 1972; Garwood & Kendall, 1985). The young snails grow rapidly at first, reaching a shell height of about 15 mm at the end of one year and about 22 mm after two, at which time they start to reproduce.

Genus JUJUBINUS Monterosato, 1884

Jujubinus exasperatus (Pennant, 1777)
(Fig. 37)

Trochus exasperatus Pennant, 1777
Cantharidus exasperatus (Pennant, 1777)
Trochus exiguus Pulteney, 1799

Diagnostic characters
Shell pyramidal, flat-sided, with pointed apex, taller than broad; ornament of nodose spiral ridges and fine growth lines; last whorl with sharp keel formed from enlarged ridge, four smaller ridges between keel and adapical suture and 6–7 on base. Columella with low bulge, umbilicus absent. Usually with irregular but mainly axial pattern of carmine red.

Other characters
The 6–8 whorls meet at sutures themselves almost invisible but located by the broad spiral ridge on their adapical side. The growth lines form tubercles on the spiral ridges and scale-like projections in the intervening furrows. The colour pattern varies: the background may be pale brown, reddish or greenish; there may be white areas on the spiral ridges and nearly always there are blotches of carmine, often arranged so as to form alternating red and white stripes on the whole shell. Up to 11 mm high, 6 mm broad; last whorl occupies 45–50% of shell height, aperture one third.

The body of the animal is much as in the species of *Gibbula* (p. 104–113). It lacks the mid-ventral fold on the snout; the margins of the cephalic lappets are lobed, the three pairs of epipodial tentacles have each a basal sense organ. The snout and the sides of the foot are brown, its sole is cream, like the neck lobes and the cephalic lappets. A line of red borders the ventral edge of the foot, occurs on all tentacles and on the grooved area under the operculum.

J. exasperatus has only a limited distribution in Britain, recorded rarely from south west England and the Clyde Sea Area. It may be found on weeds (on which it browses), mainly *Zostera*, from L.W.S.T. to depths of 200 m. Abroad it is found south to the Mediterranean and the Azores. Breeding occurs April to August. The eggs are uncoloured, fertilized externally and embedded in jelly, forming a zigzag, flattened string, oval in section, which is entangled with and attached to weeds or stones. Juveniles hatch in about 4–5 days (Robert, 1902).

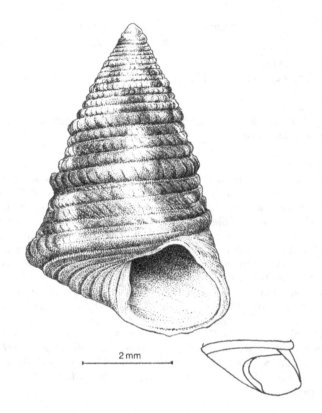

2 mm

Fig. 37. *Jujubinus exasperatus*.

Jujubinus striatus (Linné, 1758)
(Fig. 38)

Trochus striatus Linné, 1758
Cantharidus striatus (Linné, 1758)

Diagnostic characters
Much like *J. exasperatus* but with 8–10 spiral ridges between keel and adapical suture on the last whorl and 7–9 on its base.

Other characters
The apical angle of this shell is greater than in *J. exasperatus* (p. 116) and the growth lines are usually less obvious. The last whorl and the aperture are both relatively larger. The colour pattern is extremely variable but usually consists of coffee to chocolate brown with cream or white areas of different sizes, the apical region being a carmine red. The brown and white may form broad alternating bands or a chequered pattern. Some shells are nearly all red. Up to 10 mm high, 8 mm broad; last whorl occupies two thirds of shell height, aperture 40–45%

The animal is like *J. exasperatus* but lacks the red markings of that species.

J. striatus is confined to the Channel Islands (where it is not uncommon), south west England, and perhaps some southern and western Irish localities (in all of which it is rare). The species ranges south to the Mediterranean. The animals live on weeds, especially *Zostera*, and on stones with *Ulva* and *Codium* at L.W.S.T. and to depths of 200 m. The eggs and spawn of this species are said (Robert, 1902) to be indistinguishable from those of *J. exasperatus*.

2 mm

Fig. 38. *Jujubinus striatus*.

Jujubinus montagui (W. Wood, 1828)

(Fig. 39)

Trochus montagui W. Wood, 1828
Cantharidus montagui (W. Wood, 1828)

Diagnostic characters
Shell more or less pyramidal but spire a little dome-shaped; whorls gently
swollen, last whorl bluntly keeled; spiral ridges narrow, not enlarged at the
periphery; columella rather thick, with bulge. Cream to pale brown with
dark spots set along the spiral ridges, darker towards the apex.

Other characters
Whilst in general like other *Jujubinus* species, *montagui* is distinct in shape
and colour. The spiral ridges appear as flat-topped threads; there are seven
on the base of the last whorl, 6–7 between its periphery and the adapical
suture. Growth lines are fine and numerous. There is no umbilicus in adult
shells. Up to 8 mm high, 6 mm broad; last whorl occupies half the shell height,
aperture 40–50%.

The animal hardly differs in its external features from the other species.
The flesh is cream, flecked with brown and white; a dark Y-shaped mark
lies on the snout. The neck lobes and the cephalic lappets are white and
the epipodial sense organs are black.

J. montagui is not found intertidally but may be collected from stony and
gravelly bottoms 10–200 m deep. It is the most widespread of the species
of *Jujubinus* in the British Isles, occurring as far north as the Northern Isles,
more frequently on western than on eastern shores. It is found off the Channel
Islands and extends to the Mediterranean.

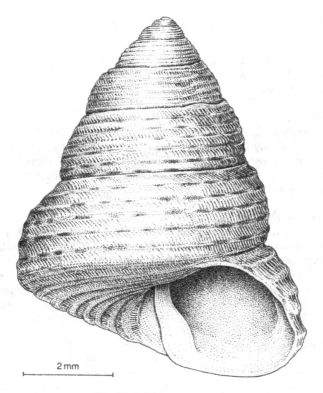

Fig. 39. *Jujubinus montagui*.

Jujubinus clelandi (W. Wood, 1828)
(Fig. 40)

Trochus clelandi W. Wood, 1828
Cantharidus clelandi (W. Wood, 1828)
Trochus millegranus Philippi, 1844

Diagnostic characters
Shell a sharply pointed, nearly regular cone, about as broad as high, with flat sides and nearly flat base; last whorl sharply keeled by enlarged spiral band, repeated above each suture in the spire; 5–6 spiral ridges lie between this band and the suture on the last whorl and also on each whorl of the spire; all spiral ridges nodose.

Other characters
There are 7–8 whorls. In addition to the spiral ridges the shell has many prosocline growth lines which contribute to the tubercles on the ridges and form small elevations in the grooves between. The apertural plane is more oblique (prosocline) to the shell axis than in other species. The columella leans to the left, its tooth very slight. White or pinkish, refringent, with a pink apex and often with brownish pink spots on the spiral ridges. Up to 12 mm high, 11 mm broad; last whorl occupies 50–55% of shell height, aperture about 30%.

The animal is like that of the other *Jujubinus* species. The lip, however, is not split mid-ventrally, the cephalic lappets are papillated marginally and linked to the eye stalks ventral to the tentacle base, that on the right joined in turn to the neck lobe; the foot is rather narrow, bearing on each side three epipodial tentacles with associated sense organs. The flesh is cream with brown marks on the side of the foot, some olive-green colour on the dorsal surface of the snout, a dark Y-shaped mark between the tentacle bases, and one or more dark lines along all tentacles.

This species is always sublittoral, living on stony and gravelly bottoms 35–800 m deep. It is not rare. The animals live between the Mediterranean and northern Norway and they have been found off all British and Irish coasts except those of the eastern Channel and southern North Sea.

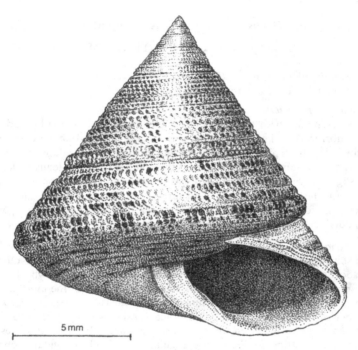

5 mm

Fig. 40. *Jujubinus clelandi.*

Genus CALLIOSTOMA Swainson, 1840

Calliostoma zizyphinum (Linné, 1758)
(Fig. 41)

Trochus zizyphinus Linné, 1758
Trochus conuloides Lamarck, 1822

Diagnostic characters

Shell pyramidal, with sharply pointed apex and nearly flat sides and base; last whorl sharply keeled by thick spiral band repeated above each suture in the spire; 6–9 spiral ridges lie between periphery and adapical suture on the last whorl and on each whorl of the spire. Columella leans to left and has slight bulge. Umbilicus closed except in the smallest shells. Protoconch with reticulate pattern. Snout with terminal fringe of papillae and tubular extension from mid-ventral lip curving back on right. Four or five pairs of epipodial tentacles.

Other characters

The spire is occasionally slightly concave in profile near the apex. Numerous prosocline growth lines occur which, on the four most apical whorls, cause the spiral ridges to be beaded, and this is also a feature of the most adapical ridge on younger whorls. The number of spiral ridges depends upon the number of minor ones in the furrows; each ridge is steeper on its adapical side. There are usually ten whorls. Most shells are yellowish, with chestnut, red, or purple stripes on the whorls, regular on the keels, irregular elsewhere, reduced on the base, the apex often with a lilac tint; wholly red, purple, or white shells may occur, the last sometimes believed to be a distinct species, *C. lyonsi*. Up to 30 mm high, 30 mm broad; last whorl occupies 40–45% of shell height, aperture 33%.

The cephalic tentacles are long and setose; the eye stalks connect at their base with small cephalic lappets and posteriorly with smooth-edged neck lobes. The anterior edge of the foot is double; there are no sense organs at the base of the epipodial tentacles. The ridged area on the dorsal surface of the foot posteriorly is usually kept with its edges upraised and meeting, its secretion being apparently used by the foot to wipe the shell free of attaching organisms (Jones, 1984). Yellowish with reddish brown streaks or blotches; tentacles with dark central line.

C. zizyphinum occurs, fairly commonly, amongst weeds and under stones on rocky shores a little above L.W.S.T. and to 300 m depth. It has been found on all suitable British and Irish shores, in the Mediterranean, and on western European shores north to the Lofoten Islands. It is said to eat small coelenterates (Salvini-Plawen, 1972) but also takes vegetable matter.

Yellowish eggs are laid in a rope of jelly about 3–4 mm in diameter and up to 35 mm long which is fastened at intervals by the foot of the female to stones and weed; from this young snails emerge in 7–10 days.

Vernacular name: painted top shell.

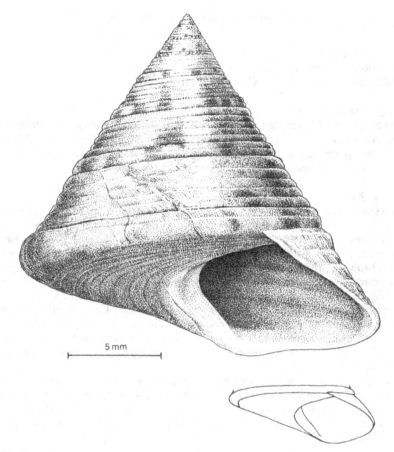

Fig. 41. *Calliostoma zizyphinum*. Inset, outline of a more common shape of the aperture.

Calliostoma granulatum (Born, 1778)
(Fig. 42)

Trochus granulatus Born, 1778
Trochus papillosus da Costa, 1778
Calliostoma papillosum (da Costa, 1778)

Diagnostic characters
In general like *C. zizyphinum* (p. 124) but the profile of the spire is concave, the last whorl inflated and less sharply keeled; there are 7–12 spiral ridges between periphery and suture on the last whorl; all major ridges bear tubercles. Snout heavily papillated; three pairs of epipodial tentacles.

Other characters
The shell has 7–9 whorls. The smaller spiral ridges (between the larger ones) are smooth; the two ridges below each suture are more prominent than the others and those on the base of the shell are less tuberculated. The columella slopes to the left, with only a slight indication of a bulge. Cream or yellowish with brown or reddish spots and streaks, the peripheral alternation of light and dark less obvious than in *C. zizyphinum*. Up to 35 mm high, 35 mm broad; last whorl occupies 60% of shell height, aperture 40–50%.

The external features of the animal are like those of *C. zizyphinum* but the snout is more papillose, especially near the tip, as is also the left neck lobe. The flesh is cream with brown specklings; the cephalic tentacles have a dark central line, while eye stalks, neck lobes, epipodial tentacles and the posterior end of the foot are all white.

C. granulatum is not found intertidally but is occasionally dredged from depths of 7–300 m on gravelly or soft bottoms. Its main area of distribution is south of the British Isles to the Azores and the Mediterranean. It has been found off south west England, in the Irish Sea, and off the west coast of Scotland.

Fig. 42. *Calliostoma granulatum*. A, shell; B, head and anterior end of foot in ventral view. (Fretter & Graham, 1962).

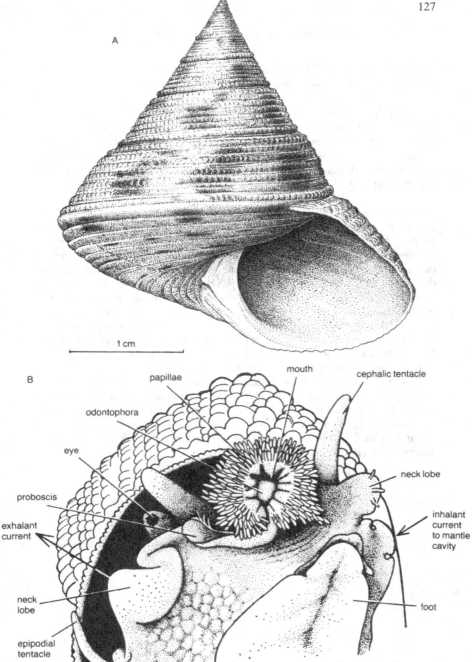

Calliostoma occidentale (Mighels & Adams, 1842)
(Fig. 43)

Trochus occidentalis Mighels & Adams, 1842
Trochus alabstrum Lovén, 1846

Diagnostic characters
Shell a straight-sided cone with reddish opalescent lustre; last whorl keeled; ornament of prominent, rather well-spaced spiral ridges of which the most adapical on the last whorl is nodose, though higher in the spire nodosities occur on other ridges too. Outer lip rather acutely angulated. Cephalic lappets minute or absent; three pairs of epipodial tentacles.

Other characters
The shell is slightly transparent and has seven whorls. The spiral ridges are narrow, that at the periphery of the last whorl sharp-edged and forming the keel; there are four or five on the last whorl above the keel, four on the next two, gradually reducing up the spire. They are less prominent on the base and may be absent there except marginally and centrally. Prosocline growth lines are present but are not obvious. There is little or no bulge on the columella; a small umbilicus is present. The protoconch has a reticulated pattern but this is commonly lost by erosion. The spiral ridges are often yellowish. Up to 11–12 mm high, 10–11 mm broad; last whorl occupies 55% of shell height, aperture 40%.

The body is cream with purplish brown markings.

C. occidentale is a northern species found from the British Isles and Scandinavia to Iceland and also off eastern American coasts. In the British Isles it has been recorded from the Moray Firth, off the north coast of Scotland, the Northern Isles and Rockall, and, more recently, from the southern Irish Sea and Celtic Sea. It lives on stones 19–1000 m deep and feeds on alcyonarians and hydroids (Perron & Turner, 1978).

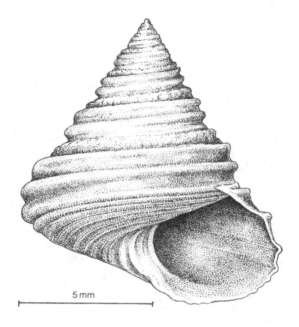

5 mm

Fig. 43. *Calliostoma occidentale*.

Family SKENEIDAE Clark, 1851

Genus SKENEA Fleming, 1825

Small animals with white or colourless shells comprising only a few whorls. They lack nacre and are either smooth or have spiral ridges; the aperture is circular and the umbilicus is prominent. The operculum is horny and has only a few spiral coils. The head lacks cephalic lappets but does possess neck lobes. There are three pairs of epipodial tentacles on the foot.

Key to British species of Skenea

1. Shell with delicate spiral lines over most or all of last whorl; peristome circular, slightly angulated at beginning of outer lip **2**

   Shell with spiral lines limited to umbilical area or absent **3**

2. Last whorl with spiral lines all over; outer lip with sinuses; growth lines flexuous; umbilicus narrow; apical angle about 100° ....
   .... *Skenea cutleriana* (p. 134)

   Last whorl without spiral lines on adapical half; outer lip without sinus; growth lines straight; umbilicus wide; apical angle 140–145°
   .... *Skenea serpuloides* (p. 132)

3. Outer lip with two sinuses, one above and one below periphery; spiral lines inconspicuous or absent; umbilicus large; surface glossy
   .... *Skenea nitens* (p. 136)

   Outer lip without sinuses; opercular stop (= groove) along columellar area; umbilicus wide and round, the surrounding spiral lines clear; junction of outer and columellar lips rounded ....
   .... *Skenea basistriata* (p. 138)

Skenea serpuloides (Montagu, 1808)
(Fig. 44)

Helix serpuloides Montagu, 1808
Cyclostrema serpuloides (Montagu, 1808)
Skenea divisa (Adams, 1797)?

Diagnostic characters
Shell not more than 2 mm high, spire very depressed (apical angle 140–145°), last whorl large, its basal half with numerous spiral lines, smooth or with only delicate spiral lines elsewhere, but with growth lines. Aperture with peristome, nearly circular; outer lip nearly straight in side view; umbilicus large and round, exposing many whorls.

Other characters
The shell is nearly transparent, with 3–4 swollen whorls. There are about ten very fine spiral ridges on the penult whorl. The aperture lies in a gently prosocline plane, its inner lip not out-turned over the umbilicus. In some shells the peristome separates from the last whorl. White or colourless. Up to 2 mm high, 1.5 mm broad; last whorl occupies 90–95% of shell height, aperture about 75%.

The animal has a broad depressed snout, slightly bifid at the tip. The cephalic tentacles are fringed on each side with sensory papillae, which are also present on the mantle edge. Cephalic lappets are absent. The right neck lobe has a fringed edge and joins the right epipodial fold; the left is plain and separate from the epipodium. There are three pairs of epipodial tentacles on the foot and an extra single one behind the right cephalic tentacle. The flesh is whitish with a dense white line on each cephalic tentacle; some red on the snout comes from the buccal mass.

S. serpuloides occurs on rocky shores from L.W.S.T. to about 50 m depth on and amongst weeds and stones and is also found on gravelly bottoms. It is a detritus feeder and is probably commoner than records suggest, being overlooked because of its size, or regarded as a juvenile of some other species. It occurs throughout Europe from the British Isles southwards; absent from much of the North Sea and eastern part of the Channel.

Eggs are laid in jelly masses attached to weed. There is probably no free larval stage.

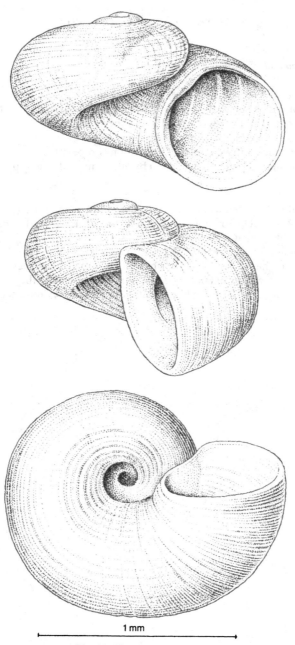

1 mm

Fig. 44. *Skenea serpuloides*.

Skenea cutleriana (Clark, 1849)
(Fig. 45)

Cyclostrema cutlerianum Clark, 1849

Diagnostic characters
Like the previous species but with a higher spire (apical angle about 100°)
and ornamented all over with spiral lines; aperture nearly circular, the outer
lip showing three shallow sinuses, one at the periphery, one at the base,
one where it meets the columella; growth lines repeat this shape; umbilicus
narrow, comma-shaped, partly hidden by out-turning of inner lip.

Other characters
There are three swollen whorls, the last with about 30 spiral ridges, the penult
with ten. White. Up to 1 mm high, 1 mm broad; last whorl occupies 95%
of shell height, aperture 60–65%.

The animal is like *S. serpuloides* (p. 132). White.

The species has a southern distribution. In the British Isles it has been
found in south west England, in some places in Ireland, the Isle of Man,
and the Hebrides, in the same kind of situation as the previous species.

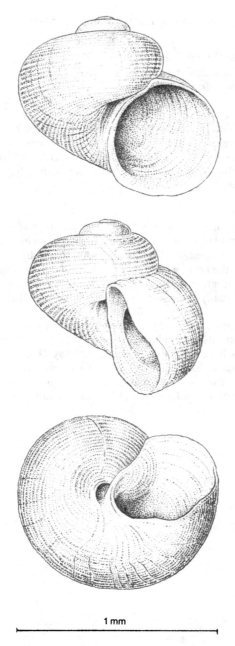

1 mm

Fig. 45. *Skenea cutleriana*.

Skenea nitens (Philippi, 1844)
(Fig. 46)

Delphinula nitens Philippi, 1844
Cyclostrema nitens (Philippi, 1844)
Trochus pusillus Jeffreys, 1848

Diagnostic characters
Like *S. serpuloides* (p. 132) but with a slightly higher and more pointed
spire (apical angle 110–115°); surface glossy and without ornament except
perhaps for a few spiral lines near the umbilicus, which is slightly blocked
by an out-turning of the inner lip; outer lip with a bay at the periphery
and another basally.

Other characters
There are four rather swollen whorls. The aperture is ear-shaped and the
peristome is not plane. The umbilicus is smaller than in *serpuloides* and oval
or comma-shaped, margined on its abapertural side by a spiral ridge which
originates from the base of the aperture. The growth lines are sinuous, showing the same curvature as the outer lip. White or brownish. Up to 1.5 mm
high, 1.5 mm broad; last whorl occupies 90–95% of shell height, aperture
70%.

The animal presents much the same external features as *S. serpuloides*
but is without an extra epipodial tentacle on the right. White.

S. nitens has been recorded between Norway (Höisaeter, 1968a,b) and
the Mediterranean. In the British Isles it has been found on most shores
except those of the eastern Channel and the North Sea. It lives amongst
weeds and in pools on rocky shores at L.W.S.T. and to depths of 100 m.

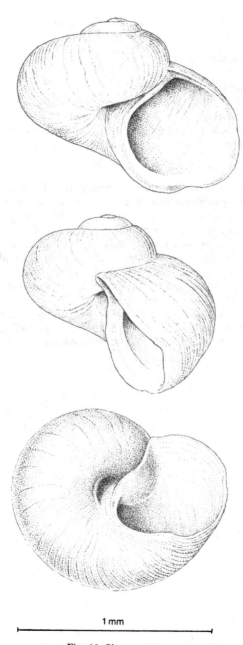

1 mm

Fig. 46. *Skenea nitens*.

Skenea basistriata (Jeffreys, 1877)
(Fig. 47)

Cyclostrema basistriatum Jeffreys, 1877

Diagnostic characters
Distinguished from other *Skenea* species by a relatively high spire (apical angle 105–110°), more swollen whorls, ornament of spiral ridges usually confined to the area round the deep umbilicus; aperture nearly circular, outer lip without marked sinuses and hardly turned out basally.

Other characters
The shell has 3–4 whorls with deep sutures placed below the periphery of the upper whorl, especially near the aperture, so that the last whorl has a rather flattened subsutural shelf. The spiral ridges may be found away from the umbilical region, but rarely at the periphery of the last whorl. There are also delicate growth lines, simply curved because of the absence of sinuses on the outer lip. The inner lip slightly covers the umbilicus and may, between that point and the origin of the outer lip, lie a little away from the surface of the last whorl. White. Up to 2 mm high, a little over 2 mm broad; last whorl occupies 95% of shell height, aperture 70%.

The animal has the same external features as *S. serpuloides* (p. 132).

S. basistriata has been recorded from deep water (90–2400 m) off the coasts of western Europe between Spain and Norway, on soft bottoms.

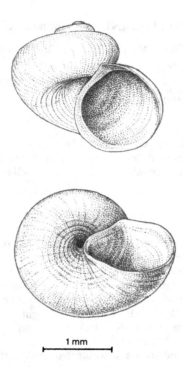

1 mm

Fig. 47. *Skenea basistriata.*

Family TURBINIDAE Rafinesque, 1815

Genus MOELLERIA Jeffreys, 1865
Moelleria costulata (Möller, 1842)
(Fig. 48)

Margarita costulata Möller, 1842

Diagnostic characters

Shell opaque with depressed spire and large, round umbilicus; operculum calcareous with depressed centre. Ornament of many slightly prosocline costellae with flattened summits and (on base of shell only) a few well-spaced spiral ridges.

Other characters

The shell has 3–4 rapidly expanding whorls with deep sutures. The costellae number 40–50 on the last whorl. The aperture is surrounded by a peristome which is approximately axial and not oblique in orientation, and is nearly circular. White or yellowish. Up to 1.5 mm high, 2 mm broad; last whorl occupies about 90% of the shell height, the aperture about two thirds.

The animal has a broad, depressed snout, setose cephalic tentacles, each with a lateral eye stalk at its base. There are right and left neck lobes with smooth margins, each linked posteriorly to an epipodial fold with (?) four tentacles on each side of the foot.

M. costulata is found over the whole North Atlantic from 7 to 2000 m, on soft bottoms, living at the greater depths in lower latitudes. Its way of life is not known. British specimens are usually held to be subfossil.

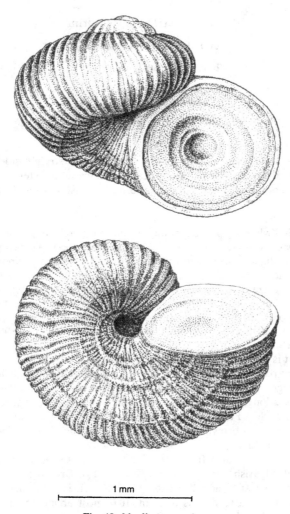

1 mm

Fig. 48. *Moelleria costulata*.

Family TRICOLIIDAE Woodring, 1928

Genus TRICOLIA Risso, 1826

Tricolia pullus (Linné, 1758)
(Fig. 49)

Turbo pullus Linné, 1758
Phasianella pulla (Linné, 1758)

Diagnostic characters
Shell glossy, solid, with conical spire more or less straight-sided; whorls slightly tumid; operculum hemispherical and calcareous externally. Colour very variable but usually with erratic pattern of reddish spots and streaks.

Other characters
There are 5–6 whorls, apparently smooth but with delicate growth lines and spiral lines more obvious near the apex. The aperture lies in a gently prosocline plane and is slightly oval. The outer lip arises below the periphery of the last whorl and is thin, a little turned out at the base; the inner lip usually forms only a glaze over the last whorl. The colour pattern is extremely variable: the shell may be pale cream, golden, carmine red, or chocolate brown, all without pattern, but usually there is a mottling or striping of red and cream. Up to 10 mm high, 5 mm broad; last whorl occupies 80% of shell height, aperture about 50%.

The snout is rather small, the cephalic tentacles long, slender, and fringed with sensory papillae, each with a lateral eye stalk. Both neck lobes have a fringed edge. The foot is narrow, its anterior and lateral edges double, and there is a longitudinal groove along the sole. There are three pairs of epipodial tentacles. Flesh yellowish, but greenish on eye stalks and mantle edge; there are red-brown lines along the head, the sides of the cephalic tentacles and the foot.

T. pullus extends north from the Mediterranean to the western parts of the Channel, all Irish coasts and the west coast of Britain to the Orkneys; absent from the east Channel and the North Sea. It is found near L.W.S.T. on rocky shores and to depths of 35 m. It is most abundant in rock pools and on the small red weeds on which it feeds. Breeding may occur all the year round. Fertilization is external as the orange eggs escape from the mantle cavity. A larva hatches after about 10–12 hours, settles after 2–5 days at a shell height of about 1.1 mm, mainly on the small red weeds on which adults live (Fretter, 1955; Fretter & Manly, 1977b).

Vernacular name: pheasant shell.

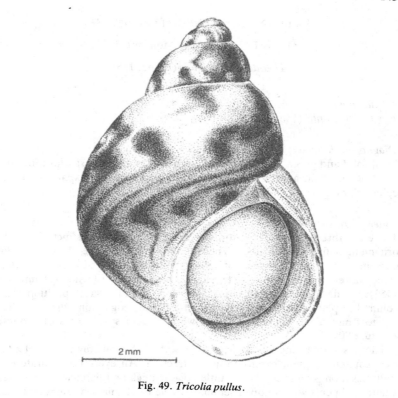

2 mm

Fig. 49. *Tricolia pullus*.

Family NERITIDAE Rafinesque, 1815

Genus THEODOXUS Montfort, 1810

Theodoxus fluviatilis (Linné, 1758)
(Fig. 50)

Nerita fluviatilis Linné, 1758
Neritina fluviatilis (Linné, 1758)

Diagnostic characters
Shell solid and glossy, hemispherical, with low spire, the last whorl forming most of the shell; aperture D-shaped, half blocked by an expansion of the columellar region.

Other characters
There are three whorls, the sutures of variable depth but distinct. The only ornament is fine growth lines. There is no umbilicus. The shell is usually dark in colour with a pattern of yellow-white streaks, and there are often three darker areas separated by two lighter ones on the last whorl (Neumann, 1959); the dark areas are subsutural, peripheral and basal in position. The columellar plate is white or yellow with a darker edge along the inner lip. Up to 8 mm high, 8 mm broad; last whorl occupies 95% of shell height, aperture 90%.

There is a broad snout, its edge lobulate. At its base lie the cephalic tentacles, capable of great extension and retraction. An eye on a separate eye stalk lies alongside each. The mantle edge is thick and papillated. In males a flattened penis with an open seminal groove lies behind the right tentacle. The foot has an epipodial fold on each side but this carries no tentacles; that on the left is joined to the eye stalk. The operculum, partly calcified, is D-shaped and has an internal peg. The flesh is white-yellow, with many black flecks, particularly on the snout, the mantle edge, the foot.

T. fluviatilis lives mainly in rivers, hiding under stones or wood by day, foraging by night on plants and detritus (Frömming, 1956). It also occurs in canals and the littoral region of lakes where there is movement in the water. The animals are limited to hard waters, preferring a calcium content of 20–30 mg.l^{-1}. They tolerate a moderately high salinity, up to 17‰ in the Loch of Stenness, Orkney (Nicol, 1938). The species occurs generally throughout Europe except for the most northern parts, the Iberian peninsula and the central Danube basin. In the British Isles it is absent from Devon and Cornwall in England, occurs only in Glamorgan in Wales, in the Lochs of Stenness and Harray (Orkney) in Scotland (Boycott, 1936), and in the limestone districts of Ireland.

Breeding begins late spring to early summer with spawning from early summer to early autumn. Females lay capsules on wood or stones or the shells of other animals. Each capsule consists of two halves, approximately

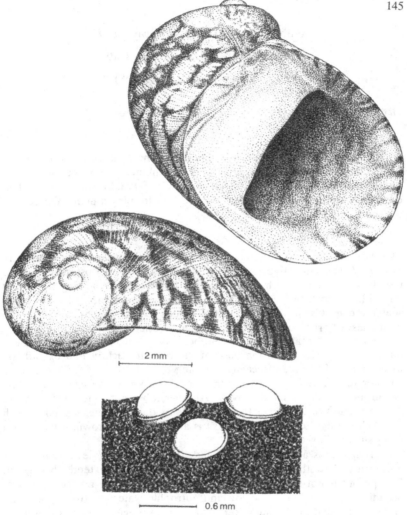

Fig. 50. *Theodoxus fluviatilis*. Bottom, three egg capsules.

hemispherical, the lower attached to the substratum, the upper fitting over
its mouth. They are made of mucous material and the upper is hardened
by a coat of sand grains, diatom cases and similar objects derived from the
faeces and stored in a so-called crystal sac until used (Fretter, 1946). Each
capsule may contain up to about 150 eggs, but only one develops successfully,
hatching as a juvenile snail, the others being used as its food. In brackish
water the number of eggs in a capsule drops to 50–60 (Bondesen, 1940).

Vernacular name: freshwater nerite.

Family VIVIPARIDAE Gray, 1847

Genus VIVIPARUS Montfort, 1810

Viviparus viviparus (Linné, 1758)
(Fig. 51)

Helix vivipara Linné, 1758

Diagnostic characters

Shell rather solid and opaque, with blunt tip; whorls moderately tumid, not flattened subsuturally, sutures not very deep; umbilicus minute or none. Aperture elongated and a little angulated adapically. Spiral brown bands on shell conspicuous. Right tentacle of males enlarged forming a penis. Operculum with concentric rings.

Other characters

The shell has a greenish periostracum but this often wears away. There are 5–6 whorls marked with prosocline growth lines. If the periostracum persists the shell is greenish with three chestnut brown bands showing through it on the last whorl, two on the others; where the periostracum is eroded the shell is cream. Usually 26–30 mm high (though recorded up to 40), 23–28 mm broad; last whorl occupies 80% of shell height, aperture 50–55%.

The head has a tapering snout and bears two tentacles, each with an eye on a lateral bulge. From the base of the right tentacle a fold extends on to the floor of the mantle cavity, forming the left edge of a food groove; another fold, forming its right edge, ends in a scroll-like exhalant siphon at the mouth of the mantle cavity. Another lobe, not rolled into a scroll, lies on the left of the neck. In females the oviduct becomes a brood pouch in which the eggs develop. The flesh is a dark velvety brownish black with many small yellow or black points.

V. viviparus is absent from Scotland, Ireland, west Wales and the most northern and southern parts of England. Elsewhere it extends throughout Europe but is not found in Iberia, southern Italy or the northern parts of Scandinavia. It is limited to lowland eutrophic waters in rivers and lakes and is more fond of moving water than the next species, though often living more or less buried in the bottom sediment (Frömming, 1956). It can eat plant material and detritus, using the radula, but also collects suspended particles on the ctenidium and transports them to the mouth via the groove on the floor of the mantle cavity (Cook, 1949; Starmühlner, 1952). The animals copulate in spring and summer, though not frequently (Sylvest, 1949). The eggs develop in the terminal part of the oviduct and are born as miniature adults, being shed in small groups throughout the summer, each with a shell about 4 mm high.

Vernacular name: river snail.

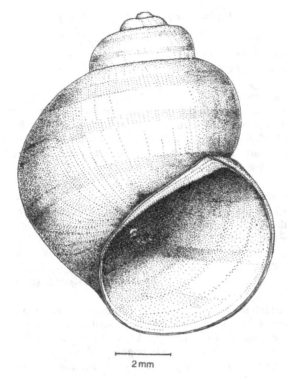

2 mm

Fig. 51. *Viviparus viviparus*.

Viviparus contectus (Millet, 1813)
(Fig. 52)

Cyclostoma contectum Millet, 1813
Viviparus fasciatus of authors

Diagnostic characters

Differs from *V. viviparus* in the following: the shell is semitransparent, delicate, retains its periostracum and is therefore a dark olive green; the brown spiral bands less obvious; spire sharply pointed; whorls very tumid, flattened subsuturally, sutures very deep. Aperture rounded, not angulated adapically; umbilicus conspicuous.

Other characters

There are 6–7 whorls. Young shells (up to three whorls) have a row of periostracal bristles at the periphery of each whorl: these become eroded but the scars remain visible. Up to 26–30 mm high, 23–28 mm broad, but shells up to 50 mm high have been found; last whorl occupies 80% of shell height, aperture 50–60%.

There are no significant differences from *V. viviparus* (p. 146) so far as the body of the animal is concerned.

V. contectus inhabits rivers, lakes and ditches, often living in quieter waters than *viviparus*. It has a more restricted distribution than that species, being limited to the English Midlands and Yorkshire in the British Isles, but it has a slightly more extensive range in continental Europe. The reproduction of *V. contectus* is as in *V. viviparus*, but the animals are said to copulate frequently (Dembski, 1968).

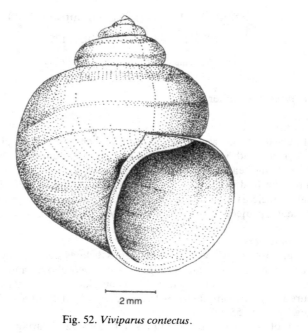

2 mm

Fig. 52. *Viviparus contectus*.

Family VALVATIDAE Thompson, 1840

Genus VALVATA Müller, 1774

Valvata piscinalis (Müller, 1774)
(Fig. 53)

Nerita piscinalis Müller, 1774

Diagnostic characters
Shell delicate, transparent, glossy if clear of encrustations; spire moderately high, cyrtoconoid, whorls tumid and sutures deep; umbilicus large; aperture circular, lying nearly wholly below penult whorl. Cephalic tentacles linked basally by a fold, a penis (in every individual) behind the right one. Right pallial tentacle as long as the cephalic ones and held erect in life. Ctenidium bipectinate, projecting from mantle cavity in living animals.

Other characters
There are 4–5 whorls with irregular growth lines and usually covered with plant growth. The aperture is slightly angulated above, and though a peristome is usually present, it may be incomplete over the last whorl. Horn coloured. Up to 10 mm high, 3–4 broad; last whorl occupies 90–95% of shell height, aperture 55–60%.

The foot is broad, with a sinuous anterior edge bearing recurved lateral points; it narrows in the middle but expands to a rounded, very thin, posterior end. The snout is rather long; the eyes lie median to the tentacle bases, behind the linking fold. Flesh yellowish, with dark and light points; tentacles pale.

V. piscinalis is widespread in Europe and is found throughout the British Isles except for Cornwall, south west Wales and north east Scotland. It occurs in fresh waters, both hard and soft, mainly where there is a gentle current, in ponds, ditches and streams, less commonly in lakes. The snails are detritus feeders but also rasp plants. They are hermaphrodite.

Breeding occurs from April to September (Cleland, 1954; Russell Hunter, 1961). Eggs are laid in spherical capsules measuring 1–2 mm across and fastened to weed, sometimes to stones; there are commonly five to about forty greenish eggs in a capsule. These develop to juveniles which hatch in 15–30 days.

Vernacular name: valve shell.

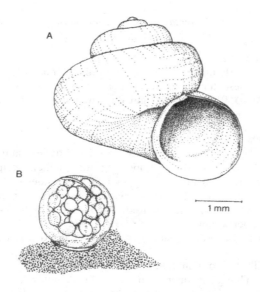

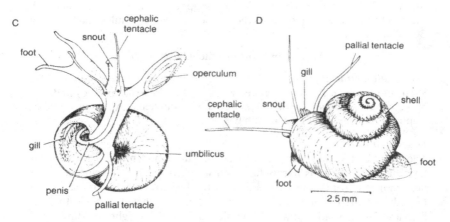

Fig. 53. *Valvata piscinalis*. A, shell; B, an egg capsule; C, animal extended from shell; D, animal crawling. (Fretter & Graham, 1962.)

Valvata cristata (Müller, 1774)
(Fig. 54)

Diagnostic characters
Shell with all whorls in one plane, looking like a small *Planorbis* but provided with an operculum; underside of shell deeply concave. Animal as in *V. piscinalis* (p. 150) though the gill does not ordinarily project and the pallial tentacle is relatively short.

Other characters
The shell is glossy, marked with growth lines, and occasionally with distinct spiral lines. There are 4–5 whorls meeting only at their periphery so that the sutures are deep on both the upper and lower sides of the shell. The aperture is circular, flattened a little against the surface of the last whorl, with a thin peristome. Horn-coloured. About 1 mm high, 3–4 mm across; last whorl and aperture both equal the shell height.

The animal is like *V. piscinalis* but is darker in colour, dark brown or black.

V. cristata likes a combination of muddy substratum, weeds, and gently flowing water. Though Palaearctic in general distribution like *V. piscinalis*, it is more local and less common than that species in the British Isles. Eggs are laid, 3–4 at a time, reddish, in vase-shaped capsules (about 3 mm high, 0.5 mm broad) attached to weeds (Nekrassow, 1929). Juveniles hatch in about 30–40 days.

Valvata macrostoma Steenbuch, 1847

Diagnostic characters
Shell intermediate in shape between that of *V. piscinalis* (p. 150) and that of *V. cristata* (above), with low spire, deep umbilicus. Aperture circular. Cephalic tentacles share a common base.

Other characters
There are 3–4 swollen whorls with deep sutures and marked with growth lines. The peristome is hardly flattened where it touches the surface of the last whorl and it may turn out a little at the base of the aperture and along the edge of the umbilicus. Horn-coloured, often rather milky. Up to 5 mm high and 5 mm broad; last whorl occupies 75% of shell height, aperture 60%.

The animal has much the same structure as other *Valvata* species. The tentacular bases adjoin, with a D-shaped eye set across each, and the pallial tentacle is stout. The head is dark brown, paler below, the tentacular tips pale; the foot is brown, but white on the sole and the opercular lobes.

V. macrostoma is the rarest of the British species and is confined to East Anglia and south east England, though living in similar places and feeding in similar fashion to the other species. Abroad it occurs in central and northern parts of Europe and is absent from Mediterranean countries (Frömming, 1956).

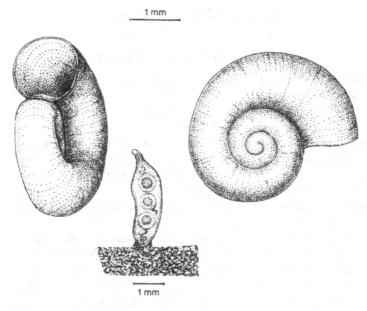

1 mm

1 mm

Fig. 54. *Valvata cristata.* Left, apertural view of shell, apex to left; right, apical view; below, an egg capsule.

Family LACUNIDAE Gray, 1857

Key to British species of Lacuna

1. Shell without umbilicus; columella not grooved; periostracum thick; last whorl occupies less than 80% of shell height; yellowish ....
   .... *Lacuna crassior* (p. 156)

   Shell with umbilicus; columella grooved **2**

2. Spire high, 5–7 whorls; last whorl occupies less than 90% of shell height; umbilical groove smooth **3**

   Spire low, 3–4 whorls; last whorl occupies more than 90% of shell height; umbilical groove ridged **4**

3. Shell turreted; periostracum thick and transversely ridged; without brown spiral bands on last whorl .... *Lacuna crassior* (p. 156)

   Shell not turreted; periostracum thin, smooth; usually 4 brown spiral bands on last whorl .... *Lacuna vincta* (p. 154)

4. Spire extremely low; umbilical groove very broad; greenish, unbanded .... *Lacuna pallidula* (p. 160)

   Spire low; umbilical groove moderately broad; usually 3 brown spiral bands on last whorl .... *Lacuna parva* (p. 158)

Genus LACUNA Turton, 1827

Lacuna vincta (Montagu, 1803)
(Fig. 55)

Turbo vinctus Montagu, 1803
Lacuna divaricata (Fabricius, 1780)

Diagnostic characters
Shell apparently smooth, with well developed spire, sharp apex, tumid whorls. Columella with groove leading to umbilicus. Aperture wide. Four orange-brown spiral bands on last whorl. Foot with two tentacles projecting from under operculum.

Other characters
The shell is delicate, covered with a thin periostracum. There are 5–6 whorls which, though tumid, are often a little flattened at the periphery. There is no ornament visible to the naked eye apart from markedly prosocline growth lines, but fine spiral ridges and grooves lie under the periostracum. The aperture is large, the outer lip curved above the periphery, rather straight towards the base where it everts a little. At the base of the columella it curves to form the abapertural edge of a groove leading to the umbilicus,

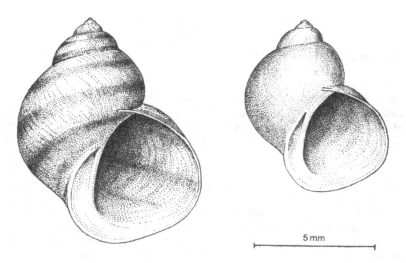

5 mm

Fig. 55. *Lacuna vincta*.

the adapertural edge of which is derived from the inner lip. Shelly material white but appears horn-coloured from the overlying periostracum. The four bands on the last whorl may fuse to give three, two, or a single one; apex of shell homogeneous purple-brown. Up to 10 mm high, 5 mm broad; last whorl 80–85% of shell height, aperture 55–60%.

The head has a long, broad snout with terminal mouth. At its base arise two long tentacles each with a basal eye. In males the penis arises behind the right eye and curves back within the mantle cavity. The foot is long, truncated anteriorly and rounded posteriorly. Body yellow, in some animals a little pinkish, with grey on the sides of the foot.

L. vincta has a northern range, the British Isles being near its southern limit. During most of the year the snails live sublittorally, but they migrate upshore to breed and then occur commonly, sometimes abundantly, at about L.W.S.T. on all British and Irish shores where there is a good growth of fucoids, particularly *Fucus serratus*, or of red weeds, on which they feed preferring sporelings (Fretter & Manly, 1977b). The animals are often very destructive. They breed over a long period in spring, laying between 1000 and 1500 capsules, each with a single egg. The capsules are embedded in jelly and laid on weed, mainly *F. serratus*, the whole spawn forming a ring or short spiral about 3 mm in diameter. The young hatch as veliger larvae which settle 2–3 months later. The animals are annuals, dying after spawning.

Vernacular name: chink shell.

Lacuna crassior (Montagu, 1803)
(Fig. 56)

Turbo crassior Montagu, 1803

Diagnostic characters
Shell covered with thick periostracum marked with prosocline ridges; whorls tumid and sutures deep. Some shells show slight spiral ridges at periphery of last whorl under periostracum. Aperture large, squarish. Umbilicus usually absent but sometimes a chink is present. Animal with two tentacles under operculum.

Other characters
The shell is rather solid, with 6–7 whorls forming a high spire with a stepped profile. If there is no umbilicus a columellar groove is not obvious, though it is when an umbilical chink occurs. The base of the aperture is drawn out to form a distinct spout. Yellow or brownish. Up to 14 mm high, 10 mm broad; last whorl occupies 75% of shell height, aperture 45–50%.

The body of the animal is like that of *L. vincta* (p. 154), is pale yellow, dark on the dorsal surface of the snout. The metapodial tentacles are white.

L. crassior is uncommon. It is usually found on soft bottoms from L.W.S.T. to depths of 90 m, and has been recorded off many British and Irish coasts; commoner in the north since its main distribution is arctic. It is said to feed on the bryozoan *Alcyonidium gelatinosum*; its life-history is not known. It has sometimes been regarded as merely a deep water form of *L. vincta*.

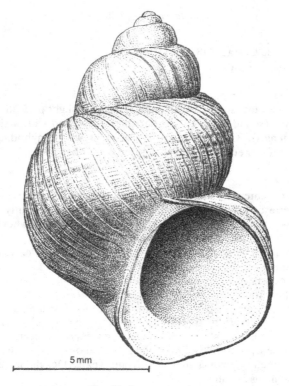

5 mm

Fig. 56. *Lacuna crassior*.

Lacuna parva (da Costa, 1778)
(Fig. 57)

Cochlea parva da Costa, 1778
Lacuna puteolus (Turton, 1819)

Diagnostic characters
Shell small, with low spire (apical angle 110–115°), slight periostracum, and large aperture (round in males, higher than broad and with a slight basal spout in females). Columellar groove marked with longitudinal ridges. Brownish, apex sometimes purplish; without colour bands.

Other characters
The shell is delicate, has three tumid whorls meeting at deep sutures below which they are a little flattened; the last whorl is often slightly angulated below the periphery. Up to 4 mm high, 4 mm broad; last whorl occupies about 90% of shell height, aperture about 70%.

The animal resembles *L. vincta* (p. 154). It is cream, the head pinkish with white tentacles, the sides of the foot grey.

L. parva occurs in the same situations as *L. vincta* but often somewhat higher on the shore, on *Fucus serratus, F. vesiculosus* and on red weeds, which, with their epiphytic growths, form its food. The animals are found on European shores from the Arctic Ocean to Iberia and on all suitable shores in the British Isles. The spawn is a hemispherical, clear mass of jelly fastened to weeds, preferably *Phyllophora*; it is about 2 mm in diameter and contains about ten eggs which hatch as juvenile snails (Ockelmann & Nielsen, 1981). The animals are annuals.

Vernacular name: chink shell.

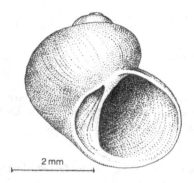

Fig. 57. *Lacuna parva.*

Lacuna pallidula (da Costa, 1778)
(Fig. 58)

Nerita pallidula da Costa, 1778

Diagnostic characters

Shell delicate, somewhat globular with hardly any spire (apical angle 150°), the last whorl very large. Aperture expanded and D-shaped. Columellar groove very broad and umbilicus wide. Greenish.

Other characters

The shell has 3–4 rapidly expanding whorls the last of which constitutes most of its bulk. Female shells may grow to a height of 12 mm and a breadth of 6 mm, whereas male shells are only half that size. The last whorl occupies nearly the whole of the shell height, the aperture 80–85%.

The animal is like that of other species. White, with grey on the sides of the foot.

L. pallidula is found on the same type of shore as other *Lacuna* species, on weeds, almost always *Fucus serratus*, at L.W.S.T. and to depths of 70 m (Smith, 1973). It is widespread in the British Isles and occurs in Europe from France to the Arctic as well as in Greenland and south to Connecticut. Breeding may occur throughout the year but is maximal February to May. Eggs are laid in jelly masses on the fronds of *Fucus serratus*; each mass contains a variable number of eggs (15–113), each in its own capsule. The mass is oval or circular, not kidney-shaped, in outline, slightly domed and tightly fastened to the substratum. In it the egg capsules lie in rough concentric rings, pressed together so that their outlines are angular (Goodwin, 1979). Development is direct and young snails emerge. They grow rapidly from May to October; thereafter the rate of growth of the sexes differs: females grow by 0.91 mm shell height per month from October until February then by 0.27 mm per month; males grow slowly and so reach a much smaller final height than females. Growth slows abruptly at spawning, and all the snails are dead by June or July (Smith, 1973).

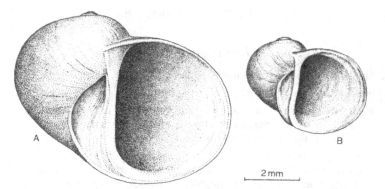

Fig. 58. *Lacuna pallidula*. A, female; B, male.

Family LITTORINIDAE Gray, 1840

Key to British species of Littorina

1. Shell with low spire, apex flattened **2**

   Shell with well developed spire, apex pointed **3**

2. Shell with distinct spire; outer lip arises below level of apex; aperture wide, throat not much constricted; ovipositor pigmented; penis with short tip and about 30 glands in 2–3 rows ....
   .... *Littorina obtusata* (p. 170)

   Shell with hardly any spire; outer lip arises nearly level with apex; aperture narrow, throat constricted; ovipositor not pigmented; penis tip long, about twelve glands in one row ....
   .... *Littorina mariae* (p. 172)

3. Whorls of spire almost flat in profile, sutures shallow; outer lip arises tangential to last whorl **4**

   Whorls of spire tumid, sutures deep; outer lip arises approximately at right angles to last whorl **5**

4. Adult shell large, usually with spiral ridges and grooves; without a periostracal flap along outer lip; columella white; transverse black stripes on cephalic tentacles .... *Littorina littorea* (p. 166)

   Adult shell small; periostracal flap along outer lip; columella dark; longitudinal black stripes on tentacles; usually very high on beach
   .... *Littorina neritoides* (p. 168)

5. Shell small, nearly globular, spire short; a peripheral spiral brown band bounded above and below by white enters the aperture; floor of mantle cavity dark; penis with long tip and 3–4 glands; ovoviviparous, so female duct not glandular . *Littorina neglecta* (p. 176)

   Shell with well-marked spire; whorls usually with clear spiral ridges and grooves (but may be eroded); colour pattern as above absent; floor of mantle cavity with diagnonal white band; oviparous or ovoviviparous **6**

6. Animals on weeds in sheltered brackish habitats commonly not exposed at low tides; shell dark, thin and usually smooth ....
   .... *Littorina tenebrosa* (p. 181)

   Animals live on upper part of beach in crevices or under stones in fully saline conditions, exposed at low tides; shell solid, usually with spiral ridges **7**

7. Shell with strap-like spiral ridges, often with central furrow; grooves often darkly pigmented; penis with very short tip and up to twelve (usually fewer) glands; oviparous (female duct glandular) ....
.... *Littorina nigrolineata* (p. 172)

Shell with spiral ridges, usually rising to a crest but sometimes strap-like; penis tip moderately long; oviparous or ovoviviparous **8**

8. Oviparous, female duct with glands; in male prostate not reaching level of anus, with only small ciliated field alongside, not red in living animals (not wholly reliable character); penis with tapering tip, glands reaching to or round its curved base ....
.... *Littorina arcana* (p. 180)

Ovoviviparous, female duct without conspicuous glands; prostate reaches beyond anus, ciliated field large and red in living animals; penis with blunt tip, its glands not reaching its base **9**

9. Spiral ridges on shell with crests placed centrally; aperture everted only slightly at base; throat dark; in crevice at top of beach ....
.... *Littorina saxatilis saxatilis* (p. 178)

Spiral ridges crested along their adapical edge; aperture with marked basal eversion; throat commonly not dark; among weeds, under stones with *Pelvetia* at top of beach ...... *Littorina saxatilis rudis* (p. 178)

Identification of Littorina *species*

There is usually little difficulty in identifying *L. littorea* (p. 166). Young specimens have clearly ridged shells and may be mistaken for *L. saxatilis* (p. 178), but may be distinguished by their white columella. *L. neritoides* (p. 168) is instantly recognizable by the periostracal flap along the outer lip, and *L. obtusata* (p. 170) and *L. mariae* (p. 172) by the flat spire. *L. mariae* may be picked out by the greater flattening of the apical region of the shell, by the constricted throat and by the long back-turned tip of the penis; juvenile shells lack a nick where the outer lip joins the surface of the last whorl when seen in apical view.

All the other species, though originally described as separate, have a sufficiently similar appearance to make understandable the attempt to unite them into a single species, sometimes called *saxatilis*, sometimes *rudis*, as was first done by Jeffreys (1862–69). Recent work, however (Heller, 1975; Raffaelli, 1982), has shown this situation to be untenable and up to six possible species have been regarded as occurring on the shores of the British Isles: *nigrolineata, neglecta, arcana, tenebrosa, rudis* and *saxatilis*, though it is not yet safe to assume that the last two or three represent more than ecological forms.

Identification of species in the group, the '*saxatilis* complex', often involves difficulty because of the scarcity of undoubtedly absolute differences. For certainty, shell, body, and habitat should all be known; even then single specimens may be troublesome and the student should make a reference collection of named shells with which comparison may be made. Identification of *nigrolineata* (p. 172) and *neglecta* (p. 176) is straightforward: doubts arise mainly with *arcana, rudis, saxatilis* and *tenebrosa*.

L. arcana (p. 180) can be identified at once by its oviparous habit, revealed by the glandular nature of the oviduct, whereas the other three species are ovoviviparous. *L. arcana* lacks pigment in the right part of the mantle skirt which the others possess (though this is not a reliable character), and certainly prefers more exposed shores. In the male the relationship of prostate and anus and the shape of the penis are helpful.

L. saxatilis (or *L. saxatilis saxatilis*) (p. 178) usually has a patterned shell with a dark throat, and lives high, in crevices. The last whorl and aperture are large (breadth of last whorl/breadth of penult = 2 or more), and the spout below the columella is little developed. In *L. rudis* (or *L. s. rudis*) (p. 178), by contrast, the shell normally lacks pattern, the throat has no different colour from the outer surface, the apertural area is less, the last whorl smaller (breadth of last whorl/breadth of penult = 2 or less), and the apertural spout is usually prominent. It lives at a slightly lower level, under stones. *L. tenebrosa*, thinner and smoother, usually dark, is commonly limited to brackish water, is often permanently submerged and shows a fondness for weedy substrata.

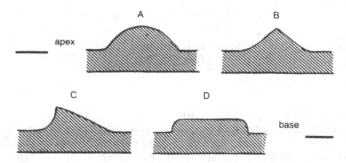

Fig. 59. Spiral ridges of shells of *Littorina* species in transverse section. A, *littorea*; B, *saxatilis saxatilis*; C, *saxatilis rudis*; D, *nigrolineata*.

Genus LITTORINA Férussac, 1822

Littorina littorea (Linné, 1758)
(Figs 59A, 60)

Turbo littoreus Linné, 1758

Diagnostic characters (see also p. 164)
Shell solid with moderately high, pointed spire; whorls nearly flat in profile, with numerous low spiral ridges (Fig. 59A, but may be eroded). Outer lip approaches last whorl tangentially and forms short spout at base of columella. Umbilicus absent. Reddish brown to black, columella white. Tentacles with transverse black stripes. Penis with 16–36 glandular papillae in several rows. Oviparous, female duct glandular.

Other characters
The shell is not glossy, has 5–6 whorls and its profile is often concave towards the apex. The spiral ridges are often worn except on the base, but are always clear on young shells; they are strap-shaped on the last whorl, narrower and sharper on older ones. Prosocline growth lines also occur, often irregularly enlarged. The aperture is pointed adapically and the first part of the outer lip is frequently a little concave; in young shells the lip is thin and crenulated by the spiral ridges. The columella is short. The apical region may be eroded and ashen in colour. Up to 30 mm high, 25 mm broad; last whorl occupies 80% of shell height, aperture 60–70%.

The snout is broad, the tentacles a little flattened, each with an eye on a lateral bulge basally. Males have a penis behind the right tentacle, sickle-shaped and tapering distally. Its glandular papillae lie on the left (anterior) edge; its right side carries a seminal groove. In the same position in females is an ovipositor, a glandular field with raised margin. Both are reduced or lost outside the breeding season (Grahame, 1970). The foot is shield-shaped. Yellowish with numerous black lines, the penis and ovipositor paler.

L. littorea is abundant throughout the British Isles (though rare in the Scillies and Channel Islands), on rocky shores and any others sufficiently firm to give some attachment. Winkles live between H.W.S.T. and L.W.S.T. and, in high latitudes, to depths of 60 m. Their upper limit on a shore depends on the degree of shelter and shade, both raising it. They eat diatoms, young and older algae, and are particularly attracted to *Ulva*, *Enteromorpha*, avoiding tougher fucoids. At the higher levels on a beach vegetable detritus becomes an important food (Watson & Norton, 1985). The species ranges in Europe from Spain to the White Sea; it is also found in Greenland and south to New Jersey in America.

Breeding is concentrated into the months February to May but may occur at almost any time (Smith & Newell, 1955; Williams, 1964; Fish, 1972; Grahame, 1975). Females lay lens-shaped capsules with an equatorial rim

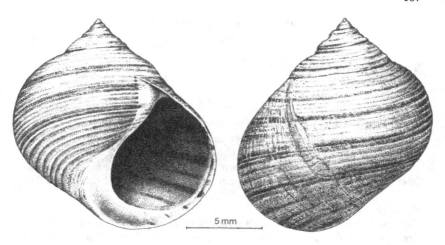

Fig. 60. *Littorina littorea.*

which contain 1–5 colourless eggs (Tattersall, 1920), several bursts of laying occurring during the breeding season (Daguzan, 1976b). The capsules are planktonic and after 5–6 days veliger larvae hatch which have a free life of 4–7 weeks, depending on locality. They settle in waves, giving a polymodal population, either sublittorally (Smith & Newell, 1955; Daguzan, 1976b) or in intertidal crevices or amongst barnacles. Growth, at first rapid, slows if the temperature falls below 8°C, but by one year the animals have a shell height of 8–10 mm, at 2 years 13–16 mm, 3 years 15.5+ mm; most then die (Williams, 1964; Cousin, 1975).

Vernacular name: edible winkle.

Littorina neritoides (Linné, 1758)
(Fig. 61)

Turbo neritoides Linné, 1758
Melaraphe neritoides (Linné, 1758)

Diagnostic characters (see also p. 164)
Shell with flexible periostracal flap along outer lip, of which the origin is tangential to the last whorl. Surface smooth, whorls only slightly tumid and sutures shallow. Outer lip everted at base of columella. Dark bluish or brownish black, spire usually eroded to fawn. Tentacles with only longitudinal black lines. Penial glands in more than one row. Oviparous, female duct glandular.

Other characters
The shell is solid, the spire with a nearly straight profile. Growth lines are present, sometimes wrinkling the surface of the whorls. The aperture is pointed at its apical end. Though the general appearance suggests no colour pattern there is often a white spiral band on the base of the shell and a dark one round the periphery of the last whorl; dark and light stripes may also run across this whorl. Up to 9 mm high, 7 mm broad; last whorl occupies about three quarters of the shell height, the aperture more than half.

The animal is in general like *L. littorea* (p. 166). The ovipositor is linked to the female opening by a ciliated groove. The body is grey with dark lines along the tentacles, white round each eye and on the sole of the foot.

In most places on British and Irish coasts where it occurs *L. neritoides* is the highest of the *Littorina* species. It is to be found in cracks in, and on the surface of, rocks from the *Pelvetia* level to heights determined by exposure: the higher the splash zone the higher the winkles can survive. Sometimes they are encountered as low as M.H.W.N.T. (Moyse & Nelson-Smith, 1963) and this is usually explained as due to high humidity or to a downward breeding migration. Their food is predominantly black lichens (Daguzan, 1976a), but they also eat detritus. They are found on all coasts except in the southern half of the North Sea, where suitable habitats are absent; they are, however, local and are often not present even where conditions seem suitable. Outside the British Isles *L. neritoides* has been recorded from Scandinavia to the Mediterranean and Black Seas.

Breeding occurs from autumn to spring, each egg being laid in a disk-shaped capsule shed into the plankton (Lebour, 1935c); this may be achieved by the downward migration referred to above, but capsules may be deposited in pools at the level at which the adults live. Veliger larvae hatch from the capsules and spend about three weeks in the plankton before settling at a much lower level on the shore than their final home, which is reached by upward migration (Fretter & Manly, 1977a). Breeding rates are very variable and many animals apparently fail to reproduce in any given year, perhaps depending on weather conditions (Hughes & Roberts, 1980). The rate of

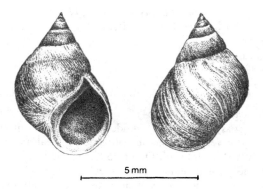

5 mm

Fig. 61. *Littorina neritoides*.

growth, for the same reasons, is equally variable, but animals with shells 7 mm high are probably about five years old (Lysaght, 1941; Hughes & Roberts, 1980).

Littorina obtusata (Linné, 1758)
(Fig. 62A,B,C,D)

Nerita obtusata Linné, 1758
Littorina littoralis of authors
Littorina aestuarii Jeffreys, 1869

Diagnostic characters (see also p. 164)
Shell solid, not glossy; last whorl rounded, sometimes a little flattened at the periphery, spire very low and smooth-sided (apical angle about 120°). Surface apparently unsculptured. Outer lip arises not at right angles to shell axis. In apical view (of juvenile shells especially) a notch where outer lip and last whorl meet. Throat not markedly constricted. Colour variable. Penis flat with short tip and many glands in 2–3 rows. Ovipositor with some pigment. Oviparous, female duct with glands.

Other characters
There are 4–5 whorls with many fine prosocline growth lines and, less constantly, delicate spiral lines. The first part of the outer lip is usually straight or concave in apertural view, and the inner lip spreads outwards alongside the columella and blocks the umbilicus. The shell may be of one colour (orange to red, yellow, brown-black, light or yellowish green), or it may have a dark reticulated pattern, or spiral bands overlying the ground colour (Dautzenberg & Fischer, 1914). The plain yellow, green and brown shells and those with a reticulated pattern are the common morphs in the British Isles (Smith, 1976). Up to 15 mm high, 17 mm broad; last whorl occupies about 90% of shell height, aperture 75–80%.

The animal resembles *L. littorea* (p. 166) but has only longitudinal stripes on the tentacles, and the penis and ovipositor are much reduced. It is yellowish with dark lines the degree of development of which parallels the amount of pigment in the shell.

L. obtusata occurs abundantly throughout the British Isles and extends south to the Mediterranean and north to north Norway; it is also found in the western parts of the Baltic, in eastern Canada, and south to New Jersey. It nearly always lives in association with fucoids, particularly *Fucus vesiculosus* and *Ascophyllum nodosum* growing along the edge of tidal pools, whose floats the shells mimic. It centres on the lower half of the beach, H.W.S.T. to L.W.S.T., avoiding the most exposed shores and tolerating some brackishness (Daguzan, 1976b; Goodwin & Fish, 1977).

The animals are capable of breeding throughout the year, usually less actively in summer. Eggs, each in a capsule, are aggregated into sheets of jelly attached to damp fronds of *Fucus* species, less frequently to stones. The spawn forms flat rather than domed masses, round, oval or kidney-shaped, in which the egg capsules are packed loosely so that their outlines are round, not angulated. The spawn is usually rather easily lifted off the

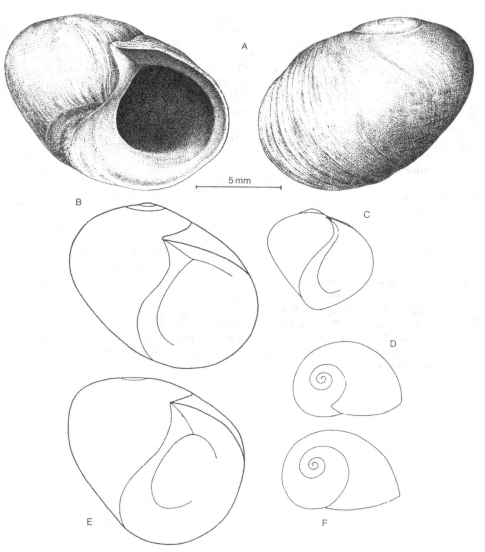

Fig. 62. A, *Littorina obtusata*; B, C, D, *Littorina obtusata*; E, F, *Littorina mariae*.
B, and E, apertural views of adult shells; C, apertural view of juvenile shell; D, and
F, apical views of juvenile shells.

substratum (Goodwin, 1979). The eggs hatch in three or four weeks as young
snails (Goodwin, 1978).

Vernacular name: flat winkle.

Littorina mariae Sacchi & Rastelli, 1966
(Fig. 62E,F)

Diagnostic characters (see also p. 164)

Closely similar to *L. obtusata* (p. 170) but with a shorter, even totally flat spire, the outer lip arising higher on the last whorl and nearly normal to the shell axis. In apical view (in young shells especially) there is no notch where outer lip and last whorl meet. Throat constricted. Penis with long tapering tip held under glandular part, which has about twelve glands in one row. Ovipositor not pigmented. Oviparous, hence female duct glandular.

Other characters

The same colour morphs occur in *mariae* as in *obtusata*, yellow and dark reticulated shells being the commonest. Up to 11 mm high, 12 mm broad; last whorl occupies 95–100% of shell height, aperture 85–90%

L. *mariae* has the same habits as *L. obtusata* but is often found at a slightly lower level on the shore, with *Fucus serratus*; it also prefers more exposed situations (Goodwin & Fish, 1977). Its full distribution within the British Isles is not certain since its recognition as a distinct species is relatively recent. It has, however, been found in Kent (Shellness), and on most southern and western shores as well as in Jersey, and on northerly east coast ones to Orkney and Shetland (McMillan, 1981).

The spawn of this species is, in general, like that of *L. obtusata*. It is almost always laid on *Fucus serratus* and is more commonly kidney-shaped rather than oval in outline and is not circular (Goodwin, 1975; Goodwin & Fish, 1977).

Littorina nigrolineata Gray, 1839
(Figs 59D, 63)

Littorina rudis var. *compressa* Jeffreys, 1866
Littorina saxatilis of authors (part)

Diagnostic characters (see also p. 164)
Shell with moderately high spire with pointed apex; whorls swollen with strap-shaped spiral ridges (Fig. 59D), each often with a small central groove. Aperture oval, outer lip arising nearly at right angles to last whorl and everted at base of columella to form a short spout. White, grey, yellow or red, the spiral grooves often dark, but never with a reticulated pattern. Penis rather flat with short tip and up to twelve glands close to tip. Oviparous, hence female duct glandular.

Other characters
The shell has six whorls. There are 10–21 ridges on the last whorl, 3–7 on the penult, 3–4 on the most apical; the intervening grooves are narrow. Up to 20 mm high, 18 mm broad; last whorl occupies 75–90% of shell height, aperture two thirds to three quarters.

The animal is like *L. littorea* (p. 166) but has only longitudinal stripes on its tentacles. The flesh is yellowish; the penis is white, and from its base a white streak passes diagonally backwards on the floor of the mantle cavity.

L. nigrolineata is found amongst weeds and stones on rocky shores, at levels related to the degree of exposure but commonly at about M.T.L., creeping actively after both emersion and immersion (Petpiroon & Morgan, 1983). It eats weeds, microphytes and detritus (Sacchi, Testard & Voltalina, 1977). The animals have been recorded from localities on the western sides of the British Isles between Scilly and Lewis as well as on the northern shores of the North Sea.

These winkles breed throughout the year, with a summer maximum (Sacchi, 1975), laying eggs in capsules embedded in jelly masses which are fastened to the underside of the stones amongst which they live. The eggs are pink and the capsules in which they lie are angular in outline. The young hatch as juvenile snails with shells 0.5 mm high in 4–7 weeks (North Wales) and are estimated to take 6–9 months to grow to 4 mm, with a winter pause in growth (Hughes, 1980), and probably a further year to reach maturity.

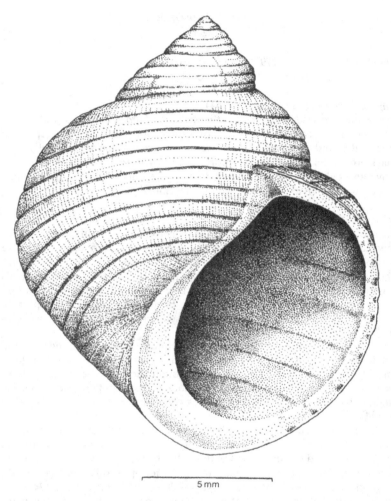

Fig. 63. *Littorina nigrolineata*.

Littorina neglecta Bean, 1844
(Fig. 64)

Littorina rudis (Maton, 1797) of authors (part)
Littorina saxatilis (Olivi, 1792) of authors (part)

Diagnostic characters (see also p. 164)
Shell rather globose, smaller than that of other *Littorina* species, with short, blunt spire (apical angle 90–100°); whorls tumid, apparently smooth. Aperture large and round, pointed apically; columellar region narrow. Usually with a dark brown spiral band on last whorl which runs into aperture and is bounded above and below by white. Penis with elongated tip with 3–4 large papillae in one row. Ovoviviparous, with young in terminal part of female duct.

Other characters
The shell is thin and has 3–4 whorls slightly flattened in the subsutural region. There are prosocline growth lines and some fine spiral striae, of which one or two at the periphery are the clearest. Though spiral coloured bands are regularly found in adults they tend to be the sole pattern of young animals and are partly replaced by a tessellated pattern in old ones (Hannaford Ellis, 1984). The columella and the throat are brown, the base of the aperture white. Up to 5 mm high, 4.5 mm broad, usually smaller; last whorl occupies 85–90% of shell height, aperture 60–65%.

The animal has the same general appearance as other littorinids but the female duct ends in a brood pouch. The flesh is dark grey; two dark lines run along each tentacle, there is white round the eye, the snout is dark, the sides of the foot rather pale. The floor of the mantle cavity lacks the diagnonal white streak of other species once united with it in *L. saxatilis* (p. 164).

L. neglecta is found on rocky shores, most commonly on those which are exposed. It occurs from near the top of the barnacle zone downwards, in small crevices and in *Laminaria* holdfasts, but is most reliably found in empty barnacle shells, the largest animals towards the top of this range (Raffaelli, 1978). It eats mainly unicellular algae. Like *L. nigrolineata* (p. 174) it is only recently that its specific status has been recognized and many specimens must have been dismissed as juveniles of other species. Its occurrence within the British Isles is widespread and it is likely that it lives wherever shores provide suitable habitats. It has also been found in New England, where it is associated with *Zostera* (Robertson & Mann, 1982), but its wider distribution is uncertain. Breeding has been described by Raffaelli (1976): it is greatest in spring. Young are born with shells about 0.5 mm high and mature at a shell height of about 2 mm.

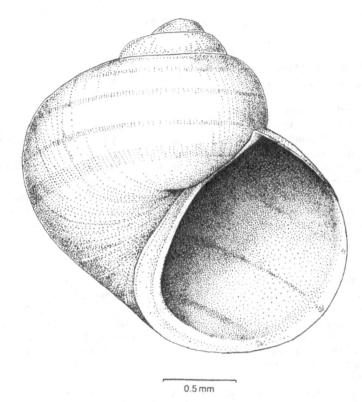

0.5 mm

Fig. 64. *Littorina neglecta.*

Littorina saxatilis (Olivi, 1792)
(Figs 59B,C; 65)

Turbo saxatilis Olive, 1792
Littorina rudis of authors

Diagnostic characters (see also p. 164)
Shell solid, opaque, with pointed spire, swollen whorls and deep sutures; ornament of numerous spiral ridges which are usually symmetrically V-shaped in section (Fig. 59B; examine those not on shell base, and see below), sometimes reduced and often eroded. Aperture round, outer lip arising from last whorl almost at right angles to shell axis. Columella short, apertural edge turned out below it slightly. Colour variable, but usually with a tesselated pattern, throat darker than outer surface. Tip of penis often blunt but with a point, its glands small and not reaching the curved base. Ovoviviparous, terminal part of the female duct containing embryos.

Other characters
The shell has 4–5 whorls. The ratio breadth of last whorl/breadth of penult is two or more (breadth = greatest measurement at right angles to shell axis; the breadth of the last whorl measured just above the origin of the outer lip). The area of mantle skirt between the genital duct and the columellar muscle may show red colour in live animals (it is sometimes difficult to be sure of this) and in males the prostate gland extends anterior to the anus. Up to 15 mm high, 10 mm broad; last whorl occupies 75–80% of shell height, aperture 50–60%.

L. saxatilis, the highest littorinid apart from *L. neritoides* (p. 168), is typically found in crevices of bedrock at and above the *Pelvetia* zone. The species is spread throughout western Europe, including the western parts of the Baltic, and is also found on north eastern and north western coasts of North America. A dwarf form with large aperture, *L. s. scotia*, has been described from algal tufts in Rockall, and another with everted lips (*L. s. patula*) also from Rockall and Eddystone Rocks probably belongs here.

Some authorities, most recently J. E. Smith (1981), regard the form *rudis* Maton, 1797 as a species distinct from *saxatilis*, though most treat it only as an ecological variant. The differences by which *rudis* may be recognized are: the spiral ridges are asymmetrical in section, their adapical sides the steeper (Fig. 59C); the out-turning of the lip below the columella is large; the shell usually shows no tessellated pattern; the throat is not dark; the ratio breadth of last whorl/breadth of penult is 2 or less; the animals occur about the level of *Pelvetia canaliculata* but prefer damp areas where they may be found under stones; if on vertical surfaces they are not usually in crevices.

Vernacular name: rough winkle.

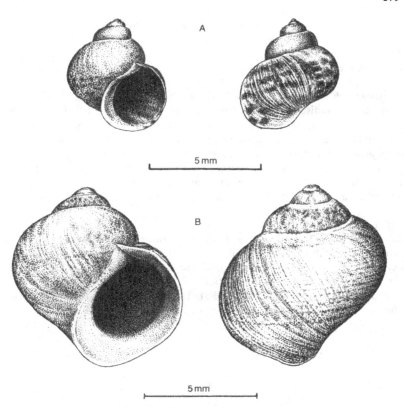

A

B

5 mm

5 mm

Fig. 65. A, *Littorina saxatilis saxatilis*; B, *L. saxatilis rudis*.

Littorina arcana Hannaford Ellis, 1978

Littorina saxatilis of authors (part)
Littorina rudis of authors (part)

Diagnostic characters (see also p. 164)
Shell almost indistinguishable from that of *L. saxatilis rudis* (p. 178 and Fig. 65B), but spire relatively smaller, eversion of aperture at base of columella more marked, and shell breadth (in apical view) parallel to apertural edge greater than that at right angles to it. Penis with tapering tip, its glands variable in number and arrangement but usually reaching its curved base. Red colour absent from right part of mantle cavity wall. Prostate does not extend as far as the anus. Oviparous, terminal part of female duct glandular.

Other characters
L. arcana often lives alongside *L. saxatilis rudis* but commonly extends a little lower on the beach and prefers more exposed shores. At breeding periods, however, it migrates downshore to a level where its egg masses are not subject to serious desiccation and the possibility of cross-breeding is eliminated (Hannaford Ellis, 1985). Its distribution is still incompletely known but it has apparently the same range in the British Isles as *L. s. rudis*.

Littorina tenebrosa (Montagu, 1803)

Turbo tenebrosus Montagu, 1803
Littorina rudis var. *tenebrosa* of authors

Diagnostic characters (see also p. 164)
Shell like that of *L. saxatilis rudis* (p. 178) in general shape but thinner and with reduced spiral ridging; whorls slightly flattened peripherally; columella narrow and aperture little everted below it. Usually dark brown or black. Ovoviviparous.

Other characters
The shell has 5–6 whorls and appears smooth to the naked eye. It sometimes has a tessellated pattern and sometimes a dark spiral band below the periphery of the last whorl. The columella and the throat are dark. Last whorl occupies about 80% of shell height, aperture about 55%.

L. *tenebrosa* is to be found in sheltered situations which are frequently brackish. It tends to prefer a weedy to a stony substratum and often lives where it is permanently submerged. The distribution is not well known.

The specific standing of this winkle is perhaps less certain than that of others and it may well prove to be only a variety of *L. saxatilis*.

Family POMATIASIDAE Gray, 1852

Genus POMATIAS Studer, 1789

Pomatias elegans (Müller, 1774)
(Fig. 66)

Nerita elegans Müller, 1774
Cyclostoma elegans (Müller, 1774)

Diagnostic characters
Shell with cyrtoconoid spire, tumid whorls and deep sutures; ornament of many fine spiral ridges and grooves. Aperture with nearly circular peristome with small adapical spout. Operculum thick, calcified externally.

Other characters
The shell is opaque and dull. There are 4–5 whorls, the last with about 35 spiral ridges, the penult with 17–18. There are also numerous prosocline growth lines, not usually visible except in the spiral grooves. A thickening in the throat, clearest on the columellar side, acts as an opercular stop. There is a deep umbilical groove leading to a distinct umbilicus. White to tawny or pinkish, sometimes with darker blotches and often with a purplish cast at the apex. Up to 15 mm high, 13 mm broad; last whorl occupies 75–80% of shell height, aperture 45–50%.

The head is extended into a long, broad, depressed snout with the mouth at its tip. The tentacles are thick and short, not exceeding the snout in length, with slightly bulbous tips. An eye lies on a bulge at the base of each tentacle laterally. In males a dorsoventrally flattened penis arises behind the right tentacle, rather far within the mantle cavity. The foot is oval, rounded anteriorly, and clearly divided into right and left halves by a median groove. The animal creeps by moving the two halves forward alternately. Dark brownish grey with many fine white and yellow points, the tentacles black, pale at the tips, eyes on pale areas.

P. elegans is a markedly calciphile species, living in hedge bottoms, among leaf litter and plant roots, often subterraneously, especially in cold weather. It may creep on the surface when temperature and humidity are both high, eating vegetable matter. The eggs are laid singly in spherical capsules buried in the soil, 2 mm in diameter, though the egg measures only 140 μm across. A young snail hatches in about three months (Creek, 1951).

P. elegans is a southern species found all round the Mediterranean and extending north to southern Germany, through France to Denmark and Britain, where it is limited to parts of England south of a line Lancaster to Scarborough; absent from Wales, much of the Midlands, East Anglia, Devon and Cornwall, though occasionally found there in sand dune areas.

Vernacular name: round-mouthed snail.

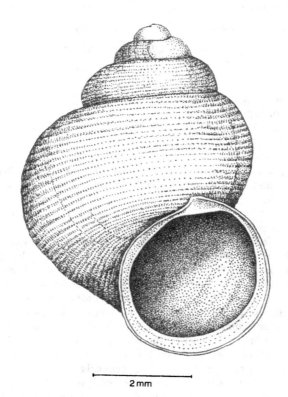

2 mm

Fig. 66. *Pomatias elegans*.

Family ACICULIDAE Gray, 1850

Genus ACICULA Hartmann, 1822

Acicula fusca (Montagu, 1803)
(Fig. 67)

Turbo fuscus Montagu, 1803
Acme fusca of authors
Acicula lineata of authors

Diagnostic characters

Shell small, columnar, with blunt tip; whorls flat-sided each with subsutural spiral ridge and axial grooves. Confined to leaf litter in woodland on calcareous soils.

Other characters

A glossy, semitransparent shell with narrow, slightly cyrtoconoid spire. There are 5–6 whorls, the last of which occupies about 40% of the total height. The aperture is square to lozenge-shaped, angulated at its apex, with a prominent peristome. The columellar lip all but obliterates an umbilical groove and an umbilicus is absent. The parietal lip is rather prominent and nearly straight. The operculum is thin, with few turns. Yellowish brown, sometimes redder. Up to 2 mm high, 1 mm broad; aperture occupies 20–25% of shell height.

The head extends to a narrow snout with terminal mouth. The tentacles are set far apart, each with an eye behind its base, and a tentaculiform penis arises behind the right eye in males; its edges curl to form a groove connected across the floor of the mantle cavity to the male pore. There is no ctenidium, and the mantle cavity is commonly filled with water containing a bubble of air, the water circulated by cilia on the walls of the cavity (Creek, 1953). The foot is narrow, broader anteriorly, pointed posteriorly. The flesh is nearly transparent with many fine brown spots on the skin, but black at the base of the tentacles and on the mantle skirt.

A. fusca is found under stones and pieces of wood amongst leaf litter in woods, especially beech woods on calcareous soils. It prefers damp, shady places and avoids light. It is generally distributed in the British Isles where these conditions can be met. Abroad, the species ranges eastwards as far as Greece, except Spain and Portugal, and is absent from most parts of Europe north of the Alps.

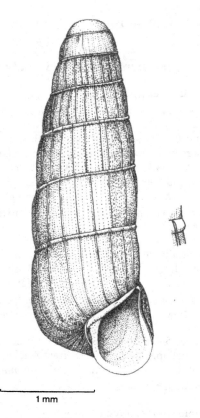

1 mm

Fig. 67. *Acicula fusca*.

Family HYDROBIIDAE Stimpson, 1865

Key to British genera and species of Hydrobiidae

1. Animals from fresh water **2**

   Animals from brackish or sea-water **4**

2. Spire rather high; 6 whorls, tumid, rarely with spiral keel or tufts
   of bristles; aperture pointed adapically, its maximal breadth near
   base; umbilicus usually blocked; white spot behind eye, snout
   evenly dark but white near tip; males rare ....
   .... *Potamopyrgus jenkinsi* (p. 194)

   Spire rather low; 4–6 whorls, not keeled or bristly; umbilicus
   obvious; yellow spot behind eye **3**

3. 5–6 moderately tumid whorls; apex pointed; aperture angled apically; in Sussex, East Anglian and southern Irish river estuaries
   only .... *Pseudamnicola confusa* (p. 196)

   4 distinctly tumid whorls, the last large; apex blunt; aperture round;
   in canals near Manchester only .... *Marstoniopsis scholtzi* (p. 198)

4. Umbilicus obvious; 5–6 whorls, moderately tumid; apex pointed;
   aperture angled adapically; in Sussex, East Anglian and southern
   Irish river estuaries only .... *Pseudamnicola confusa* (p. 196)

   Umbilicus closed or at most a minute chink **5**

5. 6 moderately tumid whorls, often with spiral keel or tufts of bristles;
   aperture with maximal breadth close to base; white transverse
   mark at apex of snout, the rest evenly dark ....
   .... *Potamopyrgus jenkinsi* (p. 194)

   Shell and animal not like this **6**

6. 6–7 whorls, usually nearly flat-sided, apex rather rounded; maximal
   apertural breadth near mid-height; snout with dark transverse
   bar behind apex; tentacles with black bar near tip; penis with
   blunt tip and smoothly curved profile on right .. *Hydrobia ulvae* (p. 188)

   Shell and animal not like this **7**

7. 5–6 markedly tumid whorls with deep sutures, apex pointed; maximal apertural breadth near mid-height; snout white dorsally, dark
   laterally, the bands converging between the eyes; penis with long
   flagellum .... *Hydrobia ventrosa* (p. 190)

   Shell and animal not like this **8**

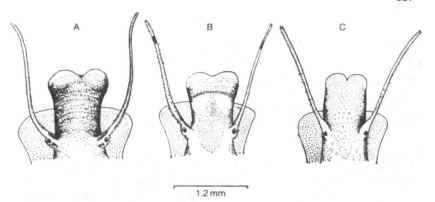

Fig. 68. Dorsal view of head of A, *Potamopyrgus jenkinsi*; B, *Hydrobia ulvae*; C, *Hydrobia ventrosa*, to show pigmentation pattern. (Fretter & Graham, 1962.)

8. Whorls moderately tumid, last whorl large; spire not noticeably cyrtoconoid; maximal apertural breadth near base; males rare
.... *Potamopyrgus jenkinsi* (p. 194)
Whorls slightly tumid, last whorl not large; spire distinctly cyrtoconoid; maximal apertural breadth at mid-height; penis with blunt tip and angulated profile on right .... *Hydrobia neglecta* (p. 192)

Genus HYDROBIA Hartmann, 1821

Hydrobia ulvae (Pennant, 1777)
(Figs 68B, 69)

Turbo ulvae Pennant, 1777
Rissoa ulvae (Pennant, 1777)
Peringia ulvae (Pennant, 1777)
Sabanaea ulvae (Pennant, 1777)

Diagnostic characters
Shell small, spire moderately high, whorls nearly flat-sided in profile, with brown periostracum; prosocline growth lines the only clear ornament. Aperture pointed apically. Snout with dark transverse line immediately behind a pale tip and (usually) dark lateral lines and (sometimes) a mid-dorsal one. A black transverse one on each cephalic tentacle a little behind its tip.

Other characters
The spire has 6–7 whorls and is a little cyrtoconoid. There is an umbilical chink in most shells. Up to 6 mm high, 2.5–3 mm broad; last whorl occupies 60–70% of shell height, aperture about 40%.

The head has a long snout, bifid at its tip. The left tentacle is thicker than the right and shows different ciliary currents. The mantle skirt has a single pallial tentacle on the right. In the mantle cavity of males, there is a large penis shaped like a question mark; its tip is blunt and it shows a smoothly curved profile on the right anteriorly. The flesh is yellowish grey with black and bright yellow speckles. In addition to the colour pattern on the snout mentioned above there is a backwardly projecting, black, triangular area with its base between the tentacles. The eyes lie in a pale patch usually bordered anteriorly by a black streak.

H. ulvae occurs, often in enormous numbers, on wet banks of sand or mud, preferring estuarine conditions but occasionally on open coasts and liking habitats with moving tidal waters (Cherrill & James, 1985). It extends upwards from sublittoral levels and is commonest on the uppermost third of the beach. It also occurs on weeds and in salt marshes where the snails are often large (Chatfield, 1972). It eats diatoms collected from the surface of sand grains (Fenchel, 1975; Fenchel, Kofoed & Lappalainen, 1976), some bacteria (Jensen & Siegismund, 1980), silt, and such algae as *Ulva* and *Enteromorpha*, preferring the latter (Nicol, 1935). It feeds best on particles 40–80 μm in size and cannot apparently take in pieces greater than 210 μm (Lopez & Kofoed, 1980). According to Newell (1962) the animals exhibit a behaviour cycle related to feeding: at ebb tide the snails migrate seawards, feeding as they go; when the tide starts to flow they float to the surface and are carried back to their starting level. Vader (1964) and Barnes (1981a,b) have not confirmed this as a regular cycle of activity. The animals can live and breed between salinities 5–40‰. The species occurs throughout the British

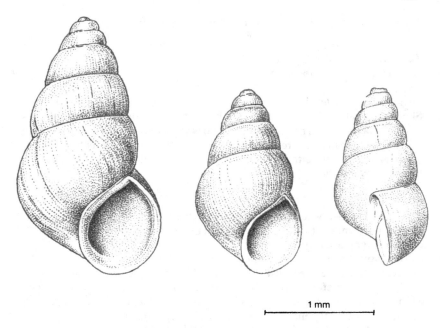

Fig. 69. *Hydrobia ulvae.*

Isles and elsewhere between the north of Norway, all of the Baltic except the most eastern parts, the Atlantic coasts of Europe and of Africa to Senegal, and in the Mediterranean.

Breeding occurs in spring and summer, sometimes with early and late peaks. The eggs are laid in jelly masses attached to sand grains or other objects, mainly to the shells of other adults. The mass is hemispherical, contains commonly 3–7 colourless eggs although up to 70 have been recorded and gets covered with sand grains. Veliger larvae hatch in 2–3 weeks, but seem to vary from place to place in their length of life before settling (Pilkington, 1971; Fish & Fish, 1977). Adults may live up to 5 years (Quick, 1924; Anderson, 1971).

Vernacular name: mud snail.

Hydrobia ventrosa (Montagu, 1803)
(Fig. 68C, 70)

Turbo ventrosus Montagu, 1803
Rissoa ventrosa (Montagu, 1803)

Diagnostic characters
Shell like that of *H. ulvae* (p. 188) but with more slender spire which is
more cyrtoconoid and has markedly tumid whorls and deep sutures. Aperture
rounded adapically, its greatest breadth about the middle of its height. The
snout is without an anterior dark transverse bar and the tentacles have no
black transverse mark. Penis ends in a long flagellum.

Other characters
The shell has 5–7 whorls, its apical angle averaging 36°; it may be a dark
or a pale brown. Up to 4.5 mm high, 2 mm broad; last whorl occupies 60–70%
of shell height, aperture 35–40%.

The snout has dark lateral lines which converge posteriorly between the
tentacles. There may be a dusky longitudinal line near the tip of each tentacle.

H. ventrosa prefers lower salinities (6–25‰) than *H. ulvae* (Nicol, 1936;
Muus, 1963) and more sheltered localities, free from tidal movement (Cherrill
& James, 1985); it tends to occur in drainage channels and lagoons with
little direct contact with the sea and with soft bottoms and vegetation, preferring the latter. In East Anglian lagoons, it is known over a salinity range
of 1–36‰ (Cherrill & James 1985; Barnes, 1987). Like *H. ulvae* it feeds
on particles rasped from the substratum, but takes smaller ones, feeding
best on those less than 40 μm in size, and not able to ingest those greater
than 120 μm (Fenchel, 1975; Lopez & Kofoed, 1980). Its general geographical
range is similar to that of *ulvae* but its occurrence is much more sporadic.

Eggs are laid April to July, each one in a capsule which is fastened to
sand grains or the shells of other snails. It becomes obscured by adherent
sand grains. A young snail emerges at hatching.

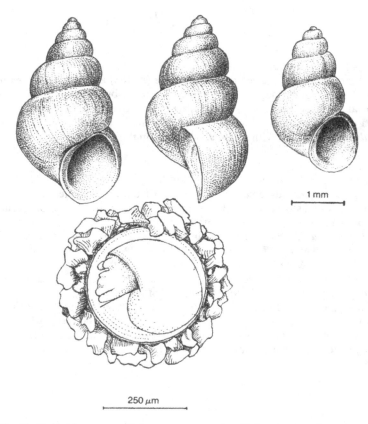

1 mm

250 μm

Fig. 70. *Hydrobia ventrosa*. Below, an egg capsule with its covering of sand grains.

Hydrobia neglecta Muus, 1963
(Fig. 71)

Diagnostic characters
Shell intermediate between that of *H. ulvae* (p. 188) and that of *H. ventrosa* (p. 190) as regards degree of swelling of whorls, but narrower than either (apical angle averages 32°); apex rather blunt. The tentacles have a conical area of black pigment close to the tip. The penis is small, rather like that of *ulvae* in shape but with an angular outline on the right anteriorly.

Other characters
The empty shell is brownish or grey, but living ones seem nearly black. The snout has a dark transverse bar as in *ulvae* with dark lateral ones and (usually) a median one as well. Up to 3–4 mm high, 2 mm broad; last whorl occupies about two thirds of the shell height, the aperture 35–40%.

H. neglecta inhabits the same kind of habitat as *H. ventrosa* but prefers more saline water (25‰ optimal, but it lives between 10‰ and 35‰: Muus, 1963; Cherrill & James, 1985). Its feeding is like that of *H. ulvae*, optimal with particles 40–80 μm in size, but like that of *H. ventrosa* in that it cannot take those greater than 120 μm (Lopez & Kofoed, 1980).

The animals have been found in Guernsey, along North Sea coasts, and the west coasts of Scotland and Ireland. It may well be more widespread as it has not been distinguished from the other species until relatively recently recognized in Denmark (Bishop, 1976; Cherrill & James, 1985).

1 mm

Fig. 71. *Hydrobia neglecta*.

Genus POTAMOPYRGUS Stimpson, 1865

Potamopyrgus jenkinsi (Smith, 1889)
(Figs 68A, 72)
Hydrobia jenkinsi Smith, 1889

Diagnostic characters
Shell in general like that of *Hydrobia ventrosa* but with less swollen whorls and usually a larger last whorl. Aperture rounded adapically, its broadest part below the middle of its height. In some shells a keel, occasionally with projecting periostracal bristles (or bristles alone), runs spirally round some or all whorls. No right pallial tentacle. Snout with anterior white transverse band, rest dark.

Other characters
The apical angle averages about 55°. The umbilicus is usually closed. The shell has a reddish horn colour, but is often more or less completely blackened by deposits. Up to 5 mm high, 2–3 mm broad; last whorl occupies 65–70% of shell height, aperture about 40%.

The tentacles are white basally, anterior to the eye, blue-grey elsewhere, though the white sometimes extends along them as a narrow central line. The sides of the foot are grey, with white along the edge of the sole.

P. jenkinsi lives in water of salinity of 0–16‰ and may be found alongside *Hydrobia ventrosa* (p. 190) in brackish situations. It is also to be found in all kinds of freshwater habitats, preferring running water to stagnant, but not apparently affected by hardness. It lives under stones, on the substratum, and on such plants as *Lemna, Elodea* and *Potamogeton*, on the epiphytic growths of which it feeds. It can take larger particles than any of the *Hydrobia* species, feeding best on particles 80–160 μm in size (Lopez & Kofoed, 1980).

Most populations of this snail contain only females, which reproduce parthenogenetically and ovoviviparously. In some places (North Wales for example) males occur to the extent of 3–11% of the total population (Wallace, 1986), but their role is still not definite. Keeled shells are more frequent in brackish than in freshwater populations, but no adequate explanation, genetical or environmental, of the occurrence of a keel has yet been offered.

P. jenkinsi is found throughout Europe except for some Mediterranean and Balkan areas. In the British Isles it is scarce or absent in mid-Wales and in much of Scotland outside the central valley.

Fig. 72. *Potamopyrgus jenkinsi.*

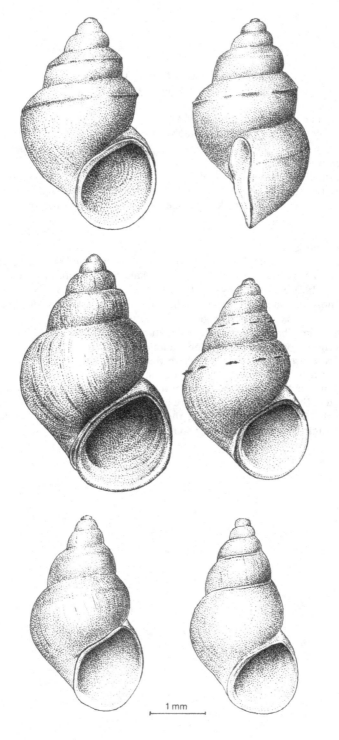

1 mm

Genus PSEUDAMNICOLA Paulucci, 1878

Pseudamnicola confusa (Frauenfeld, 1863)
(Fig. 73)

Hydrobia confusa Frauenfeld, 1863
Amnicola confusa (Frauenfeld, 1863)

Diagnostic characters

Shell small, semitransparent, glossy, with blunt tip; whorls tumid and flattened below the sutures which are deep and gutter-like; last whorl large. Pallial tentacle absent. Snout dark with pale tip; an orange-yellow spot behind each eye.

Other characters

There are 5–6 whorls forming a rather broad, squat shell with an apical angle usually about 66°, so much greater than that of species of *Hydrobia* (p. 188–192) or *Potamopyrgus* (p. 194). There is a well marked umbilicus. Pale horn colour. Up to 4 mm high, 2 mm broad; last whorl occupies about three quarters of the shell height, the aperture about half.

The animal has the same external features as most hydrobiids. The tentacles are long and pale except for a central longitudinal brown band.

P. confusa occurs in places where the water is brackish but nearly fresh, mainly in the upper tidal reaches of quiet-running rivers. It may be found on plants or the muddy edges or bottoms of water channels. Though widespread in Iberia and France it is rare in the British Isles but has been recorded from the River Arun in Sussex, the Thames estuary (Harris, 1985), some East Anglian rivers and those of southern Ireland.

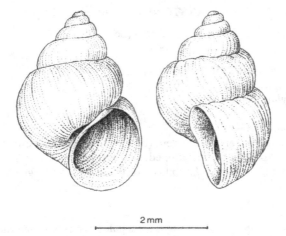

2 mm

Fig. 73. *Pseudamnicola confusa*.

Genus MARSTONIOPSIS Altena, 1936

Marstoniopsis scholtzi (Schmidt, 1856)
(Fig. 74)

Hydrobia scholtzi Schmidt, 1856
Bithynella scholtzi (Schmidt, 1856)
Amnicola taylori (Smith, 1901)

Diagnostic characters
Shell small, delicate, semitransparent (but surface usually covered by encrusting deposits). Apex blunt and flattened, whorls tumid and sutures deep. Ornament of growth lines only. Umbilicus distinct. Aperture egg-shaped, with peristome.

Other characters
The shell has 4–5 whorls. Clean shells are a pale horn colour, but their encrustations usually make them dark. Up to 2.5 mm high, 2 mm broad; last whorl occupies 65–70% of shell height, aperture 40–45%.

The animal is like other hydrobiids, with a tapering, bifid snout, long, very contractile tentacles each with a basal eye, but it lacks a pallial tentacle. In males a penis arises behind the right tentacle and curves back into the mantle cavity, ending in a long flagellum. The foot is approximately shield-shaped, pinched a little in the middle, rounded posteriorly. The flesh is grey; a yellow spot lies over each eye.

M. scholtzi is found on plants such as *Glyceria* and amongst filamentous algae to a depth of 2–3 m in lakes, canals, and rivers throughout central and northern Europe, except Scandinavia. In the British Isles, however, it is limited to a few canal sites in Cheshire, around Manchester.

Breeding occurs May-July (Jackson & Taylor, 1904) or perhaps for longer (Dussart, 1977). Egg capsules are attached singly to the surface of the plants on which the snails live. Each is more or less hemispherical, fastened by the flat base and with a keel across its summit, and contains one egg, which hatches as a juvenile in about six weeks.

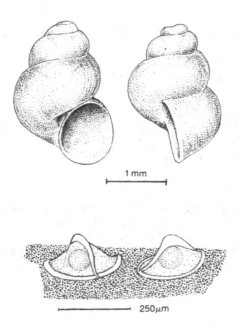

1 mm

250μm

Fig. 74. *Marstoniopsis scholtzi*. Below, two egg capsules.

Family TRUNCATELLIDAE Gray, 1840

Genus TRUNCATELLA Risso, 1826

Truncatella subcylindrica (Linné, 1767)
(Fig. 75)

Helix subcylindrica Linné, 1767
Acmea subcylindrica (Linné, 1767)
Truncatella truncatula (Draparnaud, 1801)
Truncatella montagui Lowe, 1829

Diagnostic characters
Adult shell nearly cylindrical, with 3–4 whorls, with a blunt tip marked with
a spiral turn but not a true protoconch. Ornament of many narrow costae,
though occasional smooth shells occur. Aperture rather small, oval, with
well marked peristome and labial varix. Young shells (found with adults)
have a narrow spire tapering to a point and their aperture is more pointed
apically.

Other characters
The whorls are slightly tumid though distinctly flattened peripherally, and
the sutures are rather deep. The inner lip expands to form a triangular plate
blocking the umbilicus. Buff. Up to 5 mm high, 2 mm broad; last whorl occupies half the shell height or a little more, aperture about one third.
    The animal has a cylindrical snout ending in a round, expanded disk on
which the mouth lies; at its base are the tentacles, short and rapidly tapering,
each with a kidney-shaped eye set across the base dorsally. Each eye has
a white lens on its anterior edge. There are no pallial tentacles. On the
right a ciliated groove runs from the mouth of the mantle cavity to the base
of the snout. In the mantle cavity the male has a long, recurved, tapering
penis. The foot is small, the sole rather square; a narrow operculum is set
across its dorsal surface posteriorly. Flesh white or grey, the edge of the
snout brown, with the pink buccal mass showing by transparency.
    T. subcylindrica lives under a canopy of such plants as Sea Blite (*Suaeda
maritima*) and Sea Purslane (*Halimione portulacoides*), high on the shore.
It tends to hide under stones or pieces of wood, and is found less frequently
on the soil, eating vegetable detritus and, to a lesser extent, the living plants.
It used to occur sporadically on the south coast of England from about Beachy
Head westwards, but has suffered from the destruction of many of its sites
and now seems limited to a part of its original range in Dorset. Elsewhere
it occurs in France, Spain, Portugal and throughout the Mediterranean.
    Young animals, which live with the adults, have a shell of 6–7 whorls,
the two topmost smooth, an aperture with a thin peristome and a narrow
umbilicus. At maturity the topmost whorls are broken off, the hole sealed,

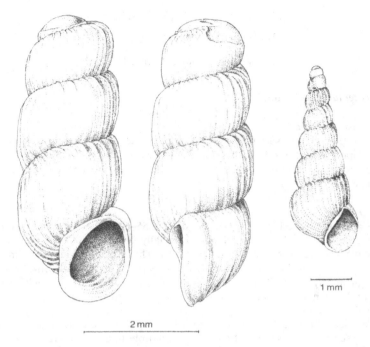

1 mm

2 mm

Fig. 75. *Truncatella subcylindrica*. Left and centre, two adult shells; right, a juvenile shell.

and, though further whorls are added, their breadth remains the same, producing the cylinder of the adult shell.

Egg capsules are more or less spherical and measure 700–800 μm in diameter, with thick walls, and each enclosing a single egg (Fretter & Graham, 1978). The capsules are fastened to detritus and the eggs hatch as small snails.

The animals are known as looping snails because of their method of locomotion (Morton, 1964): with the body held firmly by the foot the snout is stretched forwards and attached to the ground. The foot is now released and drawn forwards until it touches the snout. The cycle is now repeated.

Family BITHYNIIDAE Troschel, 1857

Genus BITHYNIA Leach, 1818

Bithynia tentaculata (Linné, 1758)
(Fig. 76)

Helix tentaculata Linné, 1758

Diagnostic characters
Shell rather glossy, semitransparent under a thin periostracum; spire well-developed, whorls moderately swollen, the last one large. Umbilicus absent. Aperture oval, pointed adapically. Operculum with concentric rings, partly calcified.

Other characters
The apex is pointed and the apical angle is 55–60°. There are 5–6 whorls. The outer lip is a little turned out, especially basally, and the inner lip is concave where it rests against the last whorl. Though there is no open umbilicus a slight groove lies alongside the columellar lip. The shell is horn-coloured though nearly white if the periostracum is lost. Up to 10 mm high, 6 mm broad; last whorl occupies about three quarters of the shell height, the aperture about 60%.

The head bears a long snout with a bifid tip. The cephalic tentacles are long, each with a basal eye. Males have a double-tipped penis recurved into the mantle cavity. All animals have a ciliated groove on the floor of the mantle cavity, its right edge expanding anteriorly to form an exhalant siphon. The foot is broad anteriorly with recurved lateral points. The flesh is grey with yellow speckles.

B. tentaculata is common in hard water areas of the British Isles (absent from Cornwall, Devon, North Wales and northern Scotland), especially in the gentler reaches of rivers, where it creeps over the vegetation or on the bottom, feeding primarily on epiphytic growths but also filtering suspended particles (Schäfer, 1952, 1953a,b; Tsikhon-Likanina, 1961; Meier-Brook & Kim, 1977). The species is generally spread throughout the western Palaearctic region.

Breeding occurs late spring to early summer (Lilly, 1953). Egg capsules are fastened to the underside of leaves, side by side, to form 2–4 rows with 10–20 capsules in all (Nekrassow, 1929). There is a single reddish egg in each capsule which hatches to a small snail in 2–3 weeks.

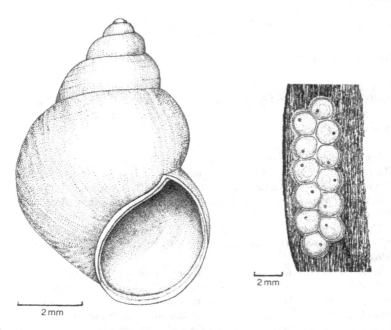

Fig. 76. *Bithynia tentaculata*. Right, a spawn mass on weed.

Bithynia leachi (Sheppard, 1823)
(Fig. 77)

Turbo leachi Sheppard, 1823

Diagnostic characters
Shell like that of a small *B. tentaculata* (p. 202) but glossier, with more tumid whorls, deeper sutures, an umbilicus, a rounder aperture which is not angulated adapically.

Other characters
The apex is blunter than that of *B. tentaculata*, and there are only 4–5 whorls. The inner lip is convex against the last whorl. Horn-coloured, but often greyer than that of *tentaculata*. Up to 5 mm high, 4.5 mm broad; last whorl occupies 80% of shell height, aperture 50%.

The body of the animal is like that of the previous species but paler in colour, grey or white with black spots, the tentacles and the foot nearly devoid of pigment.

B. leachi occurs in the same situations as *B. tentaculata*, with which it is often found, though usually less abundant. It has a more restricted distribution in the British Isles, tending to be limited to the south and east. The animals breed in early summer, laying capsules which are indistinguishable from those of *B. tentaculata* though a little smaller. Young snails emerge in about three weeks. They are probably annuals (de Wit, 1965) though Beer, Korolova & Lifshits (1969) have suggested a life span of 3–4 years.

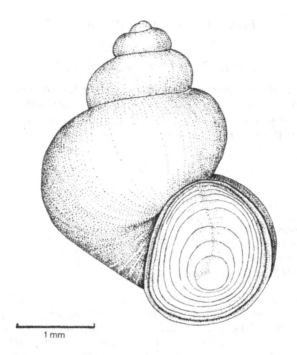

1 mm

Fig. 77. *Bithynia leachi*.

Family IRAVADIIDAE Thiele, 1928

Key to British genera of Iravadiidae

Spiral ridges distinct; shell whitish; spire more or less columnar with
blunt apex .... *Ceratia proxima* (p. 208)
Spiral ridges very fine or absent; adaptical suture of last whorl very
deep; shell thin; spire cyrtoconoid with pointed apex . *Hyala vitrea* (p. 206)

Genus HYALA H. & A. Adams, 1852

Hyala vitrea (Montagu, 1803)
(Fig. 78)

Turbo vitreus Montagu, 1803
Rissoa vitrea (Montagu, 1803)
Onoba vitrea (Montagu, 1803)
Cingula vitrea (Montagu, 1803)

Diagnostic characters
Shell tall, narrow, distinctly cyrtoconoid, semitransparent, with flattened,
blunt apex; suture between penult and last whorl very deep, often cutting
into shell profile closer to shell axis than that between antepenult and penult
whorls. Without costae and usually also without apparent spiral ridges. Posterior end of foot cleft medially.

Other characters
The sides of the basal half of the shell are nearly parallel (discounting the
curvature of the whorls), but bend to the rather blunt apex. If any spiral
ridges occur they are extremely weakly developed. The yellow colour resides
in the periostracum, the shell below being white. Up to 3 mm high, 1.5 mm
broad; last whorl occupies 60% of shell height, aperture 35%.

   H. vitrea is distributed from the Mediterranean north to Norway but is
recently recorded only from a few Scottish and western Irish localities where
it may be locally not uncommon. It lives on muddy-sandy bottoms from
10 to 50 m deep.

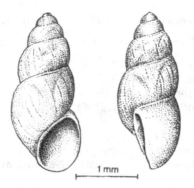

Fig. 78. *Hyala vitrea*.

Genus CERATIA H. & A. Adams, 1852

Ceratia proxima Forbes & Hanley, 1850
(Fig. 79)

Rissoa proxima Forbes & Hanley, 1850
Onoba proxima (Forbes & Hanley, 1850)
Cingula proxima (Forbes & Hanley, 1850)

Diagnostic characters
Shell tall and slender with blunt apex, tumid whorls and deep sutures; with many fine spiral ridges but lacking costae. Aperture oval. Posterior end of foot cleft in mid-line.

Other characters
The shell is semitransparent, narrow (apical angle *c*. 25°). The spiral lines are numerous – up to 40 on the last whorl, 16–18 on the penult – but low and delicate. The protoconch is large, 450 μm across. The yellow, tan or orange colour is due to the periostracum and the underlying shell is white. Up to 3 mm high, 1.25 mm broad; last whorl occupies two thirds of the total height, the aperture about one third.

The animal has neither pallial nor metapodial tentacles. The posterior cleft in the foot is visible through the operculum of a retracted animal.

C. proxima ranges from the Mediterranean north to the British Isles where it is now apparently limited to some sites between the Scillies and the Hebrides, but is nowhere common. It lives sublittorally and may be dredged from muddy bottoms at 30–50 m.

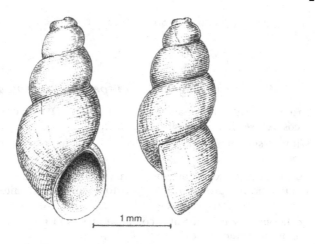

Fig. 79. *Ceratia proxima*.

Family RISSOIDAE Gray, 1847

Key to British genera and species of Rissoidae

1. Last whorl with a reticulate series of pits (Fig. 82), with or without costae elsewhere on shell **2**

   Without such pits on last whorl though other ornament may be present **4**

2. Costae absent except for labial varix; shell rather thin; apex and outer lip red-brown, rest of peristome and varix white; inner lip not extended over last whorl .... *Rissoa porifera* (p. 221)

   Costae usually present; lilac colour at apex and round aperture; inner lip extends over last whorl **3**

3. Costae on last whorl die out at periphery; total number of costae not over 24; in apertural view costae visible only on last whorl; parietal lip slight .... *Rissoa rufilabrum* (p. 220)

   Total number of costae more than 24; in apertural view costae visible on last and penult whorls; expansion of parietal lip over last whorl considerable .... *Rissoa lilacina* (p. 218)

4. Ornament on shell none, or of spiral ridges, sometimes with costae, but if present these are confined to subsutural parts of whorls and do not reach periphery; they may be on all or only some whorls **5**

   Ornament different from this **15**

5. Breadth of shell equals 50% or more of height **6**

   Breadth of shell equals less than 50% of height **12**

6. Shell globose with many fine spiral lines; colour uniform yellow white .... *'Putilla tumidula'*, *Obtusella intersecta* (p. 260)

   Shell conical, with or without colour markings **7**

7. Last whorl commonly with 3 spiral colour bands (an unbanded variety exists) but without transverse markings; 6–7 whorls; spiral ridges on last whorl; high on shore .... *Cingula trifasciata* (p. 254)

   Last whorl with numerous transverse colour bands or with rows of spots, sometimes also one spiral colour band at base of last whorl **8**

8. Shell with labial varix, lilac apex, dark comma-shaped mark on last whorl by varix; whorls slightly tumid; metapodial tentacle finger-shaped; opercular lobes black .... *Rissoa parva* (p. 214)

   Shell and animal not like this **9**

9. Whorls nearly flat-sided, with 2 rows of brown transverse markings; metapodial tentacle with 3 points; opercular lobes yellow ....
.... *Alvania semistriata* (p. 236)

Whorls tumid, sutures deep **10**

10. Shell with blunt apex, 4 whorls; umbilicus large; last whorl with 2–4 rows of brown streaks or spots; not north of Channel Islands
.... *Setia pulcherrima* (p. 228)

Shell with sharp apex, 5–6 whorls; umbilicus small or closed; transverse brown bands on last 2 whorls **11**

11. Apex bronze; a few spiral lines on last whorl; whorls more tumid, sutures deeper than in next species (these points not obvious unless specimens are available for comparison); some shells have brown spiral band on base of last whorl; metapodial tentacle flattened .... *Pusillina sarsi* (p. 226)

Apex purple; sometimes a shallow reticulation of surface of penult whorl; no basal brown spiral; metapodial tentacle filiform ....
.... *Pusillina inconspicua* (p. 224)

12. Shell with obvious spiral ridges and grooves; small costae on adapical parts of some or all whorls; sutures markedly oblique **13**

Shell smooth or with obscure spirals, no costae **14**

13. Costae on all whorls; often 2 spiral brown bands on last whorl; larval shell rather small, up to 340 µm across ....
.... *Onoba semicostata* (p. 256)

Costae few, limited to more apical whorls, occasionally absent; larval shell rather large up to 550 µm across .... *Onoba aculeus* (p. 258)

14. Shell thin, 7 or more whorls, last whorl less than half height; a small bulge sometimes on columella; peristome sometimes flared, usually from brackish water ....
.... *Rissostomia membranacea* (p. 222)

Shell not like this **15**

15. Costae or transverse ridges absent from topmost whorls of teleoconch but present on others, reaching at least the periphery of last whorl; pallial tentacle on right; metapodial tentacle single and filiform **16**

Costae on all whorls of teleoconch; pallial tentacles on right and left; metapodial tentacle triple or 3-pointed **20**

16. Shell tall; aperture large, peristome flared all round; usually a bulge on columella and a thickening in the throat level with labial varix; marine .... *Rissostomia membranacea* (p. 222)

Shell not like this **17**

17. 18–30 thin costae on last whorl, dying out at periphery; many fine spirals; apex purple; last whorl with subsutural brown streaks; peristome brown .... *Pusillina inconspicua* (p. 224)

8–25 broad costae on last whorl with microscopic spirals in the intervening spaces **18**

18. Shell usually with dark comma-shaped mark on last whorl by labial varix; apical region pink-lilac; 8–12 costae on last whorl, dying out at periphery where they join a spiral ridge ... *Rissoa parva* (p. 214)

Shell without comma mark **19**

19. Shell solid; spire may be cyrtoconoid towards apex; 10 costae on each whorl in lower half of shell, apical whorls without costae, nearly flat-sided; apex, peristome, throat, tinged lilac; southern .... *Rissoa guerini* (p. 216)

Shell semitransparent; spire straight-sided or cyrtoconoid; all whorls tumid 12–17 costae on the youngest whorl which has them, but their occurrence is irregular; apex bronze, last whorl often with transverse brown lines and brown spiral on its base; dark streak on columella .... *Pusillina sarsi* (p. 226)

20. Costae numerous in the form of fine ridges much less obvious than prominent spiral keels; spire turreted .... *Alvania carinata* (p. 230)

Costae as prominent as spirals or more so **21**

21. Costae more conspicuous than spirals which are slight except for a prominent basal keel; costae opisthocline, flexuous; aperture oblique, small, with double peristome .... *Manzonia crassa* (p. 250)

Costae and spiral ridges about equally developed **22**

22. Throat with ridges or teeth within outer lip **23**

Such teeth absent **25**

23. Reticulation of costae and spiral ridges coarse; sutures deep; aperture large with 12–13 teeth within outer lip; knoblike tooth on columella in most shells .... *Alvania cancellata* (p. 248)

Shell not like this **24**

24. 16–20 costae on last whorl; apex golden yellow .... *Alvania cimicoides* (p. 234)

35–45 costae on last whorl; apex like rest of shell in colour .... *Alvania beani* (p. 232)

25. Spire noticeably cyrtoconoid; last whorl = 70% of shell height; interstices of ornament oblong; aperture narrow, elongated; Channel Islands only .... *Alvania lactea* (p. 242)

Shell not like this **26**

26. Apex blunt; larval shell with zigzag spiral lines; spiral ridges on last whorl better developed than costae; only 1 record of living animals this century .... *Alvania jeffreysi* (p. 246)

Shell not like this **27**

27. About 14 costae on last whorl crossed by 4–5 spiral ridges with 2 more spiral ridges below the costae; reticulation with tuberculated nodes; sutures deep and whorls shouldered ....

.... *Manzonia zetlandica* (p. 252)

Many more than 14 costae and 4–5 spiral ridges on last whorl; profile of whorls rounded **28**

28. 30–40 costae on last whorl with 12–15 spiral ridges forming delicate square reticulation; 6 whorls .... *Alvania punctura* (p. 244)

19–24 costae on last whorl; spiral ridges more prominent towards the shell base; 4–5 whorls **29**

29. Costae and aperture opisthocline; whorls a little flattened, sutures deep; protoconch up to 400 μm across ...... *Alvania abyssicola* (p. 238)

Costae and aperture orthocline; whorls rounded; protoconch more than 400 μm across; all British records from Scilly area ....

.... *Alvania subsoluta* (p. 240)

Genus RISSOA Desmarest, 1814

Rissoa parva (da Costa, 1778)
(Fig. 80)

Turbo parvus da Costa, 1778
Turboella parva (da Costa, 1778)
Rissoa interrupta (Adams, 1798)

Diagnostic characters

Shell rather broad with moderately high spire. Apical 3–4 whorls smooth, others smooth or with costae. On last whorl costae (if present) end basally by joining a spiral cord. Labial varix present. Last whorl angulated in juvenile shells. Apical region with lilac tint; last whorl with subsutural dark streaks, that nearest the aperture forming 'comma-shaped' mark over varix.

Other characters

The 7–8 whorls are slightly tumid, apparently more so in costate shells. The apical angle is about 40°. All whorls have fine spiral lines, clearest on the last whorl at the periphery, below which one forms a distinct cord. Costae, when present, are slightly opisthocline, wave-like in section, the spiral lines obvious in the troughs between. They cross the whorls of the spire but end at the subperipheral spiral on the last whorl, on which there are 8–12. The angulation of the last whorl, pronounced in juvenile shells (lacking a varix), persists into some (usually smooth) mature ones. The aperture is oval to D-shaped, pointed adapically, with the outer lip everted basally. Apart from the lilac apical region the shell may be any shade from cream to dark brown, with the smooth apical whorls commonly dark and not patterned. The 'comma mark' near the aperture does not appear until the shell has four whorls. In addition to the subsutural streaks the last whorl may have on its base a brown spiral band from which streaks extend adapically. The peristome is brown. Up to 5 mm high, 3 mm broad; last whorl occupies about two thirds of shell height, aperture one third to nearly half.

The snout is bifid anteriorly and the tentacles are long and slender, each with a basal eye. There is one right pallial tentacle. Behind the right cephalic tentacle in males is the penis, whose tip forms a short filament. The foot is narrow, constricted in the middle where it can fold transversely, and has a median groove posteriorly. The body is pale yellow with purple-black streaks; tentacles pale with central white line, a yellow patch behind each eye. Opercular lobes black.

R. parva occurs in two forms, one smooth-shelled, the other with costae, the former sometimes called *R. interrupta*; there are, however, numerous and varied intermediates (see Wigham, 1975a,b). The animals are extremely abundant on rocky shores, especially in summer, in all parts of the British Isles. They are commonest about L.W.S.T., extending up to about M.T.L. and down to depths of about 15 m, on fronds and at the base of algal tufts, in laminarian holdfasts, under stones and in pools, where they often climb

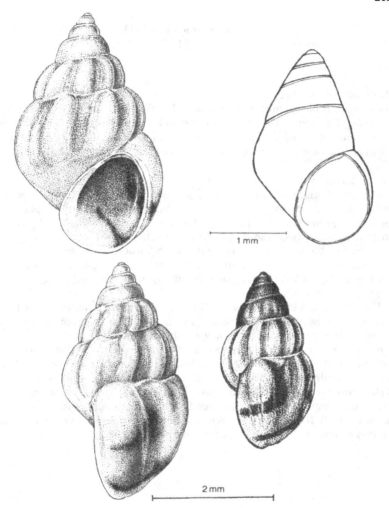

Fig. 80. *Rissoa parva*. Top right, a shell of the variety *interrupta*, sometimes regarded as a separate species.

on mucous ropes. They are epiphytic grazers and detritivores. The species ranges from the Mediterranean to the north of Norway.

Breeding may occur throughout the year but is greatest in summer and least in winter. Females deposit transparent lens-shaped capsules on weeds, each about 600–700 μm across and containing 6–50 white eggs. These hatch in about three weeks as veliger larvae which have a bilobed velum with marginal orange spots or lines, the right lobe the larger. Settlement takes place when the shell is about 0.5 mm high, usually on fine branching weeds.

Rissoa guerini Récluz, 1843
(Fig. 81)

Rissoa costulata Alder, 1844

Diagnostic characters
Shell ovoid-conic in profile with relatively tall spire, coeloconoid towards
the apex. The apical five whorls smooth and nearly flat-sided, the youngest
three tumid and with costae; all with fine spirals. Costae disappear on last
whorl near aperture but there is a thick labial varix. Brownish, costae white.

Other characters
The shell has eight whorls and there are ten costae on each of the penult
and antepenult whorls. They cross these but fade below the periphery of
the last; they are usually a little opisthocline. The spiral ridges and grooves
are clearest between costae. The aperture is oval, narrowing adapically. There
may be some brown streaks crossing the smooth whorls, and the apical region
is often slate-coloured; a similar brown-lilac colour appears in a band in
the throat. Up to 6 mm high, 3 mm broad; last whorl occupies half the shell
height, the aperture a third.
    The animal is like *R. parva* (p. 214) but is yellower. The tentacles have
a central yellow line and the snout and the sides of the foot are brown.
There is a long metapodial tentacle with a broad base.
    R. guerini is a southern species found from the Canary Islands north to
the south coast of Ireland and to the English Channel. It lives at L.W.S.T.
and sublittorally on weeds and under stones. It is locally not uncommon.
    Breeding (at Plymouth) takes place from February to April, when clear,
lens-shaped capsules are laid on weeds. They are about 1 mm in diameter
and contain 80–100 colourless eggs which develop to veliger larvae. The velum
has two prominent brown blotches on each of its two lobes.

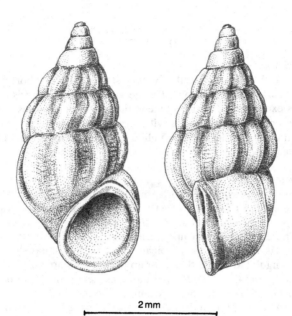

2 mm

Fig. 81. *Rissoa guerini*.

Rissoa lilacina Récluz, 1843
(Fig. 82D)

Diagnostic characters
Shell glossy, rather broad basally. Youngest part of last whorl marked by a rectangular arrangement of pits. Costae visible on two whorls of the shell when that is examined in apertural view, their total number less than thirty. Throat constricted by tuberosities within outer lip and on columella. Inner lip spreads slightly over last whorl. Yellow with brown marks and lilac colour on peristome.

Other characters
There are 6–7 gently swollen whorls, the most apical distinctly more tumid. Costae are confined to the youngest three whorls and tend to vanish near the outer lip: the last whorl has usually 7–8, the penult thirteen, the next up to ten. There are also many fine spiral ridges most visible between costae. The aperture is oval, narrower apically, but with a squarish appearance because it is angulated where the outer lip joins the columellar lip and where that joins the parietal lip. The varix and constricted throat do not develop until the shell is full grown. The brown marks on the shell lie between costae and are clearest near the sutures; they also occur on the smooth whorls. The varix is white and may be bounded by orange-brown bands which occasionally elongate to form a spiral round the base. A ring of deep violet, often incomplete, lies on the peristome. Up to 5 mm high, 2.75 mm broad; last whorl occupies 60% of shell height, aperture 35–45%.

The animal is similar to *R. parva* (p. 214). It is yellowish, its tentacles with a bright yellow central line, the snout and sides of the foot brown, with yellow under the operculum.

R. lilacina is found amongst weeds and on sandy ground at L.W.S.T. and sublittorally, not uncommonly in places. It occurs in the Channel as far east as the Isle of Wight, in the southern Irish Sea, on the south and west coasts of Ireland and the west coast of Scotland. Elsewhere it occurs on European Atlantic coasts southwards.

Little is known about the breeding of this species. One transparent lens-shaped capsule has been seen attached to the shell and containing 28 developing embryos (Lebour, 1937). It is probable that these would later have become veliger larvae, but these have not been observed.

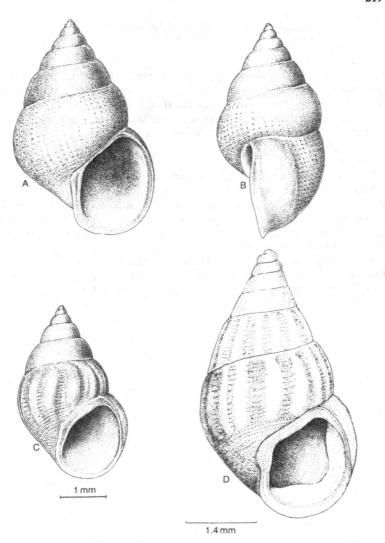

Fig. 82. A, B, *Rissoa porifera*; C, *R. rufilabrum*; D, *R. lilacina*.

Rissoa rufilabrum Alder, 1844
(Fig. 82C)

Diagnostic characters
Shell like that of *R. lilacina* (p. 218) in having a punctate appearance on the younger parts of the last whorl, but with not more than 24 costae which are visible on only one whorl when the shell is looked at in apertural view. Aperture rounder and throat less constricted than in *lilacina*; spread of inner lip over the last whorl not marked. Varix present.

Other characters
The shell has often a slightly translucent appearance and the last whorl is relatively large. The angles between outer lip and columella and between columella and parietal lip are wider than in *lilacina* so that the whole aperture looks less square. The colour pattern and size of the shell are as in that species but the last whorl occupies 63–70% of the shell height and the aperture 42–48%. Bodily features are like those of *lilacina*.

    R. rufilabrum has been recorded from the Channel Islands north to Norway; the main records come from the west coast of Scotland where it is commoner than *R. lilacina*, though living in similar places (Smith, 1970).

Rissoa porifera Lovén, 1846
(Fig. 82A,B)

Diagnostic characters
As in the two previous species the younger parts of the last whorl bear a series of squarish pits, but costae are altogether absent. Inner lip barely spreads over the last whorl; tuberosities in throat absent, but varix present.

Other characters
This is a more delicate shell than that of either *lilacina* (p. 218) or *rufilabrum* (p. 220). It is horn-coloured, the varix is white, the apex often has a lilac tint and the peristome less coloured than in the other species.

R. *porifera* has been found in west Ireland, west Scotland, the Shetlands, and Scandinavia; it inhabits the same type of habitat as other species.

The three species, along with the Mediterranean R. *violacea* Desmarest, 1814, form an aggregate, each with its characteristic geographical range, with some overlap. The production of costae and of tuberosities in the throat is maximal in the more southern *violacea* and *lilacina* and decreases in the more northerly forms.

Genus RISSOSTOMIA G. O. Sars, 1878

Rissostomia membranacea (J. Adams, 1800)
(Fig. 83)

Turbo membranaceus J. Adams, 1800
Rissoa membranacea (J. Adams, 1800)
Rissoa labiosa (Montagu, 1803)

Diagnostic characters
A very variable shell which may be thin and transparent or thick and opaque;
have a tall, slender shape or be short and broad, but which nearly always
shows an aperture with flared lips and sometimes partly blocked by internal
thickenings. Costae and labial varix present or absent. One metapodial
tentacle. In brackish or fully marine habitats.

Other characters
The appearance of the shell depends on the habitat: shells from brackish
water tend to be tall (8–9 whorls), slender (apical angle 35–45°), semitransparent, smooth or with up to ten costae on the last whorl, a thin varix or none
on the outer lip, an oval aperture, and slight or no thickenings within the
outer lip and on the columella. Shells from marine waters tend to be stubbier
(5–7 whorls, apical angle 45–55°), opaque, usually with costae (up to 18 on
the last whorl), a marked varix, an angulated aperture with the throat often
partially blocked by thickenings within the outer lip and on the columella.
In all, the most apical whorls are smooth. Solid shells may be white, yellowish
or greenish; semitransparent ones are horn-coloured, the tint in all due to
the periostracum. Longitudinal brown marks may lie between costae, and
a brown spiral band may encircle the shell base. The varix is pale, sometimes
with a brown band alongside it. Up to 9 mm high, 3 mm broad. In slender
shells the last whorl occupies 40–45% of shell height, the aperture one third;
in more squat ones the last whorl occupies two thirds and the aperture one
half of the height.
   The animal is cream, sometimes light brown; cephalic and metapodial
tentacles are white, and there is white round the eyes; the foot is dark laterally
and on the opercular lobes.
   R. membranacea is typically associated with *Zostera* but also lives on other
weeds at L.W.S.T. and below. It tolerates salinities down to 7‰. It is moderately frequent throughout the British Isles, except on the east coast of Scotland. Abroad it ranges from the Canary Islands to Norway and extends some
way into the Baltic. In the southern parts of the range the animals may
breed throughout the year, but in the north they are summer breeders. The
egg capsules are hemispherical, clear, and commonly about 1 mm in diameter;
they are fastened to weeds. The thickness of the walls varies with locality,
thin in marine, thicker in brackish localities. The number of eggs per capsule
also varies from 40–400, dependent upon the habitat. In some capsules all

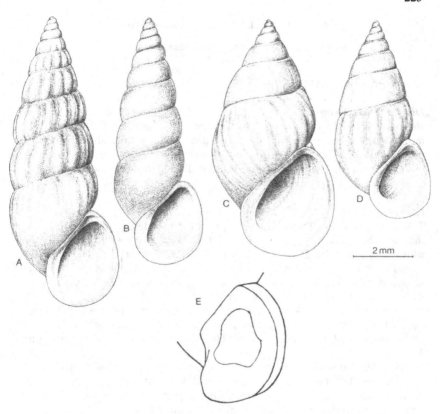

Fig. 83. *Rissostomia membranacea*. A and B, var. *octona* Nilsson, 1822, a northern form from water of salinity 12–16‰; C, var. *membranacea* J. Adams, 1800, a northern form preferring salinities of 15–20‰; D, var. *labiosa* Montagu, 1803, found in water of salinity 20–35‰; E, a common appearance of the aperture of D.

the eggs develop to free-swimming veligers, in others some eggs are eaten by the more precocious embryos, the larval stage is omitted and the young are born as juveniles (Rehfeldt, 1968; Rasmussen, 1973). The significance of this is not known.

In view of the great variation in the shell characters and reproductive behaviour it is not clear whether we are dealing with a single species which happens to be very sensitive to slight environmental changes, or whether we have an aggregate of sibling species susceptible of division when our knowledge of these is more complete.

Genus PUSILLINA Monterosato, 1884

Pusillina inconspicua (Alder, 1844)
(Fig. 84)

Rissoa inconspicua Alder, 1844

Diagnostic characters
Very similar to *Rissoa parva* (p. 214) in shape and proportions though smaller, more delicate and more transparent. Apical three whorls always smooth, others with spiral lines and nearly always numerous fine costae, less frequent on the last whorl. Umbilical groove present. Dark purple spot at apex. Metapodial tentacle finger-shaped.

Other characters
The shell is glossy. The spiral lines are most obvious at the periphery of the last whorl, where they may form a fine reticulation with the costae, of which there may be 18–30 on each of the last two whorls. The outer lip shows a clear anal sinus and the varix is not thick. Whitish or light horn-coloured, the apical whorls generally not lilac; the youngest whorls sometimes with short, light chestnut streaks on their subsutural parts though without the comma mark of *Rissoa parva*. The last whorl may have a basal spiral band of the same colour, which also appears on the peristome. Up to 3 mm high, 2 mm broad; last whorl occupies 60–70% of shell height, aperture about 45%.

The animal resembles *R. parva* but has a long metapodial tentacle and the penis is flattened and lacks a terminal filamentous part. The body is whitish with numerous yellow spots, one large one by each eye, two dark lines along each side of the foot and black opercular lobes.

P. inconspicua is frequent in low rock pools and amongst algal tufts but is more often found sublittorally on algae and sandy gravel to depths of about 100 m. The species is widespread from the Mediterranean to the Arctic and has been found in most parts of the British Isles. The animals are epiphytic grazers and detritivores. There is a brief summer breeding period (August and September at Plymouth) during which clear, hemispherical egg capsules are laid on debris or the shells of other animals, each with 6–9 eggs. Veliger larvae hatch which develop red and yellow markings on the velum and foot.

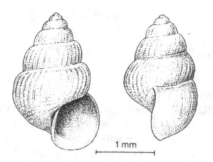

Fig. 84. *Pusillina inconspicua*.

Pusillina sarsi (Lovén, 1846)
(Fig. 85)

Rissoa sarsi Lovén, 1846
Rissoa albella Lovén, 1846

Diagnostic characters

A rather squat little conical shell (apical angle about 60°), semitransparent, with deep sutures and swollen whorls, the last often inflated. Apical three whorls smooth, the rest smooth or with costae which die away near periphery of last whorl without joining a spiral cord. Last whorl of juvenile shell not angulated. Labial varix sometimes present. Brown streaks cross adapical parts of some or all whorls between costae and may be linked basally by spiral band on last whorl; 'comma' mark absent. Metapodial tentacle flattened.

Other characters

The development of costae in this species is erratic and they may appear on any or all of the younger whorls (4–6), but usually disappear on the youngest part of the last whorl. They are slightly opisthocline, number 12–17 on the youngest whorl on which they are present, 9–16 on the next adapical. The shell is cream or horn-colour in general, the smooth apical whorls often rather dark and the apex sometimes showing some violet colour; varix usually white. The aperture is oval, the peristome turning out a little basally. There is a narrow umbilical chink. Up to 3 mm high, 2.5 mm broad; last whorl occupies two thirds to three quarters of the shell height, the aperture about half.

The animal has the same appearance as *Rissoa parva* (p. 214) but the tentacles on head and foot are relatively longer. The flesh is cream with whiter and darker areas, the latter on the sides of the foot and under the operculum.

P. sarsi is found from the Mediterranean to north Norway. It is not common on the coasts of the British Isles and has been mainly taken off the west coast of Scotland. It lives among weeds, occasionally at L.W.S.T., usually below, to depths of 15 m.

The species is a variable one in respect of shell thickness, development of costae and, to some extent, colour. Thicker and more opaque shells, usually with costae and a varix but with poorly developed colour bars on the last two whorls, are referred to the form *albella*, which has sometimes been regarded as a distinct species.

The animals breed from December to May. Their egg capsules are clear and lens-shaped, about 1 mm in diameter and contain 40–70 eggs which develop into free veliger larvae.

Fig. 85. *Pusillina sarsi*. Upper row, forma *sarsi*; lower row, forma *albella*.

Genus SETIA H. & A. Adams, 1852

Setia pulcherrima (Jeffreys, 1848)
(Fig. 86)

Rissoa pulcherrima Jeffreys, 1848
Cingula pulcherrima (Jeffreys, 1848)

Diagnostic characters
Shell minute, glossy, semitransparent, with short, blunt spire; whorls apparently smooth and tumid, sutures deep. Aperture rounded, narrower adapically, its edge everted basally and over a groove leading to a small umbilicus.
Pale horn-colour with reddish brown markings, often in spiral rows.

Other characters
The four whorls may sometimes bear a few delicate spiral lines and some
prosocline growth lines. The apical angle is 55–60°. The colour pattern varies:
on the last whorl there is often a series of subsutural streaks and two peripheral
rows of spots plus a basal row (sometimes absent); or the streaks may be
interrupted to give two rows of spots (making 5 in all); or the spots may
fuse to give chevron-like markings. The apex may be lilac. Up to 1.25 mm
high, 0.8 mm broad; last whorl occupies about three quarters of shell height,
aperture about half.

The animal is without pallial tentacles; the single metapodial one is long.
White with yellow spots, metapodial lobes darker.

S. pulcherrima is a southern species which seems to reach its northern
limit in the Channel Islands where it lives in tufts of fine weeds at L.W.S.T.
Little is known of the breeding of these animals except that they produce
hemispherical transparent capsules measuring about 800 μm across. Each
contains a few eggs (3–9) which develop to young snails before hatching.

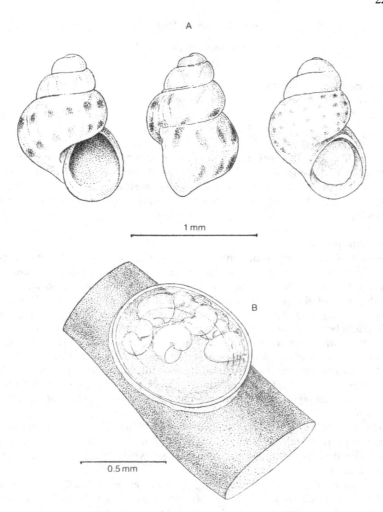

Fig. 86. *Setia pulcherrima*. A, shell; B, spawn.

Genus ALVANIA Risso, 1826

Alvania carinata (da Costa, 1778)
(Fig. 87)

Turbo carinatus da Costa, 1778
Galeodina carinata (da Costa, 1778)
Rissoa striatula (Montagu, 1803)

Diagnostic characters
Shell with large last whorl, rather small spire and markedly turreted profile. Surface with prominent spiral keels, the most adapical on each whorl the largest, the space between them marked with numerous little costae. Aperture large, pointed adapically, outer lip with thick varix. White. One metapodial tentacle.

Other characters
The shell is semitransparent, with 5–6 whorls, the most adapical part of each forming a shelf nearly at right angles to the shell axis. The 2–3 most adapical spiral ridges on each whorl are the best developed and lie far apart; those on the base of the last whorl form (1) an adapical group of 5–7, low and close together, and (2) a basal group of three higher ones lying further apart. All spaces between spiral ridges show a series of numerous, narrow costellae. The outer lip arises level with spiral 3 and is much thickened by a varix; this is absent in juvenile shells where the lip is thin and slightly scalloped by the ends of the spiral keels; some trace of this may persist in adult shells. Juveniles also show an umbilicus, closed later. Up to 4 mm high, 3 mm broad; last whorl occupies about three quarters of the shell height, the aperture about half.

The animal has the general appearance of all rissoids, but the snout is longer and narrower than usual and the neck has small lobes, one on each side. The flesh is cream, the buccal mass showing through the snout wall as a red area.

A. carinata lives under stones lying on sandy areas of beach at L.W.S.T. and to depths of about 25 m. It is found sparsely in south west England and rarely further north; probably absent from the eastern part of the Channel, the North Sea and the northern parts of Scotland. It is a southern species, extending to the Mediterranean. The reproduction is unknown: Thiriot-Quiévreux & Babio (1975) believe development to be direct.

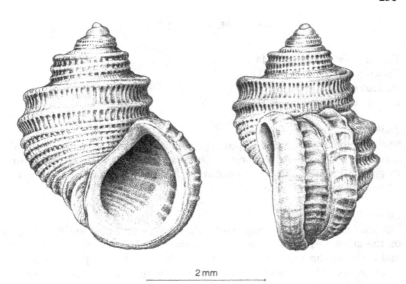

2 mm

Fig. 87. *Alvania carinata*.

Alvania beani (Thorpe, 1844)
(Fig. 88)

Cingula beani Thorpe, 1844
Rissoa beani (Thorpe, 1844)
Rissoa reticulata (Montagu, 1803)
Rissoa calathus Forbes & Hanley, 1850

Diagnostic characters
Shell with plump crytoconoid spire, apex pointed, whorls not markedly tumid, sutures in V-shaped nicks in profile. Ornament of spiral ridges crossing costae to give reticulated surface. Aperture with prominent varix; outer lip with up to twelve internal teeth.

Other characters
The shell is glossy and a little translucent when fresh. There are 35–45 costae on the last whorl, ending shortly below the periphery, 40–42 on the penult, and none on the topmost three whorls. They are low, slightly prosocline, equal in breadth to about half the intervening spaces. The spiral ridges are similar in shape and expand to tubercles as they cross the costae. There are 10–13 on the last whorl, but shells of this species fall into two groups according to the number on the penult whorl: 6–7 in *A. b. reticulata*, 4 in *A. b. calathus*. The former are usually larger shells with coarser ornament. The aperture is oval, the outer lip a little everted basally. Pale orange, cream or white, usually with two brown spiral bands on the last whorl, subsutural and subperipheral, only the former in the whorls of the spire; varix nearly white. Up to 3.5 mm high, 2 mm broad; last whorl occupies about two thirds of shell height, aperture 40–45%.

The snout is narrow, bifid distally, with long tentacles at its base, each with an eye at its origin. The foot is narrow with a lobe on its dorsal surface anteriorly and a 3-lobed metapodial tentacle under the operculum. There is a left pallial tentacle and usually also a right one. Males have a long, slender penis. The flesh is translucent white with some brown on snout and foot and yellow behind each eye.

A. beani lives about 50 m deep round the British Isles, mainly off western coasts, amongst weeds and stones, feeding on detritus. There are also a few records from L.W.S.T. The species ranges from the Mediterranean and the Azores to northern Norway. The taxonomic position of the *calathus* form is not properly understood.

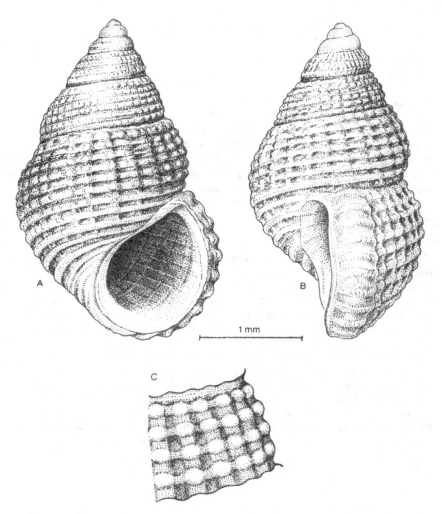

Fig. 88. *Alvania beani*. A and B, *A. beani reticulata*, whole shell; C, part of penult whorl of *A. beani calathus*. (Teeth within the aperture are not shown.)

Alvania cimicoides (Forbes, 1844)
(Fig. 89)

Rissoa cimicoides Forbes, 1844
Rissoa sculpta Philippi, 1844

Diagnostic characters
In general appearance the shell is like that of *A. beani* (p. 232) but the V-shaped nicks in which the sutures lie are wider and deeper, the lattice of ornament coarser because costae and spiral ridges are fewer; costae lost on the base of the last whorl. Aperture with varix and teeth within the outer lip. Without a tooth on columella. White or yellowish, apical whorls golden brown; a vertical brown band on the last whorl along abapertural side of varix which it crosses near the periphery, sometimes extended along the whorl subsuturally and peripherally.

Other characters
There are 6–7 whorls which are rather flat-sided and the spire is regularly pyramidal. There are 16–20 costae on the last whorl, orthocline or very slightly prosocline, 15 on the penult, 13–16 on each of the rest apart from the larval ones. Spiral ridges number 9–12 on the last whorl, 4–6 on the penult, then 2–3. The larval shell has an irregular spiral pattern. The aperture presents the same characters as in *A. beani*. Up to 4 mm high, 2.5 mm broad; last whorl occupies about two thirds of the shell height, aperture 40–45%.

    A. cimicoides has been recorded from the Mediterranean to Norway, but is probably absent from the North Sea. In the British Isles it has been found, but not recently, in a number of places between south west England and the Northern Isles; absent, apparently, in Ireland. It is uncommon and lives from a few to 500 m deep on soft bottoms. It is a detritivore. Its reproductive habits are unknown though probably not different from those of related species.

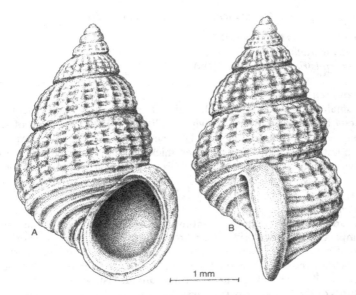

Fig. 89. *Alvania cimicoides*. Teeth within the aperture are often more distinct than is indicated in A.

Alvania semistriata (Montagu, 1808)

Turbo semistriatus Montagu, 1808
Rissoa semistriata (Montagu, 1808)
Cingula semistriata (Montagu, 1808)

Diagnostic characters

Shell small with moderately high, pointed spire (apical angle about 40°), distinctly cyrtoconoid in profile; whorls a little swollen. Ornament of spiral ridges, clearest at the periphery and on the base of the last whorl. Aperture oval, pointed adapically, with rather thick peristome turned out basally. Yellowish white with two rows of red-brown, comma-shaped marks on last whorl.

Other characters

There are 5–6 whorls which sometimes appear inflated below the sutures and impart some irregularity to the profile of the spire; the last whorl, with about 24 spiral ridges, may be a little keeled. The upper colour marks are opisthocline, the lower ones prosocline. A minute umbilical crack occurs in some shells. Up to 3 mm high, 1.5 mm broad; last whorl occupies about two thirds of shell height, aperture less than half.

The animal has a rather short snout and the tentacles are flattened. There is a short pallial tentacle on the left edge of the mantle skirt and a longer one on the right. The metapodial tentacle has three points. The flesh is cream in colour and there are yellow blotches, one behind each eye, others under the operculum.

A. semistriata occurs on the lower parts of rocky shores throughout the British Isles; not uncommon in the south, it becomes less frequent in the north. The animals seem gregarious and cluster in the base of algal tufts, under stones and in rock pools in the lower half of the beach and sublittorally. They seem to prefer silty places and are detritivores. Their recognition as a species of *Alvania* rather than *Cingula* (p. 254) is due to Ponder (1985).

Breeding (at Plymouth) occurs March–August when transparent, hemispherical capsules are laid in groups on the weeds amongst which the adults live. Each capsule has a diameter of 600 μm and contains 12–22 eggs which develop to veliger larvae.

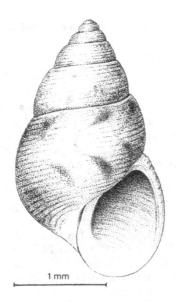

Fig. 90. *Alvania semistriata.*

Alvania abyssicola (Forbes, 1850)
(Fig. 91)

Rissoa abyssicola Forbes, 1850

Diagnostic characters
Whorls tumid, marked with lattice of spiral ridges and opisthocline costae, the latter absent from the base of the last whorl. Aperture with prominent labial varix; outer lip opisthocline and with anal sinus. Umbilicus absent. Diameter of larval shell not more than 400 μm. Cream, brownish on base and often alongside varix. One metapodial tentacle.

Other characters
The shell is glossy and has a blunt tip. The 5–6 whorls tend to be flattened peripherally and meet at deep sutures set in channelled grooves. There are 10–15 spiral ridges on the last whorl, then 5–7, 4–5, 4, up the spire, the apical 2–3 whorls smooth. The costae number 19–23 on the last whorl, 17–20 on the penult, then 15 and 6. The aperture in juvenile shells has no varix. Up to 3 mm high, 1.75 mm broad; last whorl occupies two thirds cf the shell height, the aperture rather more than one third.

The snout is deeply bifid but otherwise the body of the animal resembles that of other *alvania* species. It is white with yellow and chalk-white spots.

A. abyssicola has been found sublittorally from 15 to 100 m deep on soft bottoms, mainly off north west and north Scotland, and off the west coast of Ireland; occasional dead shells may be found elsewhere. The species ranges from northern Norway to the Mediterranean.

Capsules possibly belonging to this species have been described by Thorson (1946). They were attached to broken shells, were about 0.5 mm across and contained 6–13 eggs. These developed to veliger larvae.

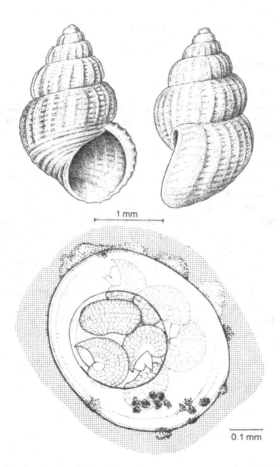

Fig. 91. *Alvania abyssicola*, shell and spawn, the latter typical of all rissoid spawn masses.

Alvania subsoluta (Aradas, 1847)
(Fig. 92)

Rissoa subsoluta Aradas, 1847

Diagnostic characters
Shell very similar to that of the preceding species but the costae and outer lip are orthocline and the diameter of the larval shell exceeds 400 μm. Whorls without peripheral flattening.

Other characters
The shell is semitransparent and has about five whorls. The numbers of costae and of spiral ridges are so similar to those of *A. abyssicola* (p. 238) that they cannot be used as diagnostic features. White or cream. Up to 3.5 mm high, 2 mm broad; last whorl occupies 66% of shell height, aperture 40%.

    A. subsoluta can usually be separated from *A. abyssicola* on the basis of the direction in which costae and outer lip lie, the angle of the latter to the shell axis in particular being very different in the two species. *A. subsoluta* is not only rare in the fauna of the British Isles but it is also restricted in distribution, all records being from the neighbourhood of the Isles of Scilly. There the animals were collected from muddy bottoms at depths of about 200 m. Elsewhere it has been found between the Mediterranean and the north of Norway. Details of its reproduction are not known.

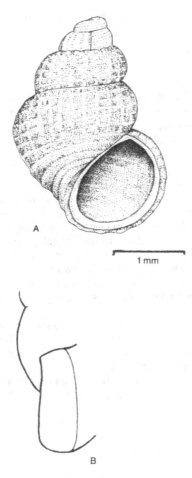

Fig. 92. *Alvania subsoluta*. A, apertural view; B, side view of apertural region showing orthocline varix (contrast with opisthocline varix of *A. abyssicola* shown in Fig. 91).

Alvania lactea (Michaud, 1830)
(Fig. 93)

Rissoa lactea Michaud, 1830

Diagnostic characters
Shell plump, with markedly cyrtoconoid outline, short spire and large last whorl. Surface, except on base of last whorl, marked by oblong reticulation, elongated along the whorls. Aperture rather narrow, pointed adapically and with peristome everted basally. Umbilicus absent. Cream.

Other characters
The shell is semitransparent and has 5–6 moderately swollen whorls, the sutures between which, in profile, lie at the base of V-shaped excavations. There are 15–17 spiral ridges on the last whorl, 7–9 on the penult, then 5–6, 2–3. The costae are prosocline and more variable in number: 16–22 or more on the last whorl, 13–18 on the penult, then 12–17, 8–15. They die out on the last whorl before reaching the base. The aperture is much narrower adapically than in the other *Alvania* species and the outer lip shows a bay near its origin. The apex is sometimes orange yellow. Up to 6 mm high, 4 mm broad; last whorl occupies three quarters of the shell height, the aperture half.

Within the region covered here *A. lactea* is found alive only in the Channel Islands, although dead shells have been recorded from further north. It lives amongst algae and under stones at L.W.S.T. and below. Abroad it extends south to Morocco and the Mediterranean.

The organization of the body of this animal and its mode of feeding are like those of other species in the genus; so probably is its mode of reproduction, but this is unknown.

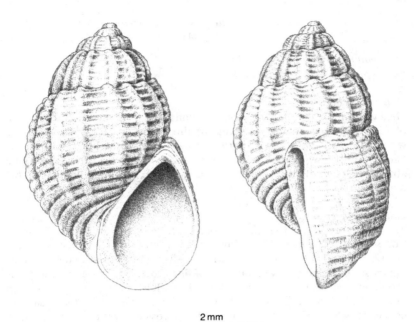

2 mm

Fig. 93. *Alvania lactea*.

Alvania punctura (Montagu, 1803)
(Fig. 94)

Turbo puncturus Montagu, 1803
Rissoa punctura (Montagu, 1803)

Diagnostic characters
Shell with sharply pointed spire; whorls tumid with deep sutures and a pattern of shallow square reticulation except on three most apical whorls. Aperture with labial varix, peristome everted at base. Umbilicus long and narrow. Cream with brown marks across whorls and sometimes a brown line along abapertural side of varix, which is white. Animal with right and left pallial tentacles, plus one metapodial.

Other characters
There are six whorls. The spire is usually a little cyrtoconoid near the apex. The sculpture is markedly shallower than in most *Alvania* species. There are 12–15 spiral ridges on the last whorl, 5–8 on the penult, then 4–5, whilst the most apical have only spiral rows of fine tubercles. There are numerous prosocline costae (35–45 or so) on the last whorl, decreasing to 28–33 on more adapical ones. Costae decrease in height or vanish on the basal half of the last whorl. Brown marks usually lie across the whorls in three series, subsutural, peripheral, and basal, the last (and sometimes also the second) visible on the last whorl only. The umbilical region is also brown. Up to 3 mm high, about 2 mm broad; last whorl occupies about two thirds of shell height, aperture a little more than a third.

The general appearance of the body of this animal is as in the other *Alvania* species. Its flesh is cream with dark lines and yellow spots; behind each eye lies a red spot and the same colour appears under the operculum.

A. punctura occurs from the Mediterranean to north Norway. It is, with *A. semistriata* (p. 236), the commonest species of the genus in British and Irish waters, though absent from the southern parts of the North Sea. The animals are not intertidal but live on sandy bottoms to 100 m deep, feeding selectively on diatoms and dinoflagellates.

The animals breed in summer laying lens-shaped capsules on weeds, each about 400 μm in diameter and containing 12–14 eggs from which veliger larvae develop. Metamorphosis and settlement take place when the larval shell is about half a millimetre in height.

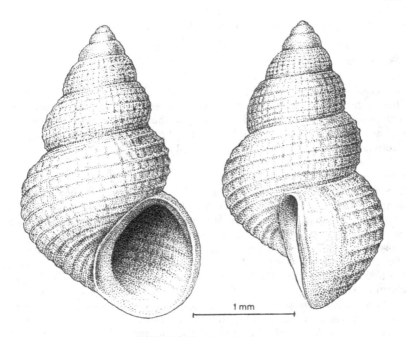

Fig. 94. *Alvania punctura.*

Alvania jeffreysi (Waller, 1864)
(Fig. 95)

Rissoa jeffreysi Waller, 1864

Diagnostic characters
Shell rather tall, slender, semitransparent, glossy; whorls tumid, decorated with lattice work of costae and spiral ridges, the latter the better developed. Costae die out on basal half of last whorl. Aperture with varix but without internal teeth. Larval shell with zigzag spiral lines.

Other characters
The shell has 5–6 whorls meeting at well marked sutures. There are nine or ten spiral ridges on the last whorl, three or four on the penult, three on the others apart from the larval shell; they sometimes decrease in size on the base of the shell and near the varix. Costae are orthocline or slightly prosocline and number about 45 on the last whorl, 40 on the penult, and about 30 on each of the others. There is an umbilical chink. White or yellowish. Up to 3 mm high, 2 mm broad; last whorl occupies 55–60% of shell height, aperture 33%.

The species ranges from the Mediterranean to the Arctic. It has, however, been found only once this century in British waters – off Uist; earlier records refer to the Scillies and to Shetland. The animals, which are said to eat foraminiferans, live on sandy bottoms between about 50 m and 600 m. Their reproduction is not known but may, on the basis of the appearance of the protoconch, include a free larval stage.

Fig. 95. *Alvania jeffreysi*. Zigzag pattern on protoconch not shown, but this is seen in the SEM photograph of a shell in apical view, ×146.

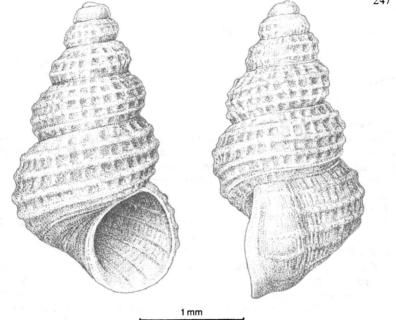

1 mm

Alvania cancellata (da Costa, 1778)
(Fig. 96)

Turbo cancellatus da Costa, 1778
Rissoa cancellata (da Costa, 1778)
Rissoa crenulata Michaud, 1832

Diagnostic characters

Shell solid, rather squatly conical; spire with sutures set in deep notches of profile. Surface with coarse lattice of costae and spiral ridges, the former running to base of last whorl. Aperture broad with labial varix and teeth within outer lip; columella broad with tubercle near its base.

Other characters

The shell is glossy and may be a little transparent. There are 6–7 whorls bearing prosocline costae and spiral ridges, 15–17 costae and seven spirals on the last whorl, 15 and 3–4 on the penult, and 13 and two respectively on the others except the larval ones which show spiral lines of points, often eroded. The aperture frequently has a small spout below the columellar tubercle; there are 11–13 teeth within the outer lip. In young shells the outer lip lacks a varix, has a thin, crenated edge, and the teeth and columellar tubercle have not developed. White or yellowish, sometimes with spiral bands; if these are present the commonest is a subsutural one on each whorl; this may be supplemented by a transverse one alongside the varix and by a sub-peripheral spiral one on the last whorl. The throat commonly shows some chestnut coloration internal to the varix which, with the peristome, is always white. Up to 4 mm high, 2.5 mm broad; last whorl occupies 70% of shell height, aperture 45%.

A. *cancellata* is a southern species at its northern limits in the British Isles. It has been recorded from isolated localities in Ireland and the British west coast as far north as Shetland, but is frequent in the south west of the area. It usually lives sublittorally to depths of 90 m, but may very occasionally be found under stones at L.W.S.T. It is a detritivore. Its reproduction is unknown but may include a free larval stage.

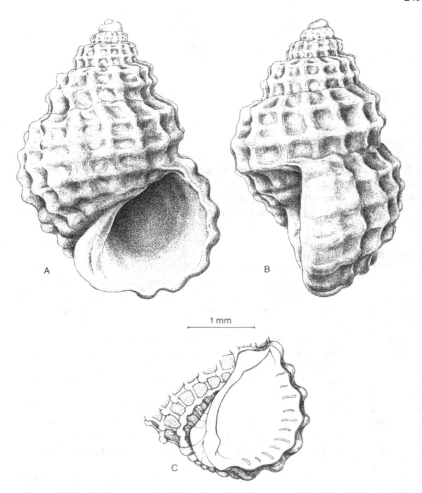

Fig. 96. *Alvania cancellata*. The shell drawn in A and B is a juvenile and lacks both internal teeth and columellar tubercle: these are shown in C.

Genus MANZONIA Brusina, 1870

Manzonia crassa (Kanmacher, 1798)
(Fig. 97)

Turbo crassus Kanmacher, 1798
Alvania crassa (Kanmacher, 1798)
Rissoa costata (J. Adams, 1796)

Diagnostic characters
Shell small, semitransparent, with long spire; whorls tumid, sutures deep. Whorls crossed by prominent sinuous costae, with numerous slight spiral ridges between and over them. Aperture small, obliquely set, with thick edge (double peristome). Umbilical groove bounded above by prominent keel. Animal has right and left pallial tentacles and a 3-pointed metapodial one.

Other characters
The shell has six whorls with a blunt tip. The most conspicuous ornament is the series of flexuous costae, mainly opisthocline, and narrower than the intervening spaces. There are ten on each whorl except the oldest. The spiral ridges are low and strap-shaped, usually becoming eroded from the costal summits; the last whorl has a strong spiral keel below the ends of the costae, separated from the apertural edge by a deep umbilical groove. There are 16–18 spiral ridges on the last whorl above the keel, 14–15 on the penult, reducing up the spire. The aperture is bounded by a thick labial varix. White. Up to 3.5 mm high, 2 mm broad; last whorl occupies 60–65% of shell height, aperture 40%.

The snout is rather deeply cleft at its tip, the tentacles long and slightly flat, each with a basal eye. The left pallial tentacle is the more obvious. The foot narrows a little in the middle and the sole has a median groove to which a gland opens. The median part of the metapodial tentacle is long. Yellowish.

M. crassa is found between the Mediterranean and Norway. It is widespread in the British Isles, occurring occasionally under stones or with weeds at L.W.S.T. and on sand to depths of 50 m. It is said (Pelseneer, 1935; Vahl, 1971) to eat coralline algae, but may be grazing growths and detritus on their surface rather than the algae themselves. The life history includes a free veliger larval stage.

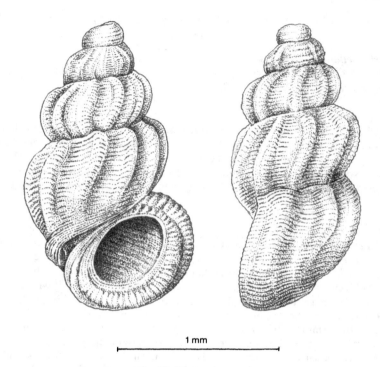

1 mm

Fig. 97. *Manzonia crassa*.

Manzonia zetlandica (Montagu, 1815)
(Fig. 98)

Turbo zetlandicus Montagu, 1815
Rissoa zetlandica (Montagu, 1815)
Alvania zetlandica (Montagu, 1815)

Diagnostic characters
Shell with rather tall spire, blunt tip, sutures (in profile) at base of deep, V-shaped notches; whorls tumid with coarse lattice of costae and spiral ridges. Two prominent, smooth, spiral keels bound a gutter on base of last whorl, the costae stopping above them. Labial varix prominent. No teeth within outer lip. White or yellowish, usually orange alongside varix and on basal keel.

Other characters
M. zetlandica looks like a coarser version of *Alvania cimicoides* (p. 234) or *A. cancellata* (p. 248), the dips to the sutures exaggerated in the profile, the reticulation coarser; the basal keels, however, identify it clearly. There are 6–7 whorls, the last with 13–16 prosocline costae, the penult with 13–14, the next with 11–12, the larval whorls smooth. There are 6 spiral ridges on the last whorl, the upper four crossing costae and tuberculated, the fifth and sixth forming smooth keels round the basal gutter. Up to 4 mm high, 2.5 mm broad; last whorl occupies 60% of shell height, aperture 40%.

M. zetlandica is an uncommon sublittoral animal in local waters and is confined to western British and Irish coasts, where it is dredged 20–350 m deep on gravelly and soft bottoms, more frequently in the south than in the north. The species is spread from the Mediterranean to Norway.

Since the protoconch seems to show two sections, Thiriot-Quiévreux & Babio (1975) have interpreted this to indicate an embryonic and then a larval stage in the life history.

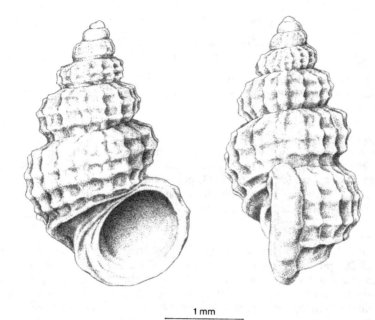

1 mm

Fig. 98. *Manzonia zetlandica*.

Genus CINGULA Fleming, 1828

Cingula trifasciata (J. Adams, 1800)
(Fig. 99)

Turbo trifasciatus J. Adams, 1800
Rissoa cingillus (Montagu, 1803)
Cingula cingillus (Montagu, 1803)

Diagnostic characters

Shell solid, conical, with well-developed pointed spire (apical angle about 50°), almost straight-sided in profile. Ornament of spiral lines on last whorl, most marked basally. Aperture oval, pointed apically, with thick peristome everted at base. Last whorl with three spiral brown bands. Pallial tentacle on right only; metapodial tentacle median, under operculum.

Other characters

The shell has 6–7 slightly swollen whorls meeting at clear sutures. There are 12–20 spiral ridges on the last whorl; in the spire 1–2 ridges at most lie on the apical side of each suture. The outer lip arises tangential to the surface of the last whorl. The colour pattern is variable: the general colour is yellow-brown with one dark spiral band below the suture in every whorl, usually one above each suture in the spire, and one near the periphery plus another near the base on the last whorl; the last two sometimes fuse. The peristome is pale. A variety *rupestris*, cream or white and unbanded, is common. Up to 4 mm high, 2 mm broad; last whorl occupies 60% of shell height, aperture 40%.

The cephalic tentacles are long and slender, the tips blunt, with an eye at the base of each. The pallial tentacle is short. The penis is long and tapering. The foot is truncated anteriorly, narrows in the middle and tapers posteriorly; the metapodial tentacle is short and triangular. The body is whitish, with a white spot behind each eye. In the snout yellow jaws and red buccal mass show by transparency, and the anterior gland in the foot appears as a V-shaped white mass.

C. trifasciata is a detritivore and a common crevice animal on all rocky shores (Glynne-Williams & Hobart, 1952; Morton, 1954), living between a few metres depth and the upper limit of the barnacle zone. It may also be found in silt in rock pools, amongst weed and under stones. *C. t. rupestris* seems to be confined to crevices and not mixed with the typical forms. The species ranges from the Bay of Biscay to Norway.

At Plymouth the breeding season lasts from March until June. Thick-walled capsules, hemispherical in shape, are attached to the walls of the crevices in which the adults live. They are about 700 μm in diameter and commonly contain 1–2 eggs which hatch as young snails.

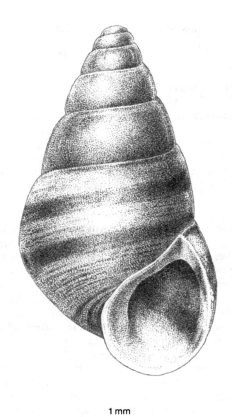

1 mm

Fig. 99. *Cingula trifasciata*.

Genus ONOBA H. & A. Adams, 1852

Onoba semicostata (Montagu, 1803)
(Fig. 100)

Turbo semicostatus Montagu, 1803
Rissoa semicostata (Montagu, 1803)
Rissoa striata (Montagu, 1803)
Onoba striata of authors

Diagnostic characters
Shell tall, somewhat columnar; whorls with numerous spiral ridges and grooves, crossed by undulate costae confined to adapical parts of each whorl.

Other characters
There are 5–6 whorls forming a shell with a cyrtoconoid spire, sometimes seemingly curved. The spiral ridges are strap-shaped, about as broad as the intervening grooves; there are 18–25 on the last whorl, 10–15 on the penult. There are about twenty costae on the last whorl, 17–20 on the penult. The aperture is oval, pointed adapically, surrounded by a slightly thickened peristome. There is no umbilicus. Usually rust-coloured, but may be yellow, black or white; the colour lies in the periostracum so in worn shells it may be restricted to the spiral grooves and the pale calcareous shell shows elsewhere. Up to 4 mm high, a little less than half that in breadth; last whorl occupies 60–65% of shell height, aperture 36–40%.

The head has a broad, depressed snout, its tip a little bilobed, and two long, setose tentacles each with a basal eye. There is always a tentacle on the left mantle edge and often one on the right as well. Males have a long slender penis lying in the mantle cavity, attached to its right floor anteriorly. the foot is rather narrow, a dorsal flap anteriorly marking the mouth of the anterior pedal gland. A flat metapodial tentacle projects backwards from under the operculum. The body is generally a shade between white and pale yellow, with darker pigment on the snout and sides of the foot; an opaque white patch lies behind each eye.

These animals are abundant on all rocky and stony shores, especially in summer; they live from near L.W.S.T. to about 100 m deep, gregarious amongst weeds, corallines, mussels, and in crevices where there is silt. They feed on detritus, licking what has been disturbed by the foot from its dorsal surface.

O. semicostata has been recorded from all round the British Isles. It ranges from the Mediterranean to southern Norway and penetrates the most saline parts of the Baltic, though absent from some eastern North Sea coasts.

1 mm

Fig. 100. *Onoba semicostata.*

Onoba aculeus (Gould, 1841)
(Fig. 101)

Cingula aculeus Gould, 1841

Diagnostic characters
This shell is extremely similar to that of *O. semicostata* (p. 256). It differs in that the costae occur only on 1–2 adapical whorls, if at all, and that the protoconch is larger (330–575 μm in diameter).

Other characters
The brown bands found on the last whorl of *semicostata* are usually absent in this species. Up to 3 mm high, 1.25 mm broad; the last whorl occupies about two thirds of the shell height, but is variable in this respect; the aperture about 40%.

    O. aculeus lives in situations very similar to those in which *O. semicostata* is found, but it avoids the siltier habitats which the latter species prefers and is commoner on algal rather than stony substrata. It is also more frequent in brackish water. It is a panarctic species and has so far been recorded locally only from the east and west coasts of Scotland and in Galway Bay; it may well prove to be much more widespread. There are recent records from southern England.

    The animals breed in Denmark (Rasmussen, 1951, 1973; Warén, 1974), the only locality for which information is available, from November to March. Capsules, usually with only one egg, are attached to weed, sand grains, or the shell of other animals. Each capsule has the usual hemispherical rissoid shape and has a diameter of 650–700 μm. A free larval stage is absent and young snails emerge from the capsules after a rather long developmental period.

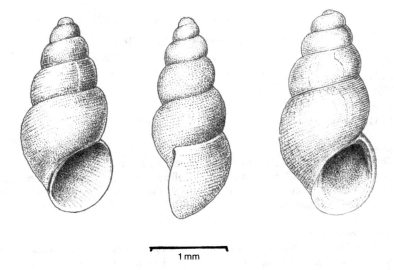

1 mm

Fig. 101. *Onoba aculeus*.

Genus OBTUSELLA Cossmann, 1921

Obtusella intersecta (Wood, 1856)
(Fig. 102)

Rissoa intersecta Wood, 1856
Rissoa soluta Philippi, 1844
Rissoa alderi Jeffreys, 1858
Cingula alderi (Jeffreys, 1858)

Diagnostic characters
Shell minute, globose, semitransparent and without coloured bands. Whorls tumid with fine sculpture of many spiral and axial lines, the former more obvious (needs much magnification to be seen). Aperture round or squarish; a small umbilicus present. White or cream.

Other characters
The shell is glossy and has 4–5 whorls, each sometimes a little flattened below the suture. The outer lip arises from the last whorl at right angles to the shell axis; the peristome by the columella is thin, a little everted. Up to 2 mm high, 1.5 mm broad; last whorl occupies about 70% of shell height, aperture 40%.

The animal is like *Cingula trifasciata* (p. 254) but more delicately built. It is cream with many white points.

O. intersecta is widespread in the British Isles, though not common, and has also been recorded from Spain to Norway. It is not an intertidal animal but is dredged to about 50 m depth, usually on sandy bottoms or amongst algae.

A rissoid, described as *Setia inflata* Monterosato, 1884, dredged on coarse gravel 13–18 m deep in Kames Bay, Isle of Cumbrae (Fretter & Patil, 1961), may prove to be either a variety of this species or a very closely related one. Its shell resembled that of *O. intersecta* closely but was a uniform greenish brown in colour, was smaller (1.25 mm high, 1 mm broad) and was without an umbilicus. The body differed from that of *O. intersecta* in that the cephalic tentacles were short, there was no right pallial tentacle and the ctenidium was reduced, partly replaced by a ciliated band. The body was cream, with the snout grey. Only the one animal has been found.

Putilla tumidula (Sars, 1878)

The local occurrence of an animal under this name is shown in Seaward's Atlas (1982) in waters between Scotland and the Faeroes. Precisely what animal this refers to must be regarded as highly doubtful. A description of an animal under this name was given by Nordsieck (1972): it was said to have a small shell of about four whorls marked with faint spiral lines. His

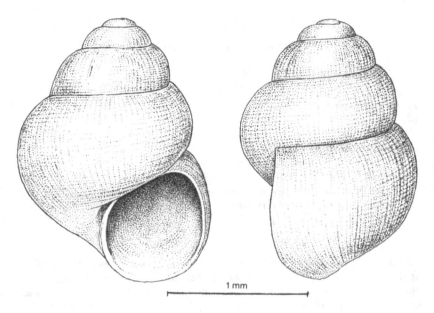

Fig. 102. *Obtusella intersecta.*

illustration (source not given) suggests a shell not unlike that of *Obtusella intersecta* (Fig. 102) but broader, with more tumid whorls and a longer aperture and more conspicuous umbilical groove. The shell is described as thin, yellowish, and up to 2.5 mm in height and 1.75 mm in breadth, the last whorl occupying about 70% of the total height and the aperture about half.

Two comments must be made. The description agrees with that of *Cingula tumidula* Sars, 1878, and may well have derived from that, but according to Warén (1974) the holotype of the species, which has now been destroyed by preservation in an acidic glass tube, was 'the only known specimen'. If this is so then the local record may rest on a misidentification. The second comment to be made is that Ponder (1985) regards *Putilla* as a skeneid rather than a rissoid, but it is not clear that both Ponder and Nordsieck are making use of the name *Putilla* in the same sense.

Family BARLEEIDAE Gray, 1857

Genus BARLEEIA W. Clark, 1855

Barleeia unifasciata (Montagu, 1803)
(Fig. 103)

Turbo unifasciatus Montagu, 1803
Barleeia rubra (J. Adams, 1797)

Diagnostic characters
A small conical shell with flat-sided profile and without obvious ornament; apical whorls usually reddish brown, younger whorls of spire with pale adapical and reddish brown abapical halves; last whorl also with basal brown spiral. Operculum crimson with silvery refringence. Animal without pallial and metapodial tentacles.

Other characters
The shell has five whorls, slightly swollen, the last sometimes slightly keeled at the periphery, and all smooth except for microscopic growth lines. The aperture is oval, pointed apically, has no varix, and there may be a small spout at the base of the columella. There is no umbilicus. The protoconch is rather globular and has a finely pitted surface. Up to 2 mm high, 1.5 mm broad; last whorl occupies 70% of shell height, aperture 40%.

The body of the animal shows much the same external features as those of any rissoid. It is yellowish, with many dark lines, and there is a bright yellow spot by each eye. The opercular lobes are black.

B. unifasciata is a southern species not found further north than the British Isles, where it is widespread except for the southern parts of the North Sea and the Channel east of the Isle of Wight. It is to be found on algae, mainly fine red weeds, and in pools near low water mark, on rocky shores, often in abundance in summer. It feeds on algae and epiphytic diatoms.

The animals breed in spring and summer, laying spherical capsules with thick walls on weeds, often on coralline algae. The capsules have a diameter of 500–600 μm and contain only a single, rather large egg. This hatches as a juvenile. Information on the biology and ecology of this species is given by Southgate (1982).

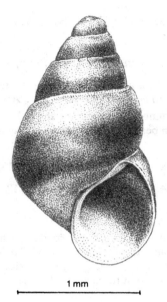

1 mm

Fig. 103. *Barleeia unifasciata*.

Family ASSIMINEIDAE H. & A. Adams, 1856

Key to British genera and species of Assimineidae

Whorls flat-sided; umbilicus absent; aperture less than half shell height;
in salt marshes Kent to Humber only .... *Assiminea grayana* (p. 264)
Whorls tumid; small umbilicus; aperture about half shell height; in
rock crevices high on beach .... *Paludinella littorina* (p. 266)

Genus ASSIMINEA Fleming, 1828

Assiminea grayana Fleming, 1828
(Fig. 104)

Diagnostic characters
Shell a rather squat cone, nearly straight-sided in profile, solid but semitransparent; umbilicus absent. Aperture rather small. The cephalic tentacles form short, rounded lobes each with a large eye at its tip. A groove on each side of the foot anteriorly from mantle cavity to the sole.

Other characters
The shell has 6–7 slightly swollen whorls which bear prosocline growth lines. The apical angle is about 60° and the last whorl is often a little angulated at the periphery. The shell is tan, often with a reddish brown spiral band at the periphery of the last whorl. Up to 5 mm high, 3 mm broad; last whorl occupies about three quarters of shell height, aperture 40–45%.

The animals have an anterior pedal gland opening on the dorsal surface of the foot anteriorly. Males have a penis with 8–12 glandular papillae. The flesh is grey with purple-brown specks, notably on the head, which is very dark; the sides of the foot are pale, but the grooves on its anterior half have black edges.

A. grayana lives in the upper parts of salt marshes, where it is wetted only by high spring tides, on the mud, amongst the vegetation, and in puddles, though it prefers emersion to submersion. It can, however, survive in normal sea water and also resist desiccation for long periods. It feeds on fresh and decaying plant material. The species has a rather limited distribution on the continent – Holland to Denmark – and also in Britain – Kent to the Humber. It may be common locally.

These snails breed April to September. Eggs are laid, each in a capsule passed down the groove from the mantle cavity to the foot, and are aggregated in heaps where the adults live. They are covered with faecal pellets to prevent drying (Sander, 1950, 1952). The eggs develop to veliger larvae which await the right conditions (prolonged immersion by spring tides or floods) for hatching (Sander & Sibrecht, 1967). The length of larval life is not known.

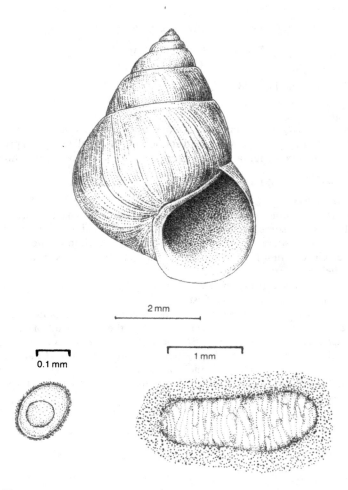

2 mm

0.1 mm

1 mm

Fig. 104. *Assiminea grayana*. Below left, an egg capsule; right, a spawn mass covered with faecal pellets.

Genus PALUDINELLA Pfeiffer, 1841

Paludinella littorina (Chiaje, 1828)
(Fig. 105)

Helix littorina Chiaje, 1828

Diagnostic characters
Shell small, glossy, semitransparent, somewhat globose, with short spire and large last whorl. Aperture oval or half-moon shaped. Small umbilicus present. Eyes close to tips of tentacles. No grooves on sides of foot. Lives high on beach.

Other characters
There are 3–4 tumid whorls meeting at deep sutures. The ornament is of growth lines and some vague spiral lines. The shell is of a pale horn colour, but occasionally somewhat orange specimens are found; in all the columella is white. Up to 2 mm high, 2 mm broad; last whorl occupies three quarters of shell height, aperture about half.

The animal has a broad snout, which is bifid at its tip, and carries short, stubby tentacles at its base. The flesh is pale grey.

P. littorina is a detrivore and a member of the crevice fauna, but may also be found under stones, or with plants, in the same kind of habitat as *Truncatella*. It appears to be limited to localities on the south coast of England west of the Isle of Wight, perhaps extending into the Bristol Channel. The animals are also found in the Channel Islands and thence south to the Mediterranean. Their biology is little known, though it is unlikely that there should be a free larval stage in their life history.

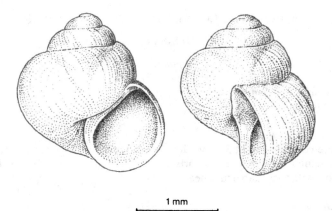

1 mm

Fig. 105. *Paludinella littorina*.

Family CINGULOPSIDAE Fretter & Patil, 1958

Genus CINGULOPSIS Fretter & Patil, 1958

Cingulopsis fulgida (J. Adams, 1797)
(Fig. 106)

Helix fulgidus J. Adams, 1797
Rissoa fulgida (J. Adams, 1797)

Diagnostic characters

A minute conical shell of 3–4 swollen whorls with a blunt tip, the only ornament irregular growth lines; whorls of spire with two spiral brown bands,
last whorl with three; aperture nearly circular.

Other characters

The shell is markedly cyrtoconoid in profile; the sutures are deep; the last
whorl is much broader than the older ones. Growth lines and the outer lip
are prosocline and there is a well-marked umbilicus. 1 mm high and 0.8 mm
broad; last whorl occupies 65–70% of shell height, aperture 40%.

The body of the animal shows most of the features of rissoid species but
has neither pallial tentacle nor penis. A broad metapodial tentacle, however,
projects backwards from under the operculum. The flesh is cream with numerous opaque white speckles.

C. fulgida is to be found in low rock pools, where it lives mainly on the
bottom in silt or weeds; it often climbs on mucous threads or hangs from
the surface film. The snails are detritivores and may be abundant in summer.
They have been recorded from most southern parts of the British Isles except
the eastern Channel, and from the west coast of Scotland and round Ireland,
but are largely absent from the North Sea. Abroad the species ranges south
to the Mediterranean.

Breeding (Plymouth, Fretter & Patil, 1961) occurs in spring and summer
when colourless, rather large eggs (150 µm across) are enclosed singly in
lens-shaped capsules fastened to coralline algae. A free larval stage is missing
and the young emerge from the capsules as small snails.

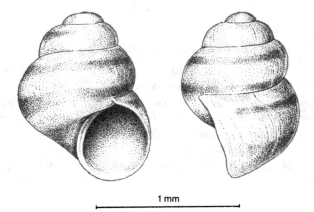

1 mm

Fig. 106. *Cingulopsis fulgida*.

Family RISSOELLIDAE Gray, 1850

Key to British species of Rissoella

1. Aperture occupies less than half of shell height, last whorl about 70%; umbilicus a chink; tentacles single ..... *Rissoella diaphana* (p. 270)

Aperture occupies more than half of shell height, last whorl more than 80%; umbilicus a chink or large; tentacles single or bifid   **2**

2. Shell taller than broad; umbilicus a chink; tentacles bifid ....
.... *Rissoella opalina* (p. 272)

Shell as broad or broader than high; umbilicus large; tentacles not bifid .... *Rissoella globularis* (p. 273)

Genus RISSOELLA Gray, 1847

Rissoella diaphana (Alder, 1848)
(Fig. 107)

Rissoa? diaphana Alder, 1848
Jeffreysia diaphana (Alder, 1848)

Diagnostic characters

A minute, glossy, milk-white, conical shell, taller than broad, with a blunt apex; there are 4–5 swollen whorls showing obscure sculpture and deep sutures; aperture oval. A conspicuous dark, oval mark lies near the anus and is visible through the semitransparent shell. Operculum with T-shaped brown mark along the columellar edge.

Other characters

Though it appears smooth to the naked eye the shell has prosocline growth lines and these may be sufficiently marked in places as to suggest slight costae; they often appear as white streaks. The last whorl is large. There is a small umbilicus. Up to 1–1.5 mm high, 0.7–1 mm broad; last whorl occupies about 70% of shell height, aperture 45%.

The tip of the snout is deeply bifid, each half appearing like a flattened, triangular tentacle. In addition the animals possess the normal cephalic tentacles, which are digitiform, each with an eye a little behind its base. The mantle edge is smooth. The animals are hermaphrodite so each has a penis behind the right tentacle. The foot has a groove along the posterior half of the sole. Cream with some darker patches.

R. diaphana is an epiphytic grazer living among red weeds, mainly in rock pools between L.W.S.T. and M.T.L., where it may be abundant in summer; it also occurs sublittorally. The animals are widespread along the coasts of

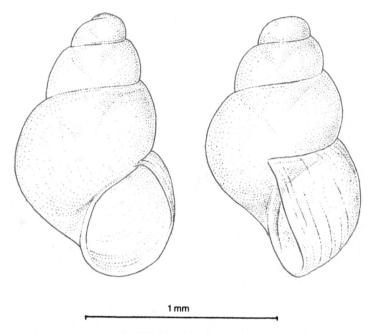

1 mm

Fig. 107. *Rissoella diaphana*.

the British Isles, predominantly in the south and west. Elsewhere they occur
from the Mediterranean to Norway.

The animals breed in spring and summer. One or two eggs are enclosed
in an ellipsoidal capsule with a flat base which is deposited on green or red
weeds. Young snails hatch from these in about two weeks.

272

Rissoella opalina (Jeffreys, 1848)

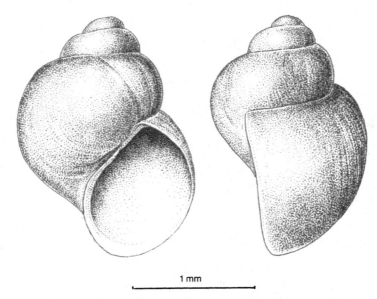

1 mm

Fig. 108. *Rissoella opalina.*

Rissoa? opalina Jeffreys, 1848
Jeffreysia opalina (Jeffreys, 1848)

Diagnostic characters
Shell minute, taller than broad, glossy and nearly transparent, with a blunt tip, swollen whorls and hardly any sculpture. Yellowish brown and refringent, sometimes with a darker peripheral spiral band. Dark mark shows through shell.

Other characters
There are 3–4 whorls which, when closely examined, show some growth lines. There is usually a small umbilicus, but it is always less clear than in *R. diaphana*. Up to 2 mm high, 1.5 mm broad; last whorl occupies 80–85% of shell height, aperture a little over half.

  R. opalina has the same general body form as *R. diaphana* except that the snout is very short and hardly bifid, whereas the cephalic tentacles are deeply divided. The animals live in the same habitat and in the same manner as *R. diaphana*; they have approximately the same distribution, though seem to be more local in their occurrence. Breeding follows the same pattern as in *R. diaphana* (Fretter, 1948).

Rissoella globularis (Forbes & Hanley, 1852)

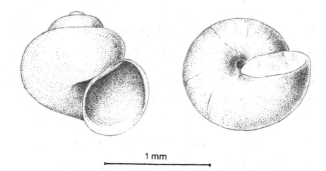

1 mm

Fig. 109. *Rissoella globularis*.

Jeffreysia globularis Forbes & Hanley, 1852

Diagnostic characters
A minute, globular shell, but with a distinct spire; broader than high, smooth, transparent and glossy; umbilicus large. Snout rather deeply bifid.

Other characters
There are 3–4 tumid whorls which meet at deep sutures. When the shell is viewed from the base the umbilicus is seen to be bounded or slightly covered on one side by the inner lip. Up to 1 mm high and about 1 mm broad; last whorl occupies 80–85% of shell height, the aperture over half.

   R. globularis has short tentacles each with an eye behind the base. The flesh is pale and has dark brown markings. The scarcity of records of this animal (mainly from Scotland) suggests that it is rarer than the other two species of the genus, though they all live in the same habitat. *R. globularis* may well prove to be commoner than believed, having been overlooked or taken to be juveniles of the other species. Breeding has not been described.

Family OMALOGYRIDAE Sars, 1878

Key to British genera and species of Omalogyridae

Shell marked only by growth lines; penult whorl meets peristome below
mid-point of inner lip .... *Omalogyra atomus* (p. 274)

Shell with spiral and transverse ridges; penult whorl meets peristome
at mid-point of inner lip .... *Ammonicera rota* (p. 276)

Genus OMALOGYRA Jeffreys, 1860

Omalogyra atomus (Philippi, 1841)
(Fig. 110)

Truncatella atomus Philippi, 1841
Skenea nitidissima (Adams, 1800)

Diagnostic characters
Shell minute, with about three whorls lying in one plane so as to form a
biconcave disk, glossy and rather transparent; nearly smooth, but with irregular, fine, scale-like rings marking growth lines. Animal lacks cephalic tentacles
but has a pair of rounded, ciliated lobes, each with a central eye, formed
from the snout.

Other characters
The whorls are more or less circular in section and are visible on both sides
of the disk. Reddish brown. Up to 0.5 mm high, 1 mm across. The animals
are hermaphrodite but the penis is not visible since it is retractile into a
pouch.

Omalogyra atomus is found, often abundantly, on fine weeds growing on
rocks and in rock pools on the lower half of the beach, and also to depths
of about 20 m. The animals seem to eat the weeds and graze epiphytic diatoms.
The species is widespread on British and Irish coasts except those of the
southern North Sea. Its further distribution extends from the Mediterranean
to Norway, Iceland, Greenland, and from Maine to Rhode Island.

These animals breed spring to autumn, laying more or less spherical egg
capsules with a wrinkled surface about 200–300 μm in diameter, on *Ulva*
and *Enteromorpha*. Usually there is only one egg in a capsule which emerges
as a young snail in 2–3 weeks. Information on biology and breeding is given
by Fretter (1948).

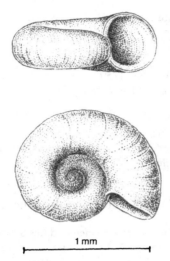

Fig. 110. *Omalogyra atomus*.

Genus AMMONICERA Vayssière, 1893

Ammonicera rota (Forbes & Hanley, 1850)
(Fig. 111)

Skenea rota Forbes & Hanley, 1850
Ammonicera tricarinata (Webster, 1856)

Diagnostic characters
Shell a minute biconcave disk, like *Omalogyra atomus* but with numerous costae forming annular thickenings on the whorls; sometimes with a peripheral keel or series of tubercles.

Other characters
There are 3–4 whorls. Reddish brown or clear yellow. 0.25 mm high, 0.5 mm across.

This is the smallest local prosobranch and has almost certainly been often overlooked on this account. It lives amongst weeds and in rock pools at L.W.S.T. and sublittorally, sometimes under stones on sand. The animals pierce algal cells and suck out the contents (Franc, 1948). They are widely spread throughout the British Isles and western Europe. The animals may well be hermaphrodites. Their egg capsules, which are egg- or lemon-shaped, have been described as fastened to the base of strands of *Cladophora*. There is a single egg in each which hatches as a young snail.

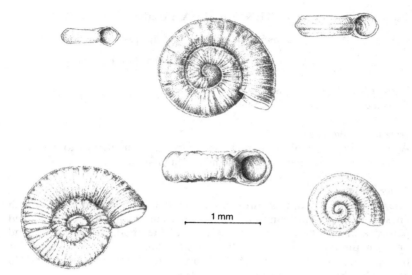

Fig. 111. *Ammonicera rota*.

Family SKENEOPSIDAE Iredale, 1915

Genus SKENEOPSIS Iredale, 1915

Skeneopsis planorbis (Fabricus, 1780)
(Fig. 112)

Turbo planorbis Fabricius, 1780
Skenea planorbis (Fabricius, 1780)

Diagnostic characters
A small, brown, semitransparent and nearly smooth shell of four whorls, with a low spire, deep umbilicus and nearly circular aperture.

Other characters
The whorls meet at deep sutures and are marked with numerous fine growth lines. The outer lip in young animals has often a projecting edge of uncalcified periostracum. In basal view the inner lip of the aperture does not bound or cover the umbilicus. The shell is chestnut brown, and often paler near the aperture. 0.75 mm high, 1.5 mm broad; last whorl occupies about 95% of shell height, aperture about three quarters.

The flesh is grey with yellow spots, with yellow around the bases of the tentacles and under the operculum.

S. planorbis may be found in pools and on weeds on the lower half of rocky shores and it also occurs to depths of 70 m. It may be extremely abundant, especially in summer, feeding on the weeds over which it crawls, and on their epiphytes (Fretter, 1948). It occurs practically everywhere in the British Isles where the shores are suitable, and ranges from the Mediterranean and the Azores to the Arctic on the eastern side of the Atlantic, through Iceland and Greenland to Arctic Canada and thence south to Florida.

Breeding may occur at any time of the year but is maximal in spring. Spherical or ovoid egg capsules are abundant on *Cladophora* or other filamentous algae; they measure 350–450 μm and contain one or two eggs which develop to young snails in 3–4 weeks (Lebour, 1937).

Note
A sinistral form of skeneopsid, *Retrotortina fuscata* Chaster, 1896, occurs south of the British Isles, though dead shells have been recorded from south west Ireland. Apart from its sinistrality it resembles *Skeneopsis planorbis* in nearly all respects though the aperture is more angulated and the spire a little lower. The shells are white and measure about 0.25 mm in height, 0.5 mm in breadth.

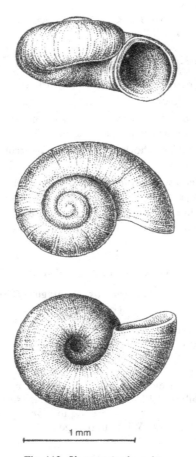

1 mm

Fig. 112. *Skeneopsis planorbis*.

Family TORNIDAE Sacco, 1896

Key to British genera and species of Tornidae

Genus TORNUS Turton & Kingston, 1830

Tornus subcarinatus (Montagu, 1803)
(Fig. 113)

Helix subcarinatus Montagu, 1803
Adeorbis subcarinatus (Montagu, 1803)

Diagnostic characters
Shell small, semitransparent, glossy, nearly circular in apical view, with low
spire and large umbilicus. Whorls with spiral keels (six on the last), crossed
by numerous small costae. Aperture lying in a very prosocline plane, with
a small anal spout adapically. White to orange.

Other characters
There are about 3.5 whorls of which 1.5 belong to the protoconch and are
without ornament. The last whorl occupies most of the shell, its six spiral
keels placed (1) alongside the suture; (2) halfway to the periphery; (3, 4,
5) above, at, and below the periphery; (6) round the umbilicus. Up to 1.5 mm
high, 2.5 mm broad; last whorl occupies about 80% of shell height, aperture
60%.
  The head has a long, narrow snout and carries two long tentacles, but
the animal is blind and lacks visible eyes. There are two pallial tentacles
at the mantle edge on the right and the ctenidium may be extended out
of the mantle cavity on the same side. Males have a penis. Flesh pale yellow
or pinkish.
  T. subcarinatus lives at L.W.S.T. under large boulders embedded on sandy
beaches, where the sand is yellow, not black. The animals are rarely found,

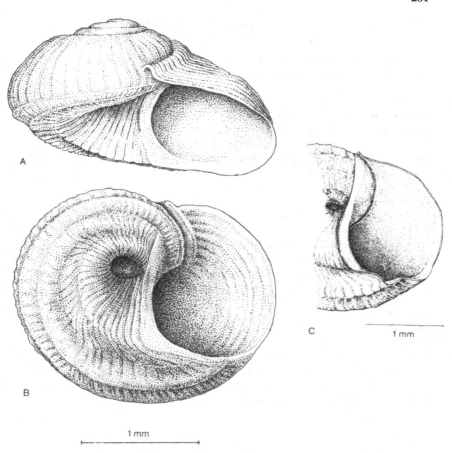

Fig. 113. *Tornus subcarinatus*. The groove shown in B at the adapical end of the outer lip is exaggerated in the shell drawn and does not develop in most, where the area appears as in C.

perhaps because the habitat is not commonly known. The species is southern in distribution, from the Mediterranean north to the British Isles, where it has been recorded from the western Channel and along western coasts.

The only reproductive stage of this species which is known is the veliger larva, which may be recognized by the bilobed velum which is dark red over its anterior half. The same colour is shown on the mesopodium though the propodium is pale.

Notes on the anatomy of this species were given by Woodward (1899) and Graham (1982).

Tornus exquisitus (Jeffreys, 1883)
(Fig. 114)

Adeorbis exquisitus Jeffreys, 1883
Adeorbis imperspicuus (Monterosato, 1895)

Diagnostic characters
Shell minute, semitransparent and glossy, with low spire, swollen whorls, deep sutures, its surface decorated with many fine prosocline costae and as many equally fine spiral ridges suggesting, at first sight, spiral rows of tubercles. Axis of aperture oblique. Cephalic tentacles bifid; eyes absent; foot without operculum.

Other characters
There are 2.5 whorls, the last much the biggest. Costae and spiral ridges cross to produce a series of fine pits. The outer lip shows an anal sinus and a peripheral bulge. The protoconch is smooth. Up to 0.8 mm high, 1.0 mm broad; last whorl occupies 90% of shell height, aperture 70%.

The head is triangular in cross section with a mid-dorsal crest and a ventro-lateral tentacle on each side in addition to the main cephalic ones. The mantle skirt has four marginal tentacles, one on the left, one median, two on the right. A ctenidium is missing. Males have a penis on the right. The sole of the foot is triangular, broad anteriorly, thin and delicate throughout. Body white.

T. exquisitus lives on fine silty bottoms and has been found from the Mediterranean to Scandinavia. It has been recorded in the British Isles off the south coast of Ireland, in the Irish Sea, and off west Scotland. It may well be more widespread and been overlooked because of its size.

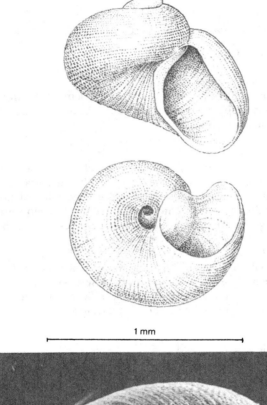

1 mm

Fig. 114. *Tornus exquisitus*.

Tornus unisulcatus (Chaster, 1897)

Adeorbis unisulcatus Chaster, 1897

Diagnostic characters

Shell minute, semitransparent, glossy, with a very depressed spire; last whorl marked by a single deep spiral groove just below the periphery, and a second, shallow one round the umbilicus.

Other characters

There are about two whorls, more or less circular in section, marked by prosocline growth lines (or low costae) as well as by the spiral grooves. The base of the shell is concave and lets the whole underside of the spire be seen. The outer lip has a deep notch at the end of the deep spiral groove. Up to 0.4 mm high, 0.99 mm broad; last whorl occupies 90% of shell height, aperture 75%.

The body of the animal has not been described. *T. unisulcatus* is clearly rare and it is doubtful whether anything more than dead shells has been found. It has been recorded in dredgings from soft bottoms off Plymouth, the north coast of Ireland and the west coast of Scotland. Abroad it ranges from the Channel coast of France to the Mediterranean.

Genus CIRCULUS Jeffreys, 1865

Circulus striatus (Philippi, 1836)
(Fig. 115)

Valvata? striata Philippi, 1836
Trochus duminyi Requien, 1848

Diagnostic characters
Shell small, flattened, semitransparent, glossy, with a low, obtuse spire and a very large umbilicus. The whorls bear spiral ridges. The aperture lies in a prosocline plane and appears nearly circular when the shell is held upright. The mantle skirt has two marginal tentacles on the right. Operculum with about 12 turns.

Other characters
There are 4–5 whorls, rounded in section, with deep sutures placed on the apical side of the periphery of each whorl. There are usually eight or ten (up to twelve) spiral ridges on the last whorl, 5–6 on the penult, 4–5 on the previous, the rest smooth. The base of the last whorl lacks ridges, but a further series of four or five appears on that part of it which lines the umbilical space. Other whorls exposed in that space also show ridges. The outer lip arises level with spiral ridge 6, 7, or 8 and shows a small peripheral bay and slight basal projection; it is crenulated by the ends of the ridges. White. Up to 1.25 mm high, 2.75 mm broad; last whorl occupies 80–85% of shell height, aperture about three quarters.

The animal has a narrow snout which is a little bifid at the tip, and two long and delicate tentacles, each with a basal eye. Males have a slender, sickle-shaped penis. The foot is narrow with recurved anterolateral points. Whitish with opaque white speckles, the snout pinkish.

C. striatus is a very rare animal at the latitude of the British Isles, where only a few shells have been found on western coasts. It is a southern species occurring from the Mediterranean north to Britain and Ireland, living on muddy bottoms at about 30 m deep. It appears to eat material adhering to sand grains (Fretter, 1956).

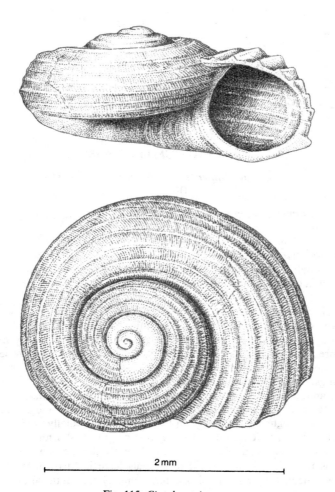

2 mm

Fig. 115. *Circulus striatus*.

Family CAECIDAE Gray, 1850

Key to British species of Caecum

Surface of tube smooth; apical plate rounded .... *Caecum glabrum* (p. 290)
Surface of tube transversely ridged; apical plate pointed ....
.... *Caecum imperforatum* (p. 288)

Genus CAECUM Fleming, 1813

Caecum imperforatum (Kanmacher, 1798)
(Fig. 116)

Dentalium imperforatum Kanmacher, 1798
Caecum trachea (Montagu, 1803)

Diagnostic characters
Shell a small slightly curved tube with numerous ring-like markings, open
at one end, closed at the other by a conical plate.

Other characters
The aperture is circular with a rather thick edge. If the animal is young
a coiled spire of 1.5–2 whorls may be attached to the end away from the
aperture, but this breaks off at maturity and the gap is sealed by secretion
of a plate of calcareous matter. Yellowish brown or reddish with some darker
markings. Up to 4 mm long, 0.5 mm in diameter.

The animal has a long narrow snout, bifid terminally. There are neither
pallial tentacles nor ctenidium. Males have a penis with some papillae. The
foot has a straight anterior edge, a little embayed medially, and a rounded
posterior end on which sits a multispiral operculum. Flesh white.

C. imperforatum is a diatom eater, moderately common on sandy, muddy
bottoms from 15–250 m deep. The animals have been recorded from the
Mediterranean to the British Isles, there mainly off southern and western
coasts as far north as Islay.

Veliger larvae of this species are common in summer plankton, the two
lobes of the velum edged with red-purple, the rest of the body colourless.
After metamorphosis the protoconch is broken off and the tight coiling of
the first part of the adult shell is replaced by a loose one to produce the
final shape.

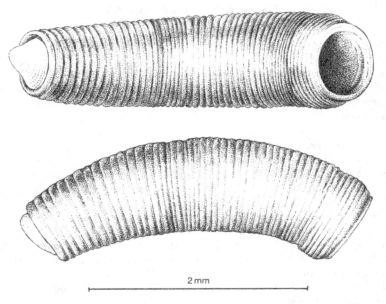

2 mm

Fig. 116. *Caecum imperforatum*.

Caecum glabrum (Montagu, 1803)
(Fig. 117)

Dentalium glabrum Montagu, 1803

Diagnostic characters

Differs from the previous species in that the tube is marked only by inconspicuous rings, in that the blind end is closed by a rounded plate, and in being smaller.

Other characters

C. glabrum has the same mode of life as *C. imperforatum*. It is more widespread, both within the British Isles and in extending to Norway and Sweden. The animals are summer breeders, laying transparent spherical capsules in the sand on which the adults live. There is one egg in each capsule which hatches to a veliger larva with a planospiral shell of 1–2 whorls. Its bilobed velum has a marginal band of purplish black. At a later stage the last whorl begins to grow away from the spire and still later, after the larva has settled and metamorphosed, the larval shell is broken off leaving only the last whorl, the upper end of which becomes sealed. With growth the oldest part of the tube is broken off, the hole again sealed so that the shell never appears to change shape (Götze, 1938).

What was once regarded as a variety of this species has recently been accepted as distinct. This is *C. armoricum* de Folin, 1869, which has been found in Brittany and Dorset. The shell may be distinguished from that of *C. glabrum* by the possession of a more prominent and asymmetrically pointed septum closing the shell.

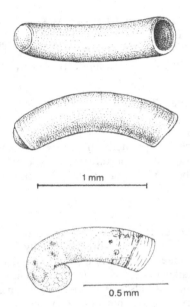

1 mm

0.5 mm

Fig. 117. *Caecum glabrum*. The bottom figure is of a young shell which still retains its initial spiral coil.

Family TURRITELLIDAE Woodward, 1851

Genus TURRITELLA Lamarck, 1799

Turritella communis Risso, 1826
(Fig. 118)

Turritella terebra (Linné, 1758)

Diagnostic characters

Shell a tall, slender, sharply pointed cone, opaque, somewhat glossy. Whorls bear spiral ridges and grooves. Aperture small, squarish, without a peristome. Mantle edge on left bears pinnate tentacles. Operculum edged with pinnate bristles.

Other characters

The apical angle is about 15°. There may be up to twenty whorls but the protoconch is always lost. Each whorl has usually three prominent spiral ridges lying at the periphery, but there may be twice that number, and there are always many smaller ones. The outer lip is almost always broken; its base is flattened in most shells and it joins the columella at right angles, often with a small spout at the junction. Brownish yellow to white, often with a lilac tinge on the base, and darker near sutures. Up to 30 mm high, 10 mm broad; last whorl occupies about 28% of shell height, aperture about 20%.

The head is drawn out into a short snout carrying short tentacles, each with a basal eye. Below the right tentacle lies the end of a deep ciliated groove extending over the floor of the mantle cavity; the right wall of the groove enlarges at the mouth of the cavity to form a scroll-like exhalant siphon. Males have no penis, and the genital tract is an open groove on the right of the mantle cavity in both sexes. The foot is small. The flesh is buff with dark spots and streaks, and there is white on the tentacles, siphon and foot.

T. communis occurs abundantly but locally in muddy sediments in shallow water, more or less buried, but maintaining contact with the water, from which the gill filters suspended particles; these are then carried forwards to the mouth in the pallial groove (Graham, 1938; Yonge, 1946). The animals are extremely inactive. They are found all round the British Isles and from the Mediterranean and North Africa to northern Norway.

The male gonad is cream, the female pink. Breeding occurs late spring and early summer. About 6–20 eggs are enclosed in a globular capsule with a thread-like process at one end which is fastened to a shell or stone; usually a number of capsules are bunched together with their stalks intertwined. Veliger larvae hatch from these. The velum is small and the larvae are weak swimmers, probably metamorphosing rather rapidly (Lebour, 1933c).

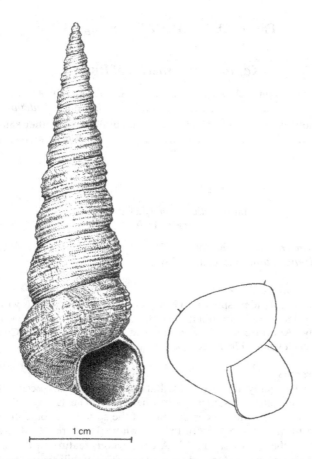

Fig. 118. *Turritella communis*. The aperture is usually more angulated, as shown to the right, than in the shell drawn.

Family CERITHIIDAE Fleming, 1822

Key to British species of Bittium

Shell with a varix on each whorl; body of animal dark; mainly on *Zostera* .... *Bittium reticulatum* (p. 294)

Shell without varices; body pale; mainly on coralline and other small algae .... *Bittium simplex* (p. 296)

Genus BITTIUM Leach, 1847

Bittium reticulatum (da Costa, 1778)
(Fig. 119)

Strombiformis reticulatus da Costa, 1778
Cerithium reticulatum (da Costa, 1778)

Diagnostic characters
Shell solid, tall, slender, spire not noticeably cyrtoconoid; whorls with costae and spiral ridges raised into blunt tubercles where they cross; base with spiral ridges only. Aperture small, with notch at base of columella. Commonly a varix on each whorl. Flesh brownish.

Other characters
The shell has usually 10–12 whorls, but up to fifteen may occur; they are a little swollen and the sutures are rather deep. The last whorl has nine or ten spiral ridges, whorls near the base of the spire have four, the number decreasing towards the apex. On the last whorl there are 12–20 costae and the whorls of the spire bear 11–12. A conspicuous feature of the ornament is the tubercles where costae and ridges cross. The umbilicus is closed. Fawn or light brown; apex, tubercles and columella pale. Up to 8 mm high, 3 mm broad; last whorl occupies 40–45% of shell height, aperture 25%.

There is an elongated snout with long slender tentacles at its base, an eye alongside each. Males are aphallic, and the genital duct is open in both sexes. The foot is long and narrow, truncated anteriorly, bluntly pointed behind. The opercular lobe projects backwards from under the operculum as a median ridge. The flesh is fawn-brown with speckles and streaks of black and white; the tentacles may be orange or reddish.

B. *reticulatum* is found on soft bottoms where *Zostera* and some other weeds grow, occasionally on stones, or in crevices; it prefers quiet places.

Fig. 119. *Bittium reticulatum*. The shell shown in A and B has less marked sculpture and a less prominent basal notch in the lip of the aperture than is usual: this is shown in C.

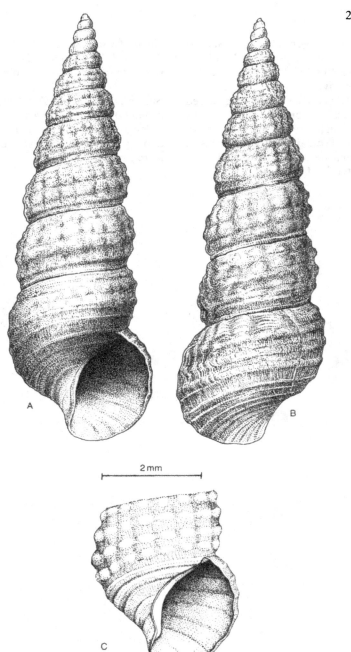

2 mm

It occurs from near L.W.S.T. to depths of 250 m, commoner sublittorally than intertidally. It is a detritivore. In the British Isles it is found everywhere except in the North Sea. Elsewhere it ranges from the Black Sea to northern Norway.

The animals are summer breeders. Spawn is attached to shells, stones or weeds and forms a cylindrical ribbon of about 3 mm diameter which is coiled in a tight spiral of 2–4 turns. Its total length is about 25 mm. In all, the spawn contains about 1000 eggs which develop rather rapidly to veliger larvae which are readily identifiable by the square tongue-shaped projection on the outer lip of their shell.

Bittium simplex (Jeffreys, 1867)

Cerithium simplex Jeffreys, 1867

Diagnostic characters
Shell like that of *B. reticulatum* but thinner and without varices; animal has pink flesh.

Other characters
This species is limited in the British Isles to Scilly, but this may only reflect the fact that it has not always been recognized as a distinct species (Bouchet, Danrigal & Huyghens, 1979).

Family APORRHAIDAE Gray, 1850

Key to British species of Aporrhais

Adapical process of outer lip not reaching level of shell apex; basal
process hooked .... *Aporrhais pespelecani* (p. 298)
Adapical process of outer lip extends beyond level of shell apex; basal
process straight; northern .... *Aporrhais serresianus* (p. 300)

Genus APORRHAIS da Costa, 1778

Aporrhais pespelecani (Linné, 1758)
(Fig. 120)

Strombus pespelecani Linné, 1758

Diagnostic characters
Shell solid, opaque, spire tall; outer lip expanded into a plate shaped like
the webbed foot of a bird, its apical point ending below the tip of the spire,
its basal part thickened and bent.

Other characters
The shell is a little glossy. There are 8–10 persistent whorls but some have
always been lost from the apex; they are swollen and bear numerous opisthocline costae. The last whorl has also two spiral bands of tubercles on its
base; these and the costae (which then shorten to tubercles) run on to the
abapertural side of the expanded outer lip. The lip has five lobes: the most
apical lies alongside the spire, 2–4 form points on the expansion, the fifth
projects basally and curves a little towards the aperture, which is an obliquely
elongated rectangle. Cream or tawny, the outer lip and edge of the aperture
darker. Up to 45 mm high (measured from tip of spire to tip of fifth lobe),
30 mm broad (measured to edge of outer lip); last whorl occupies about 55%
of shell height, aperture (measured from tip of first lobe to tip of fifth) 75–80%.
   There is a long tapering snout; tentacles arise from its base, each with
a basal eye. In males a recurved, tentaculiform penis arises behind the right
tentacle, with an open seminal groove. The foot is shield-shaped and rather
narrow, tapering to a point posteriorly. The operculum is also narrow and
lies across the foot. The flesh is mainly white, but the snout, the dorsal surface
of the tentacles and patches on the foot are carmine; in addition there are
yellow flecks on the snout and tentacles, and white ones on the foot.
   A. pespelecani ranges from the Mediterranean to northern Norway and
Iceland and occurs abundantly in almost all appropriate habitats round the
British Isles, though less common in Scilly, Channel Islands, eastern Channel
and southern North Sea. It is a detritus feeder and browser living partly

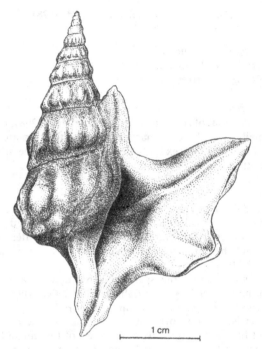

Fig. 120. *Aporrhais pespelecani.*

buried (Yonge, 1937) in muddy-sandy bottoms to depths of 180 m, never intertidally, occasionally moving from one site to another (Barnes & Bagenal, 1952). The method of locomotion is laborious (Weber, 1925). The expansion of the outer lip does not develop until the shell has about eight whorls.

The breeding period extends from March to July or August. Spherical egg capsules, each containing one egg, are attached singly or in small groups to the substratum. The egg develops to a veliger larva in about two weeks. This has a 2-lobed velum, each lobe with a red line round its margin and a red terminal blotch. Later the velum becomes first 4-, then 6-lobed. Larval life seems to be extended (Lebour, 1933c).

Aporrhais serresianus (Michaud, 1828)
(Fig. 121)

Rostellaria serresiana Michaud, 1828
Aporrhais pescarbonis (Brongniart, 1823)
Aporrhais macandreae Jeffreys, 1867

Diagnostic characters
Shell in general like that of *A. pespelecani* but smaller (p. 298), more delicate, glossier externally, with longer, sharper, more distinct points on the outer lip; the most adapical point extends beyond the tip of the spire, the most basal one is not thickened.

Other characters
There are only 7–8 whorls, and the costae are more numerous and sharper than in *A. pespelecani*. White. Up to 38 mm high (measured to tip of spire adapically), 25 mm broad (measured to outermost tip of outer lip); last whorl occupies half shell height, aperture (measured from tip of apical lobe to tip of basal lobe) 110%.

The body is as in the previous species but has a different colour pattern: the red on the head is divided into right and left halves by a median white line; the tentacles have a central white line and no red.

A. serresianus is a more northerly species than *pespelecani* and around the British Isles has been found only off the west of Ireland and the northern parts of Scotland. It lives in similar fashion to *pespelecani* but prefers finer sediments and extends to greater depths (1000 m) (Yonge, 1937). The spawn of this species is unknown, but the veliger larva has been described by Thiriot-Quiévreux (1976). The velum of older larvae has six lobes each with a reddish brown margin and distal blotch.

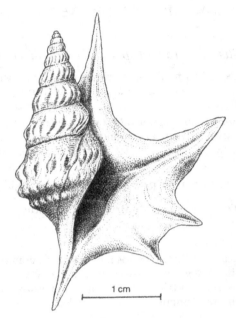

Fig. 121. *Aporrhais serresianus*.

302

Family TRICHOTROPIDAE Gray, 1850

Key to British genera and species of British Trichotropidae

Spire tall, whorls slightly tumid; spiral ridges prominent; periostracum with coarse processes along spiral ridges; aperture pear-shaped, pointed basally .... *Trichotropis borealis* (p. 302)

Spire short, whorls tumid; spiral ridges delicate; periostracum finely hirsute; aperture nearly circular, rounded basally; rare ....
.... *Torellia vestita* (p. 304)

Genus TRICHOTROPIS Broderip & Sowerby, 1829

Trichotropis borealis Broderip & Sowerby, 1829
(Fig. 122)

Diagnostic characters
Shell rather delicate, with thick periostracum drawn out into bristles or triangular plates. Spire moderately tall, with turreted profile. Whorls with prominent spiral ridges. Base ends in narrow point. Animals with grooved proboscis extending to right from mouth.

Other characters
There are about five swollen whorls meeting at deep sutures. On those of the spire there are two spiral keels, with minor ridges between the suture and the upper keel and sometimes also between the keels; on the last whorl 3–4 keels lie on the apical half and about five more (of decreasing size) below. At the base a thick ridge runs along the columellar side of the aperture, with a groove, but no umbilicus, between the two. The aperture is pear-shaped, broad above and narrow below. The calcareous shell is white on the spire, yellowish – sometimes pinkish – on the last whorl, but the periostracum masks this and imparts a general horn colour. Up to 11 mm high, 6 mm broad; last whorl occupies 70–75% of shell height, aperture 60%.

There is a short snout with the mouth at its tip. The lower lip (apparently, but actually the propodium) is drawn out into a proboscis which ends at the mouth of the mantle cavity. Males (smaller animals) have a flattened penis behind the right tentacle, with a seminal groove along it; with growth the animals become female and lose the penis. The flesh has opaque white flecks over the snout, tentacles and foot.

T. borealis is a circumboreal species found in the British Isles in the northern parts of the North Sea and off the west coast of Scotland. It lives on stones and shells on hard bottoms which are also silty, at depths from 10 to 270 m, feeding on particles collected in the mantle cavity and led to the proboscis by cilia. It is locally common.

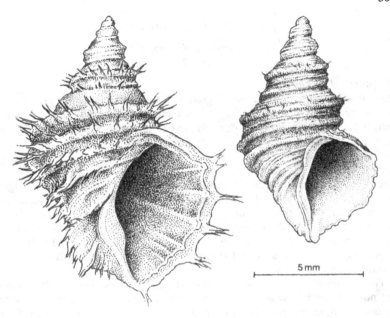

Fig. 122. *Trichotropis borealis*.

Genus TORELLIA Lovén, 1867

Torellia vestita Jeffreys, 1867
(Fig. 123)

Diagnostic characters

Shell globose, thin, completely covered with thick periostracum thrown into prosocline ridges which bear fine bristles and overlie growth lines. Animal with long proboscis, grooved dorsally, extending from ventrolateral lips of mouth.

Other characters

There is a low spire, with 5–6 tumid whorls which bear fine spiral ridges as well as growth lines, the two crossing to create a delicate reticulation of the shell surface, though this is obscured by the periostracum. In addition, in some shells, a broadly rounded keel extends from the base of the aperture to the umbilicus. The aperture is nearly circular. Yellowish. Up to 15 mm high, 15 mm broad; last whorl occupies 80–85% of shell height, aperture about two thirds.

The animal has a broad snout, extended into a proboscis as in *Trichotropis* but not necessarily turned to the right. Males have a cylindrical penis with an open seminal groove. The foot is broad and oval. Yellowish, with dark pigment on the tentacles.

T. vestita occurs widely in the North Atlantic, living on stony bottoms from 10–150 m deep off Scandinavia, and to about 2000 m in the southern parts of its range. In British waters it has been found only north and east of Shetland. Its reproduction is unknown.

Fig. 123. *Torellia vestita*.

Family CAPULIDAE Fleming, 1822

Genus CAPULUS Montfort, 1810

Capulus ungaricus (Linné, 1758)
(Fig. 124)

Patella ungarica Linné, 1758
Pileopsis hungaricus (Linné, 1758)

Diagnostic characters
Shell fragile, covered with a flounced periostracum; cap-shaped, with small spiral coil at apex. Aperture a large oval. Animal with proboscis permanently extended from mouth. Operculum absent from foot. Often living on bivalve shells.

Other characters
There is a small, spirally coiled beak of about two whorls, from which the last whorl expands rapidly to form the bulk of the shell. It bears numerous delicate ridges radiating from the apex, and there are also growth marks, some prominent, more or less parallel to the edge of the aperture. The peristome matches the curvature of the surface to which the animal clings. The shell is white, the periostracum straw-coloured. Up to 20 mm long, 20 mm broad.

The proboscis, which is formed from the propodium, is grooved dorsally as in trichotropids, and is mobile. The animals are consecutive hermaphodites: those up to about 4 mm shell length are males and have a curved penis, with open seminal groove, attached behind the right tentacle; larger animals are female and the penis is then reduced to a papilla. The head and foot are yellowish with white speckles.

C. *ungaricus* is sublittoral, and while it may be found on stones, is mainly attached to shells of living bivalves (*Modiolus*, *Chlamys*, *Pecten*) (Jones, 1949; Sharman, 1956), the gastropod *Turritella* (Thorson, 1965), and, when young, tubes of *Pomatoceros*. It uses the proboscis to collect food or the pseudofaeces of its host; it is capable of gathering particles from its own pallial currents, and those living on stones must depend upon this source (Yonge, 1938). The species is widespread off all North Atlantic coasts from the Mediterranean across to eastern America; not, however, in the southern part of the North Sea.

The eggs are held in a mass under the foot and are brooded by the female. From them hatch **echinospira larvae** (see p. 316). These have a transparent, gelatinous outer shell, almost globular and covered with minute perforations often approximately radially arranged.

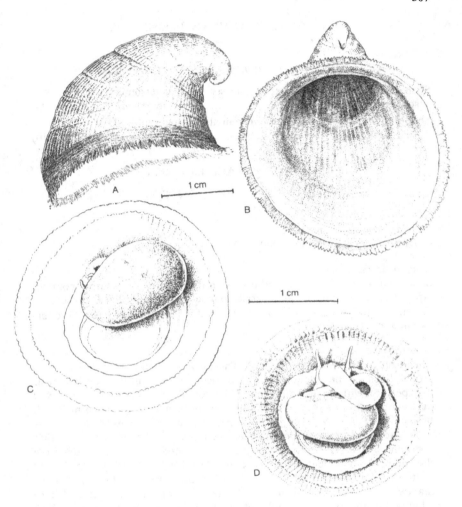

Fig. 124. *Capulus ungaricus*. A, shell, from left; B, shell in apertural view, anterior end below; C, female with egg mass held in anterior part of foot; D, female with egg mass and also showing the proboscis; its coiling is a post mortem effect.

Family CALYPTRAEIDAE Blainville, 1824

Key to British genera and species of Calyptraeidae

Shell conical with central apex; internal septum tongue-shaped; animals
solitary .... *Calyptraea chinensis* (p. 308)
Shell kidney-shaped, apex nearly terminal; internal septum D-shaped;
animals in chains or stacks .... *Crepidula fornicata* (p. 310)

Genus CALYPTRAEA Lamarck, 1799

Calyptraea chinensis (Linné, 1758)
(Fig. 125)

Patella chinensis Linné, 1758
Calyptraea sinensis (Linné, 1758)

Diagnostic characters
Shell conical, not spirally coiled except in the protoconch, sharply pointed.
Aperture nearly circular; an internal tongue-shaped partition is present. A
flattened neck lobe on each side of the head behind the tentacles. No operculum on foot.

Other characters
The protoconch is usually eroded. The only ornament takes the form of
growth lines parallel to the edge of the aperture, sometimes, especially near
the aperture, with small projections. The internal septum arises from a curved
base running from apex to apertural edge and is narrow at the former end,
broad at the latter. White or yellow. Up to 15 mm across, 5 mm high.
   The head is elongated posterior to the tentacles, the snout being rather
short and broad. The left neck lobe bears a ciliated groove which runs from
the mouth to a pouch in the thick mantle edge overlying the head; that
on the right is crossed by a ciliated groove originating on the floor of the
mantle cavity. Animals up to a shell breadth of 2 mm are male and have
a bilobed penis behind the right tentacle; larger animals are females in which
the penis becomes reduced to a knob above the tentacle (Wyatt, 1960). The
foot is large, oval in outline. The flesh is yellow, speckled with white.
   C. chinensis is a ciliary feeder, trapping particles on a mucous sheet on
the ctenidium; this is then rolled up and passed along the groove on the
floor of the mantle cavity to the mouth (Werner, 1952, 1953, 1959). The
animals may be found on shells or under stones on sheltered shores in south-
west England and sublittorally in the western Channel, off the Isle of Man
and in the Firth of Clyde.
   Eggs are laid in soft capsules, 12–25 in each, attached to stones and brooded
under the foot of the female. Development is direct (Lebour, 1936).

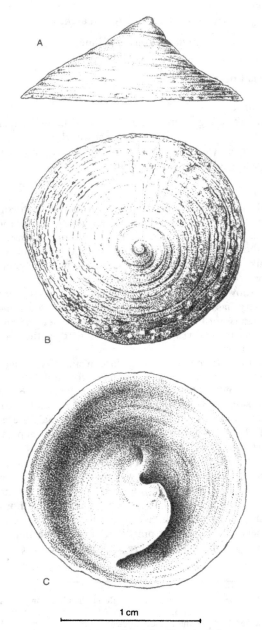

Fig. 125. *Calyptraea chinensis*. A, from the right; B, apical view, anterior end below; C, apertural view, anterior end above.

Genus CREPIDULA Lamarck, 1799

Crepidula fornicata (Linné, 1758)
(Fig. 126)

Patella fornicata Linné, 1758

Diagnostic characters

Animals usually adhering to one another to form chains or stacks. Individual shell with small depressed spire and large last whorl. Aperture large, kidney-shaped, half blocked (in empty shell) by large partition near spire. Animal with neck lobes as in *Calyptraea*. Operculum absent.

Other characters

There are only two whorls, one of which forms the protoconch the other the rest of the shell, which is solid and rather glossy, and without ornament apart from occasional growth lines. Yellow or red-brown, mottled with short dark streaks; there is usually a paler band along the periphery. The internal septum is chalk-white, with dark chestnut along its attachment; the rest of the internal surface is tan. Up to 50 mm long, 25 mm high.

The animal shows the same general organization as *Calyptraea*, though the mantle cavity has become much deeper and the gill longer; the penis of the male stage is not bilobed and it disappears when the animal becomes female. The flesh is yellowish, with dark pigment on the snout, tentacles, mantle edge and penis.

C. fornicata lives in chains of up to fifteen animals, usually fewer. The smaller animals at one end are male, some in the middle are changing sex, and the large animals at the opposite end are female; the males fertilize females in the same chain. Chains persist by the addition of small males at one end while females die at the other. The animals are ciliary feeders (Orton, 1912). They are normally sublittoral, living to depths of 10 m, but are often thrown on to beaches after storms. The species is centred on the Atlantic coast of North America but animals were introduced, with imported oysters, to Essex in 1887–90 and have now spread along the south coast of England; they have also been found in the Bristol Channel and been reported from Northumberland, Belfast Lough and Kerry, though it is doubtful whether they can form established populations in all these places. In Europe the species has spread from an original Dutch importation to the Kattegat and Skagerrak, and to the French Atlantic coast. *C. fornicata* often

Fig. 126. *Crepidula fornicata*. A and B, a single shell; C, a chain or stack of shells. The marginal sketches indicate the sexual state of the component animals. The base (bottom left) is a dead empty shell; then clockwise are three functional females, each brooding egg capsules and lacking a penis; next is an animal changing sex, followed by four functional males, each with a penis, shown black.

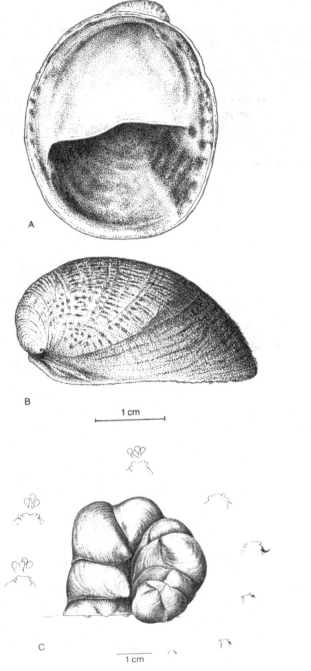

A

B

1 cm

C

1 cm

occurs in enormous numbers and is a serious pest of oyster beds (Orton, 1909; Chipperfield, 1951; Walne, 1956).

Eggs are brooded under the foot of the female; from the capsules hatch free-swimming veliger larvae which are attracted at metamorphosis to existing chains on which they settle and become small males; should they fail to find a chain on which to settle they rapidly become female and so attractive to other settling larvae, and thus the founder matriarch of a new stack. The sexual biology of this species has been discussed by Gould (1952), Coe (1953) and Hoagland (1978).

Vernacular name: slipper limpet.

Family LAMELLARIIDAE Orbigny, 1841

Key to British genera and species of Lamellariidae

1. Shell internal, mantle lobes fused over it (*Lamellaria*) **2**
   Shell external (*Velutina*) **3**
2. Shell with high spire; dorsal surface of animal dome-shaped, tuberculate; grey-purple with white markings, sometimes yellowish ....
   .... *Lamellaria perspicua* (p. 314)
   Shell with low spire; dorsal surface rather flat, smooth; yellowish with dark markings .... *Lamellaria latens* (p. 318)
3. Shell without thick periostracum; 2–3 calcified whorls, though weakly; spire distinct but short; aperture oval, very large; rare
   .... *Velutina undata* (p. 322)
   Shell with thick periostracum; very weakly calcified, even flexible **4**
4. Aperture circular .... *Velutina velutina* (p. 322)
   Aperture oval .... *Velutina plicatilis* (p. 320)

Genus LAMELLARIA Montagu, 1815

Lamellaria perspicua (Linné, 1758)
(Fig. 127)

Helix perspicua Linné, 1758
Lamellaria tentaculata (Montagu, 1811)

Diagnostic characters
Shell wholly internal with short spire, permanently covered by fused mantle lobes so that animal looks like a dorid but has smooth tentacles; the dorsal surface of the animal is dome-shaped; the mantle edge has an anterior siphonal notch, and the male has an external penis. Mantle lilac, grey or buff, often splashed with white, yellow or black. Operculum absent.

Other characters
The shell is clear and fragile, with 2–3 whorls, elevated spire, deep sutures. The aperture is large and prosocline. Up to 8 mm high, 9 mm broad; the animal may be up to 20 mm long, 12 mm across.

The head has no snout, the tentacles arising from its anterior edge; the mouth on its underside leads to an introvert. The mantle edge is thickened and folded in the mid-line to make an exhalant channel from the mantle cavity. Over the shell the mantle bears tubercles. The basal half of the penis is broad and flattened, its distal half is a narrow flagellum. The foot is shield-

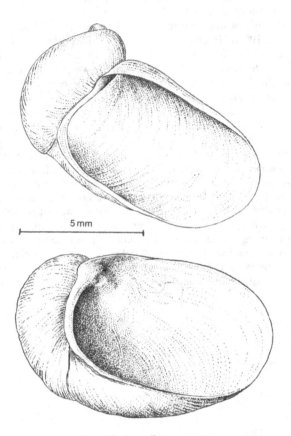

5 mm

Fig. 127. *Lamellaria perspicua*.

shaped, with an extended slit across its anterior end, the dorsal lip of which is corrugated. Markings on the dorsal mantle may mimic the appearance of tunicates, or sponges, or barnacles.

L. perspicua is not uncommon on rocky shores from about L.W.S.T. to depths of over 100 m. Intertidal animals live under stones or ledges and in rock pools, usually in association with such compound ascidians as *Botryllus*, *Polyclinum*, *Trididemnum*, and *Leptoclinum*, on which they feed. The species ranges from the Mediterranean to the north of Norway and to Iceland. It occurs on suitable substrata throughout the British Isles.

Egg capsules are vase-shaped and are buried in holes bitten out of the test of the tunicates on which the adults live (Ankel, 1935) up to the neck of the vase, the mouth still projecting a little. They contain many yellowish eggs most of which fail to develop and are used as food by those that do. From the capsule there hatch veliger larvae with a shell modified so as to comprise an inner and an outer layer with a space between the two filled with sea water: these are known as **echinospira larvae**. That of *L. perspicua* has been described by Lebour (1935a). The outer shell or **scaphoconch** has, when fully grown, 4–5 whorls coiled in one plane to form a flat disk with its margins forming very finely striated keels. The inner shell (which is the beginning of that of the adult) coiled helically and is eccentrically placed. The body of the larva is heavily pigmented. The velum of a later larva has six lobes, each with a border of dark spots.

Lamellaria latens (Müller, 1776)
(Fig. 128)

Bulla latens Müller, 1776

Diagnostic characters
Shell differs from that of *L. perspicua* (p. 314) in being flat, with hardly any spire, and with shallow sutures; dorsal surface of animal rather flat. Mantle usually buff, with few tubercles.

Other characters
This is a smaller animal than the previous, the shell being about 4.5 mm high, 9 mm broad and the animal about 10 mm long and 6 mm broad. The mantle never shows a lilac tinge. The anatomy is similar to that of *L. perspicua*. *L. latens* has the same range as *perspicua* and the same habitat and general distribution in the British Isles. The differences between this species and the last have been documented by McMillan (1939).

Breeding is like that of *L. perspicua* (Lebour, 1935a) but the echinospira larva may be distinguished by the coarse serration of the marginal keels on the outer shell, and by the lack of pigment in the animal's body.

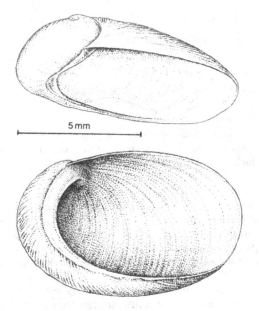

Fig. 128. *Lamellaria latens.*

Genus VELUTINA Fleming, 1822

Velutina plicatilis (Müller, 1776)
(Fig. 129)

Bulla plicatilis Müller, 1776
Velutina flexilis (Montagu, 1808)

Diagnostic characters
Shell so covered with thick periostracum that whorls and sutures are concealed, and so weakly calcified as to be flexible. Aperture an elongated oval. Mantle edge thick; foot without operculum. Light yellow.

Other characters
The periostracum is slightly hispid. When that is removed the shell is seen to have 2–3 swollen and nearly transparent whorls, the last comprising most of the shell, and the suture between it and the penult deep. Yellowish near the aperture, white towards the apex. Up to 15 mm high, 12 mm broad; last whorl occupies practically the whole of the shell height, the aperture about 85%.

The animal has no snout, the tentacles being set on a transverse ridge underneath which is the apparent mouth, leading to an introvert. The tentacles are long but can contract to blunt knobs; each has a basal eye. The thickened mantle edge restricts communication with the mantle cavity to two channels, a small left and a larger right one. The animals are hermaphrodite, all with a penis behind the right tentacle; its basal half is fleshy, its distal half a narrow flagellum. The foot sole has a figure-of-eight shape with its anterior border flattened. The flesh is white, yellow or orange.

V. plicatilis is a circumboreal species which is found in the northern parts of the North Sea and off the west coast of Scotland. The animals are not intertidal but live on hard bottoms 10–350 m deep, usually in association with compound ascidians and the hydroid *Tubularia indivisa*, on which they feed.

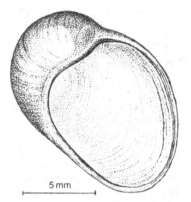

Fig. 129. *Velutina plicatilis.*

Velutina velutina (Müller, 1776)
(Fig. 130)

Bulla velutina Müller, 1776
Velutina laevigata (Pennant, 1777)

Diagnostic characters
Shell covered with periostracum concealing whorls and sutures. Spire low.
Aperture large and nearly circular. Dark brown.

Other characters
The 2–3 whorls of this shell are tumid and marked with delicate spiral and
growth lines. The periostracum is brown, the calcareous shell white or pinkish.
Up to 10 mm high, 10 mm broad; last whorl equals the whole shell height,
aperture occupies about 90% of it.

The animal has the same features as *V. plicatilis* (p. 320) but its flesh
is white or yellowish, the mantle edge with many white points.

V. velutina is a circumboreal form, more frequent in the fauna of the British
Isles than *V. plicatilis*, and occurring in most parts, though commoner in
the north. It is sometimes found at extreme low water on rocky shores but
is usually sublittoral and associated with such solitary ascidians as *Styela*,
upon which it feeds (Diehl, 1956).

The animals breed in spring and the larva has been described by Lebour
(1935a). It is an echinospira (p. 316) with a flat discoidal shell as in *Lamellaria*
species, but the outer shell, instead of being stiff and hard as in that genus,
is soft and gelatinous, lacks marginal keels, and in later stages of development
becomes rather bloated. The velum has four lobes.

A further species of this genus, *undata* Brown, 1839, has been recorded
alive from deep water off Shetland. Like *V. velutina* it has a shell with rather
pointed spire, but the aperture resembles that of *V. plicatilis* in being clearly
oval, not circular as in *velutina*. The shell is more heavily calcified and the
periostracum is thinner than in those species, and it also differs in having
more or less clear brown spiral bands.

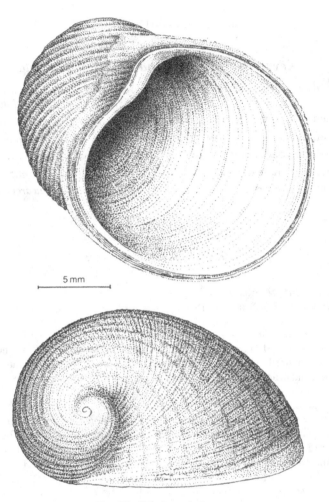

5 mm

Fig. 130. *Velutina velutina*.

Family ERATOIDAE Gill, 1871

Key to British genera and species of Eratoidae

1. Shell smooth with short spire; aperture long, narrow ....
   .... *Erato voluta* (p. 324)
   Shell not like this **2**
2. Shell with short spire, aperture round .... *Trivia* sp. (young)
   Shell convolute, ridged, a cowrie (*Trivia*) **3**
3. Shell with three dark spots .... *Trivia monacha* (p. 326)
   Shell without spots .... *Trivia arctica* (p. 328)

Genus ERATO Risso, 1826

Erato voluta (Montagu, 1803)
(Fig. 131)

Cypraea voluta Montagu, 1803
Marginella laevis (Donovan, 1804)

Diagnostic characters
Shell harp-shaped in profile, the last whorl occupying most of the shell; spire small. Aperture long, narrow, slightly sigmoid. Shell covered by mantle lobes when animal is active. No operculum on foot.

Other characters
The shell is slightly translucent and glossy. There are four whorls, those in the spire nearly flat-sided, with extremely shallow sutures; the apex is truncated. The sole ornament is a series of fine and irregular growth lines. The aperture has parallel sides and ends basally in a canal. The outer lip is thick, inturned, and has 15–18 ridge-like teeth on its inner edge. The columella is short and marked by three folds; there is no true inner lip, but it is functionally replaced by a row of about twelve tubercles on the last whorl. White, sometimes brownish basally and pink along the lip. Up to 12 mm high, 7–8 mm broad; last whorl occupies 80–85% of shell height, aperture 75–80%.

The head ends in a fold from which the tentacles arise, each with a basal eye; the mouth of the introvert lies under the fold. The mantle edge is drawn out anteriorly to form a siphon lodged in the basal canal, while laterally it forms lobes which cover the shell during activity but are withdrawn if the animal is disturbed. They bear tubercles on their outer side. The male has a penis; the foot is long and narrow. The flesh is white, generously speckled with brown, red, orange, and yellow; the mantle lobes are often dark, their

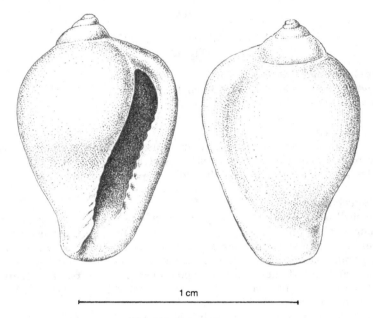

1 cm

Fig. 131. *Erato voluta.*

tubercles yellow. The tentacles have a central yellow line and are pale at their tips and base.

E. voluta occurs from the Mediterranean to Norway and is found at depths of 20–100 m off most shores of the British Isles (except between the Humber and the Isle of Wight) wherever the bottom is hard. It associates with ascidians, which form its food (Fretter, 1951a).

The larval stage is an echinospira (p. 316) with 3–4 whorls coiled helically; the coiling of the inner shell is not coaxial with that of the outer (Lebour, 1933a, 1935a).

Genus TRIVIA Gray, 1837

Trivia monacha (da Costa, 1778)
(Fig. 132)

Cypraea monacha da Costa, 1778
Cypraea europaea Montagu, 1803 (part)
Trivia europaea (Montagu, 1803) (part)

Diagnostic characters
Shell a small cowrie with three darks spots on side away from aperture; covered by mantle lobes when animal is active. Penis of male finger-like and sickle-shaped. Animal brightly coloured; mainly intertidal.

Other characters
The shell is convolute when fully grown, the last whorl growing over and hiding all older ones. Its surface is marked by numerous ridges and grooves running transverse to the long axis centrally but curving to lie along it at each end. The apertural side is flattened, the aperture long and narrow, curved at each end, the columellar region concave. White round the aperture and centrally on the convex side, brownish elsewhere. Up to 12 mm high (long), 8 mm across.

In young shells (up to 5 mm high, white in colour, about six months after settlement) the final growth of the last whorl to cover the spire has not yet occurred, and a spire of 3–4 tumid whorls is visible. The whorls show only growth lines. At this stage the aperture is a wide opening with a thin lip.

The mouth lies at the tip of a short snout; the tentacles are long, slender, each with a basal eye. The mantle edge is thickened, drawn out anteriorly into a long siphon. There is no operculum. The head, tentacles and siphon are yellow, or red-brown with yellow spots, the sides of the foot similar but paler. The mantle lobes extending over the shell have streaks of black or brown mimicking the ridges on the shell and may have red and yellow spots or streaks as well.

T. monacha is not uncommon towards L.W.S.T. on rocky shores throughout the British Isles. It lives intertidally under stones and ledges, usually where there are growths of compound ascidians, upon which it feeds, mainly *Diplosoma, Botryllus, Botrylloides, Polyclinum*. The species ranges from the Mediterranean to the British Isles.

The animals lay eggs in capsules inserted into the test of the tunicates on which they live. The capsules are vase-shaped, the mouth of the vase projecting a little out of the cavity which the female has rasped out of the test. The eggs develop into echinospira larvae (p. 316), the shells of which coil helically, the axis of both outer and inner shells the same. The velum is bilobed and the body of the larva is darkly pigmented (Lebour, 1935a).

Vernacular name: cowrie.

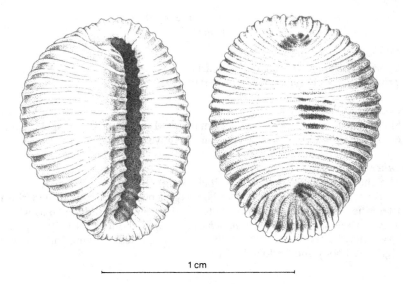

1 cm

Fig. 132. *Trivia monacha*.

Trivia arctica (Pulteney, 1799)
(Fig. 133)

Cypraea arctica Pulteney, 1799
Cypraea europaea Montagu, 1803 (part)
Trivia europaea (Montagu, 1803) (part)

Diagnostic characters
Shell as in *T. monacha* but usually smaller and without pigment spots. Penis of male leaf-like. Animal pale in colour; mainly sublittoral.

Other characters
The shell measures up to 10 mm high (long), 8 mm in breadth. *T. arctica* ranges from the Mediterranean to Norway. It occurs throughout the British Isles where substrata are appropriate for the growth of the ascidians which it eats. Breeding is as in *T. monacha* (p. 326) (Lebour, 1931b, 1935a); the larvae may be distinguished from those of that species in having only a lightly pigmented body and a 4-lobed velum.

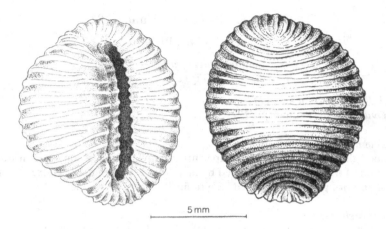

5 mm

Fig. 133. *Trivia arctica*.

Family OVULIDAE Fleming, 1822

Genus SIMNIA Risso, 1826

Simnia patula (Pennant, 1777)
(Fig. 134)

Bulla patula Pennant, 1777
Ovula patula (Pennant, 1777)

Diagnostic characters
Shell thin, glossy, convolute; aperture narrow, with lips longer than other parts of the shell. Shell covered by mantle in active animals. Dorsal surface of anterior part of foot ridged longitudinally; operculum absent.

Other characters
No spire is visible in fully grown shells. The slight ornament includes growth lines and low spiral ridges, the latter more easily seen at the extremities of the shell. The elongated aperture is drawn out at each end to a canal, the basal inhalant one broader and less distinct from the aperture than the apical exhalant one. White, yellowish, or pinkish. Up to 15 mm high, 7–8 mm broad. In young shells the spire of three tumid whorls is still exposed.

The head has a short snout ending in a suctorial disk with central mouth. Cephalic tentacles are long and have each a basal eye. The mantle edge forms an inhalant siphon anteriorly, and, laterally, two flaps which may cover the shell. Males have a long, recurved and pointed penis carrying an open seminal groove. The foot is rather large. The body is yellow with brown streaks and spots, mostly on the exposed mantle lobes; the distal half of each tentacle is white, its tip brown; front end of the foot pinkish.

S. patula is not common around the British Isles, confined to the western Channel and western coasts north to Orkney (Rendall, 1936). It is not intertidal, living 15–75 m deep on colonies of *Eunicella*, *Alcyonium* and *Tubularia* on which it feeds (Fretter, 1951a). Its further range is south to Spain.

Eggs are laid in capsules spread in a layer over *Alcyonium* colonies. They produce veliger larvae which ultimately possess a velum of four long and very narrow lobes (Lebour, 1932a).

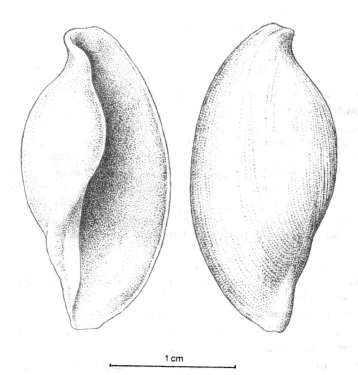

1 cm

Fig. 134. *Simnia patula*.

Family NATICIDAE Gray, 1840

Key to British genera and species of Naticidae

1. Shell without umbilicus or, at most, with minute chink **2**
Shell with distinct umbilicus **3**

2. Shell translucent; spire stepped; aperture narrow adapically; outer lip arises normal to last whorl, initially flattened; operculum horny
.... *Amauropsis islandica* (p. 346)
Shell solid, spire whorls not shouldered; aperture D-shaped; outer lip arises tangential to last whorl; operculum calcified externally
.... *Natica clausa* (p. 344)

3. Umbilicus oval, narrow adapically; outer lip arises tangential to last whorl which has 5 spiral rows of brown marks (Fig. 135A) ....
....,...... *Lunatia alderi* (p. 334)
Shell not like this **4**

4. Umbilicus round, broad apically; outer lip arises normal to last whorl which has 1 row of brown marks (Fig. 135C) ..... *Lunatia catena* (p. 336)
Shell not like this, usually without regular markings **5**

5. Umbilicus small and slit-like; northern form; rare *Lunatia pallida* (p. 342)
Umbilicus dilated **6**

6. Shell pale; a nick between last whorl and adapical end of outer lip; a groove along umbilical space produces a deep notch on the columella (Fig. 135B) .... *Lunatia montagui* (p. 338)
Shell brown; without a nick between outer lip and last whorl (Fig. 135D) .... *Lunatia fusca* (p. 340)

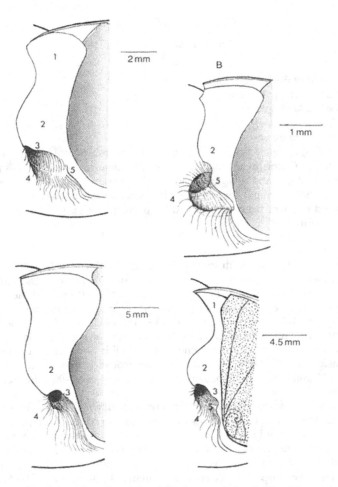

Fig. 135. Details of the columellar and umbilical areas of the shell in different species of *Lunatia*: A, *L. alderi*; B, *L. montagui*; C, *L. catena*; D, *L. fusca*. Corresponding features are similarly numbered.

Genus LUNATIA Gray, 1847

Lunatia alderi (Forbes, 1838)

(Figs 135A, 136)

Natica alderi Forbes, 1838
Natica poliana delle Chiaje, 1826
Natica nitida of authors

Diagnostic characters

Shell solid, opaque, glossy, nearly globose; spire short and blunt, with nearly flat profile. Outer lip approaches last whorl tangentially. Umbilicus present, narrow and pointed adapically. Five spiral rows of chestnut brown marks on last whorl. Animal covers shell with pedal lobes when active; propodium enlarged to cover front of shell and head except for part of tentacles. Operculum horny.

Other characters

There are 6–7 whorls with some irregular growth lines as the sole ornament. The inner lip extends to give rise to two pads lying over the last whorl. A ridge from the base of the aperture forms the left edge of the umbilical groove, and a second ridge, arising about the middle of the columella, runs into the umbilical opening (Fig. 135A). The shell is buff or yellow, paler on the base and columella. The subsutural row of markings on each whorl is more conspicuous than the others. The shell is also chestnut between the extensions of the inner lip and along the groove leading to the umbilicus. Up to 15 mm high, 12 mm broad; last whorl occupies 80–90% of shell height, aperture 75–85%.

The head has a short, broad snout at the base of which lie the two somewhat flattened tentacles, united across the mid-line by a basal fold. Eyes are insunk and usually invisible. Males have a large penis with an open seminal groove attached behind the right tentacle. A fold on the left of the propodium forms an inhalant siphon, and a ventral groove separates propodium from mesopodium. The propodium is probably inflated so as to cover the head of a creeping animal by uptake of sea water into a series of internal spaces: this has been shown for other species of the genus (Schiemenz, 1884, 1887; Russell-Hunter, 1968) but not for *alderi*. On the dorsal side of the foot lies the operculum, supported on lobes which extend to cover the back half of the shell as the animal creeps. Flesh cream or yellow with red-brown marks; tentacles dark.

L. alderi lives on sandy beaches, usually buried; it is sometimes found at L.W.S.T. but is commonly sublittoral between 10 and 50 m, and may occur as deep as 2000 m. In normal circumstances it never retracts into its shell, a process which would involve emptying sea water from the internal spaces and which would prevent normal locomotion; this restricts it to levels below those at which it would be exposed at low water. It eats bivalves, mainly

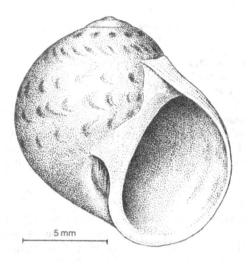

Fig. 136. *Lunatia alderi*.

tellinaceans, the shell of which it partly bores, and partly erodes by chemical means effected by an organ on the ventral lip of the mouth (Ziegelmeier, 1954; Carriker, 1981). The animals occur all round the British Isles and the species ranges from the Mediterranean to Norway.

Eggs are laid in capsules which are embedded in a collar-shaped mass of jelly, hardened and strengthened by incorporation in it of sand grains. Its formation has been described by Ziegelmeier (1961) and, though not for this species, by Giglioli (1955). Breeding occurs in spring and summer, when egg collars are produced and may be found on the sands where the adults live. They have a diameter of 25–30 mm, the central hole about 10 mm; the egg spaces lie irregularly in them in one plane, each containing only a single egg which hatches as a free-swimming veliger larva.

Lunatia catena (da Costa, 1778)
(Figs 135C, 137)

Cochlea catena da Costa, 1778
Natica catena (da Costa, 1778)
Natica monilifera Lamarck, 1822

Diagnostic characters
Shell helicoid with moderately high spire; whorls tumid and sutures distinct;
outer lip meets last whorl at right angles. Umbilicus large, usually rounded.
One subsutural row of brown marks on last whorl. Animal as in *L. alderi*
(p. 334).

Other characters
The shell of this animal is in most respects like that of *L. alderi* apart from
the points mentioned above. Although the umbilicus is usually rounded at
its adapical end it may be pointed as in *alderi*; the ridge which in that species
runs into the umbilicus from the middle of the columella is here reduced
or absent (Fig. 135C). The colour of the shell is less intense than in *alderi*
with the umbilical groove pale. Up to 30 mm high, 30 mm broad; last whorl
occupies about 90% of shell height, aperture about 70%.

The animal is like *L. alderi*.

L. catena has the same way of life as *alderi* but is restricted to shallower
water, from L.W.S.T. to 125 m deep. Its distribution in the British Isles and
abroad is also similar, but it is absent from Scandinavia.

Breeding occurs throughout spring and early summer when egg collars
may be found washed up on sandy beaches. They may be distinguished from
those of *L. alderi* first by their size, the diameter of the collar being about
75 mm, and second by the arrangement of the egg capsules within the collar:
these lie in regular lines and bulge a little on its external surfaces. Each
capsule contains many eggs (80–90) but of these only two or so develop
to veliger larvae, the rest being used by them as food.

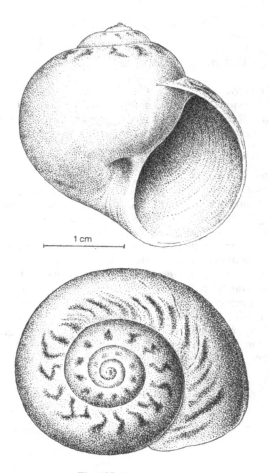

Fig. 137. *Lunatia catena.*

Lunatia montagui (Forbes, 1838)
(Figs 135B, 138)

Natica montagui Forbes, 1838

Diagnostic characters
Like the two previous species in general shell features but without colour markings; small. Recognizable by a nick on the umbilical side of the columella and by another between the origin of the outer lip and the wall of the last whorl.

Other characters
The spire is low, deep sutures giving it a moderately stepped profile. The notch on the columella marks the end of a groove which extends into the umbilicus (Fig. 135B). Buff. Up to 8 mm high, 8–9 mm broad; last whorl occupies 80–90% of shell height, aperture 75–80%.

The animal resembles that of other *Lunatia* species but is cream, with the posterior edge of the propodium brown.

L. montagui is not found intertidally but may be dredged on soft bottoms 15–200 m deep, usually on finer sediments than either *alderi* (p. 334) or *catena* (p. 336). It is a more northern species than those and most records from around the British Isles are northern. It also occurs off Northern Ireland, in the Celtic Sea, and south to the Mediterranean, though at greater depths. The egg collar is similar in dimensions to that of *L. alderi* but may be distinguished from that by the fact that the egg capsules in it lie in several planes.

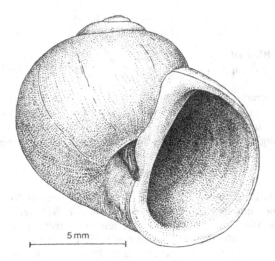

Fig. 138. *Lunatia montagui*.

Lunatia fusca (Blainville, 1825)
(Figs 135D, 139)

Natica fusca Blainville, 1825
Natica sordida Philippi, 1836

Diagnostic characters
Shell as in other *Lunatia* species but with a taller spire nearly flat-sided in profile, as in *L. alderi*. Outer lip arises tangential to last whorl, its initial part often slightly concave. Shell dark in colour, brown or chestnut, with no pattern. Umbilicus large.

Other characters
The columella is narrowed by the end of a groove emerging from the umbilicus, but is not notched as in *L. montagui* (Fig. 135D); the ridge entering the umbilicus from the basal end of the narrower section of the columella is well developed. Much of the colour resides in periostracum, and the shell is paler where that is lost; the columella is always brown, however, and the base of the aperture white. Up to 25 mm high, 20 mm broad; last whorl occupies 90% of shell height, aperture 75%.

Except for its colour (reddish brown, paler on the head and the edges of the foot) this animal is like the other species of the genus.

L. fusca is a southern species extending from the Mediterranean to the British Isles where it has been recorded from the western parts of the Channel, thence north along Irish and Scottish coasts. Not apparently found now in the North Sea and common only off west Ireland. The animals live sublittorally on muddy sand. Though presumably like those of other naticids the details of their breeding are not known.

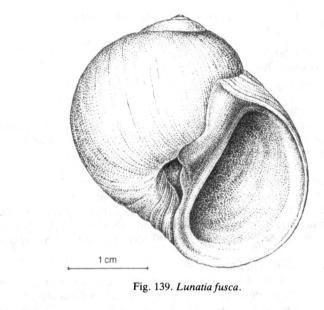

1 cm

Fig. 139. *Lunatia fusca*.

Lunatia pallida (Broderip & Sowerby, 1829)
(Fig. 140)

Natica pallida Broderip & Sowerby, 1829
Natica pusilla Gould, 1841
Natica groenlandica Möller, 1842

Diagnostic characters
Shell distinct from that of other *Lunatia* species in the small size of the umbilicus, the reduction in size of the ridges entering it, the broad columellar lip, and generally pale colour.

Other characters
The whorls are slightly tumid, the sutures well marked. The outer lip arises at right angles to the surface of the last whorl, its initial part convex. White or cream, sometimes brownish, sometimes with brown flecks. Up to 20 mm high, 18 mm broad; last whorl occupies 90–95% of shell height, aperture 75–80%.

The animal is like that of other species.

L. pallida is a circumpolar species which stretches south at increasing depths. Records from round the British Isles are few and mainly northern. The animals live on clay bottoms from 10 to 2000 m deep, the greater depths in the southern part of the range. Their spawn has been described by Thorson (1935, 1946), Lebour (1937) and MacGinitie (1959): the collar has a diameter of 30–45 mm but is recognizable by the fact that it contains rather few egg capsules which are well spaced and leave the upper and lower margins of the collar unoccupied. Each capsule contains only 1–2 large eggs which hatch as juveniles, the free larval stage being suppressed.

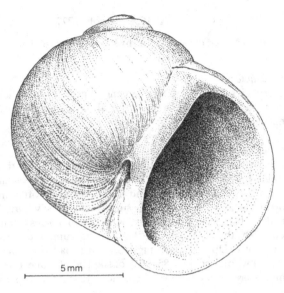

Fig. 140. *Lunatia pallida*.

Genus NATICA Scopoli, 1777

Natica clausa Broderip & Sowerby, 1829
(Fig. 141)

Diagnostic characters

Like *Lunatia* species (p. 334–342) in the general aspect of the shell but umbilicus is closed by a thick pad of shelly material. Operculum white and glossy, with external calcareous layer.

Other characters

The shell of this species is a little thinner and more translucent than in *Lunatia* species, retains more of the periostracal cover than they do and so is less glossy. There are 4–5 tumid whorls, and the short spire is often much eroded. The outer lip arises nearly normal to the surface of the last whorl. The development of shelly substance over the last whorl is much less than in *Lunatia* species. Yellowish, the colour due to the periostracum, and the shell is white where that is worn. Up to 12 mm high, 10 mm broad (up to 60 × 50 mm in Arctic seas); last whorl occupies 85–90% of shell height, aperture about 70%.

The animal is cream and shows the same external features as do *Lunatia* species.

N. clausa is distributed around the globe in Arctic waters, and extends south, at ever-increasing depths, to Portugal, the Mediterranean, North Carolina, California and Japan. In the Arctic it lives on soft substrata about 4 m deep, but reaches over 2000 m in temperate latitudes. Only a few animals have been found locally, in far northern British waters.

The egg collar of this species has been described by Thorson (1935): the egg capsules in it are arranged in only 3–4 rows, each with only one large egg in it which hatches as a juvenile.

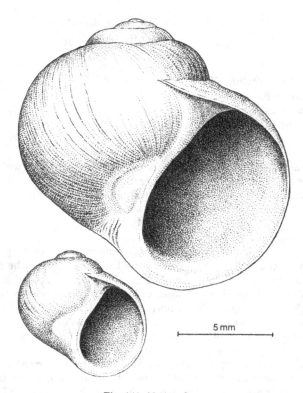

Fig. 141. *Natica clausa*.

Genus AMAUROPSIS Mörch, 1857

Amauropsis islandica (Gmelin, 1791)
(Fig. 142)

Nerita islandica Gmelin, 1791
Natica islandica (Gmelin, 1791)
Bulbus islandicus (Gmelin, 1791)
Natica helicoides Johnston, 1836

Diagnostic characters
Shell slightly translucent, not glossy, with relatively high spire with stepped profile due to flattening of subsutural area of each whorl; last whorl large; growth lines marked. Aperture large, oval, lip everted at base of columella. Umbilicus absent. Eyes hardly visible.

Other characters
The shell has six tumid whorls and deep sutures. The outer lip arises at right angles to the surface of the last whorl. The columella may be narrowed near its mid point by a sinus along its umbilical border, but there is only rarely a chink-like umbilicus. White, yellowed by the periostracum. Up to 25 mm high, 18 mm broad; last whorl occupies about 80% of shell height, aperture about two thirds.

The animal resembles other naticids. It is cream with many opaque white points.

A. islandica is a circumpolar species recorded locally only from the northern parts of the North Sea on sandy bottoms to 80 m deep; never intertidal. Its mode of life is not known. When breeding it produces an egg collar similar in general to that of other naticids (Thorson, 1935); each capsule in the collar is large and contains a single large egg which undergoes direct development.

Fig. 142. *Amauropsis islandica*.

Family CARINARIIDAE Reeve, 1841

Genus CARINARIA Lamarck, 1801

Carinaria lamarcki Péron & Lesueur, 1810
(Fig. 143)

Carinaria mediterranea Blainville, 1824
Carinaria atlantica Adams & Reeve, 1828

Diagnostic characters
Animal pelagic with expanded, nearly transparent head-foot and reduced visceral hump covered by transparent and delicate shell. Foot a longitudinally and laterally flattened keel or fin, with posterior sucker.

Other characters
The colourless and very fragile shell is cap-shaped, the 5–6 whorls lying nearly in one plane, with the last forming almost the whole shell. A thin and low keel runs along the centre of the convex side of the shell, and growth lines lie roughly parallel to the apertural edge. Up to 15 mm high, 30 mm long.

The body is much too large to be withdrawn into the shell and has the form of a slightly curved cylinder tapering at each end. The mouth lies at the anterior end and there are two tentacles, laterally placed, some way behind it, each with a prominent black eye. The fin (=foot) lies more or less centrally. Opposite it a neck connects to the visceral hump, over which the shell lies. The gill filaments tend to project anteriorly out of the mantle cavity; on the animal's right, in males, a ciliated groove runs from the cavity to a cylindrical penis. The flesh is colourless except for the eyes and some viscera. The body may be up to 100 mm long.

The animals swim freely, but not very powerfully, in open water, orientated so that the shell and visceral mass are below. They are carnivores, taking jelly fish, worms, tunicates. They are only occasional visitors to western coasts of the British Isles, being carried by currents from their more normal southern haunts: they are usually found at greater depths than they normally occupy. Animals encountered locally are unable to breed, but in their normal range they produce veliger larvae (Thiriot-Quiévreux, 1973, 1975).

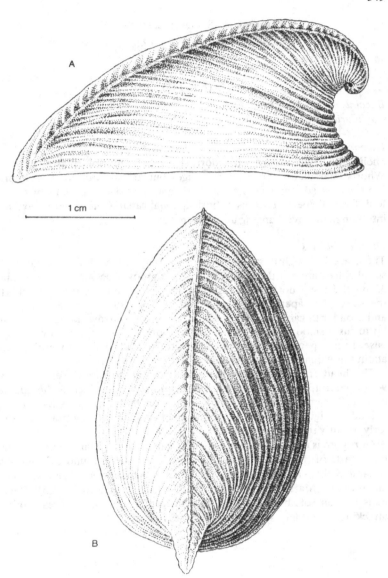

Fig. 143. *Carinaria lamarcki*. A, from the left; B, from above.

Family CASSIDAE Latreille, 1825

Genus GALEODEA Link, 1807

Galeodea rugosa (Linné, 1771)
(Fig. 144)

Buccinum rugosum Linné, 1771
Cassidaria rugosa (Linné, 1771)

Diagnostic characters
Shell large, solid, glossy, with short, sharply pointed spire; last whorl large.
Whorls bear many narrow spiral ridges but are without costae. Outer lip
is often everted, bears an external varix and is marked internally by a ridged
fold. Base of aperture forms a short siphonal canal. Columellar lip extended
into thin flange overhanging umbilical groove.

Other characters
There are 7–8 slightly swollen whorls in this shell, with deep sutures. The
spiral ridges are a little narrower than the grooves between, number about
35 on the last whorl (lying closer at its base), and 7–8 on each whorl in
the spire. The aperture is surrounded by a peristome, more or less oval,
and broader towards the adapical end; the throat narrows rapidly. The inner
lip forms a smooth sheet over the last whorl. Cream, often browner at the
base. Large specimens up to 120 mm high, 70 mm broad; last whorl occupies
about three quarters of shell height, aperture about 60%.

The head ends in a transverse ridge and has no snout. Tentacles, each
with a lateral, basal eye, arise from the ridge under which is the apparent
mouth. The mantle edge forms a siphon on the left. Males have a penis
behind the right tentacle. The foot is shield-shaped, broadly truncated anteriorly, bluntly rounded posteriorly.

G. rugosa is a southern species reaching its northern limits in the southernmost parts of the British Isles. Specimens have been found off south-west
Ireland, Scilly, and in St George's Channel, less than twenty in all, and hardly
any recently (Marshall, 1911; Massy, 1930; O'Riordan, 1984, 1985). The animals live on soft bottoms, 70–700 m deep. They are carnivores, probably
attacking echinoderms and other molluscs.

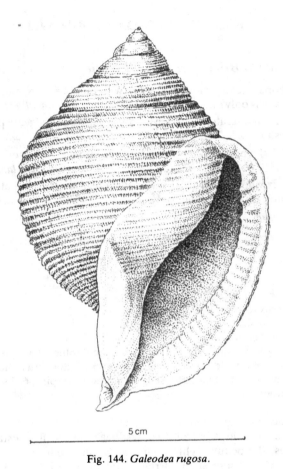

Fig. 144. *Galeodea rugosa*.

Family CYMATIIDAE Iredale, 1913

Key to British genera and species of Cymatiidae

1. 8–10 whorls, greatly swollen, marked with tubercles; siphonal canal
   long; shells only? .... *Ranella olearia* (p. 354)

   6–8 whorls, moderately tumid, with costae and spiral ridges; siphonal
   canal short **2**

2 Spire of 7–8 whorls; columella ridged; no groove within outer lip;
   rare, off S.W. coasts .... *Charonia lampas* (p. 356)

   Spire of six whorls; columella smooth; deep groove within outer lip;
   very rare, Channel Islands only .... *Cymatium cutaceum* (p. 352)

Genus CYMATIUM Röding, 1798

Cymatium cutaceum (Linné, 1767)
(Fig. 145)

Murex cutaceus Linné, 1767
Triton cutaceus (Linné, 1767)

Diagnostic characters
Shell solid, glossy, with prominent spire, the profile of which is stepped.
Ornament of costae and spiral ridges, most of the latter with a shallow groove
on their summit. Labial varix present with others on spire; 6–7 swollen teeth
within outer lip.

Other characters
The shell has six whorls of which the last bears four or five costae and eight
spiral ridges, the penult eight or nine costae and four spiral ridges; in all
whorls the two subsutural spiral ridges are reduced in size. The aperture
is oval, pointed at each end; the upper end, when viewed from the correct
angle (the columellar side) is shaped like a keyhole. Within the outer lip
is a deep groove underlying the varix. The siphonal canal is short and narrow,
the columella long; its lip turns out over a deep umbilical groove, the other
side of which is formed by a thick keel (siphonal fasciole) running up from
the tip of the siphonal canal. The inner lip bears a few adapical tubercles.
Light brown, sometimes pinkish, the throat white. Up to 60 mm high, 40 mm
broad; last whorl occupies three quarters of shell height or a little more,
aperture 60–65%.

The head is flattened with a slit-like mouth ventrally and long tentacles
at its tip, each with a basal eye. The mantle edge bears a short siphon on
the left. Males have a long penis arising behind the right tentacle, with an
open seminal groove. The front edge of the foot is axe-shaped, its posterior

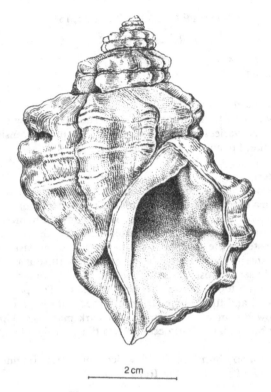

2 cm

Fig. 145. *Cymatium cutaceum*.

end broadly rounded. The flesh has dark purplish spots and white lines on a pale violet background; on the foot are reddish points outlined in white.

C. cutaceum is a southern species which is common in the Mediterranean and which extends as far north as the Channel Islands. It does not seem to have been found alive there since 1885, though finds of dead shells in 1902 and 1932 encourage hopes that it survives (Crowley, 1961). It lives sublittorally on soft bottoms, probably eating echinoderms.

Genus RANELLA Lamarck, 1816

Ranella olearia (Linné, 1758)

Turbo olearius Linné, 1758
Ranella gigantea Lamarck, 1816

Diagnostic characters

Shell large, with very tumid whorls and deep sutures, bearing widely spaced wavy and laminar varices. Aperture with small adapical extension or canal, teeth within outer lip, and long siphonal canal.

Other characters

The shell is glossy, has 8–10 whorls which form a tall, slightly cyrtoconoid spire ending in a narrow apex. The whorls are marked by prosocline costellae or enlarged growth lines and spiral ridges, which combine to produce a reticulated and tuberculated surface, often better marked at the periphery of each whorl. The costellae die out about the periphery of the last whorl though the spiral ornament continues to the base. The aperture is oval and tapers to an obliquely set, almost closed siphonal canal. The outer lip has a varix and many internal teeth, tending to occur in pairs. The parietal lip bears teeth adapically, and the columellar lip is turned out over a deep umbilical groove. Yellow brown with a scattering of dark markings. Up to 200 mm high, 100 mm broad; last whorl occupies two thirds of shell height, aperture about 45%.

R. *olearia* is a southern species of which only a few fragmentary shells have been found off south western Ireland.

Genus CHARONIA Gistel, 1848

Charonia lampas (Linné, 1758)
(Fig. 146)

Murex lampas Linné, 1758
Triton nodiferus (Linné, 1758)

Diagnostic characters
Shell large, solid, glossy, with tall pointed spire and angulated profile; last whorl large. Ornament of spiral ridges, the larger ones nodose; varices at 120° intervals. Outer lip with paired short ridges internally, columella fluted with ridges, one large one adapically.

Other characters
The shell has 7–8 whorls meeting at shallow sutures. The whole surface is covered with small spiral ridges, but up to ten on the last whorl and two on each whorl in the spire are much bigger than the others and are also nodose. Each varix consists of a prominent swelling across the whorl with a sharp, undercut edge on the side facing down the spiral. The aperture is a broad oval, pointed above and below, the siphonal canal short and open; the inner lip spreads at its base over an umbilical groove, and, more adapically, widely over the surface of the last whorl. White with brown blotches, especially near sutures, on the inner folds of the outer lip, and at the base of the columella. Up to 330 mm high, 180 mm broad; last whorl occupies about two thirds of shell height, aperture a little more than half.

The animal resembles that of *Cymatium cutaceum* (p. 352) except in its coloration, being reddish with many scattered brown spots. Each tentacle has two longitudinal black lines. *C. lampas* is a southern species ranging between the Mediterranean and the extreme southern parts of the British Isles. It is reasonably common as far north as the southern parts of the Bay of Biscay but beyond that is sporadic in its occurrence. Local finds are limited to the Channel Islands, where three or four were caught in 1825–47 and three in 1972 (Crowley, 1961; Brehaut, 1973); off Co. Kerry, where four were taken 1970–71 (O'Riordan, 1972); between Land's End and the Isles of Scilly, one specimen in 1975 (Turk, 1976). Three live specimens were found off Dover in 1986 (de Ligt, 1987, *J. Conch.*, **32**, 385). The animals probably feed on starfish and lay egg capsules from which free veliger larvae escape.

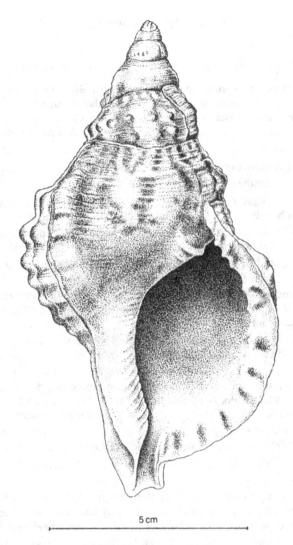

5 cm

Fig. 146. *Charonia lampas*.

Family MURICIDAE Rafinesque, 1815

Key to British genera and species of Muricidae

1. Last whorl occupies more than 80% of shell height; ornament of
   spiral ridges only (sometimes slightly tuberculated); siphonal canal
   short; a siphonal fasciole alongside umbilical area ....
   .... *Nucella lapillus* (p. 366)

   Shell not like this **2**

2. Ornament comprises costae only **3**

   Ornament of costae and spiral ridges **4**

3. 14 or more costae on last whorl, each without a shoulder spine;
   canal rather short .... *Boreotrophon truncatus* (p. 360)

   Less than 14 costae on last whorl, each with a shoulder spine; canal
   long; only subfossil in British waters .. *Boreotrophon clathratus* (p. 360)

4. Breadth of shell less than half its height; canal as long as rest of
   aperture, open; 12 or more costae on penult whorl **5**

   Breadth equal to or more than half shell height; canal shorter than
   length of rest of aperture, sometimes closed; fewer than 12 costae
   on penult whorl **6**

5. 7 shouldered whorls; costae laminar, each with a shoulder spine;
   3–4 spiral ridges and 12–14 costae on penult whorl; shell white
   .... *Trophonopsis barvicensis* (p. 364)

   7–8 whorls, spire turreted; costae rounded, spineless; 5–6 spiral
   ridges and 15–20 costae on penult whorl; shell yellow with brown
   spiral bands .... *Trophonopsis muricatus* (p. 362)

6. Breadth of shell more than half its height; 6–7 teeth within outer
   lip; canal closed in older animals; reddish . *Ocinebrina aciculata* (p. 374)

   Breadth of shell about equal to half its height; no teeth; canal open
   or closed **7**

7. Aperture occupies more than half shell height; 7–8 costae on last
   whorl with 8–9 larger spiral ridges and many fine imbricated ones;
   siphonal canal closed in older shells .... *Ocenebra erinacea* (p. 372)

   Aperture occupies less than half shell height; 10–12 costae on last
   whorl with 16–18 spiral ridges; siphonal canal open; with oysters,
   only on Kent and Essex coasts .... *Urosalpinx cinerea* (p. 370)

Genus BOREOTROPHON Fischer, 1884

Boreotrophon truncatus (Ström, 1768)
(Fig. 147A, B)

Buccinum truncatum Ström, 1768
Trophon truncatus (Ström, 1768)

Diagnostic characters

Shell with sharply pointed conical spire; whorls tumid with narrow costae shaped like a breaking wave with the concavity down the spiral; without obvious spiral ornament. Base with narrow, open siphonal canal. Outer lip smooth.

Other characters

There are seven whorls meeting at deep sutures, each with a slightly flattened subsutural area. There are 15–25 costae on each of the last two whorls, those on the last dying out towards the base. Yellow to pinkish brown. Up to 15 mm high, 7.5 mm broad; last whorl occupies about 70% of the shell height, aperture about half.

The head has no snout, divergent tentacles arising from a low ridge with the opening of a proboscis sac on its underside. Each tentacle has an eye about two thirds of its length up from the base. Males have a C-shaped penis on the mantle cavity floor on the right. The foot is shield-shaped, straight anteriorly, and carries a small operculum. Flesh cream with opaque white markings.

The species is primarily Arctic in distribution but extends to the northern parts of the North Sea and along the western coasts of the British Isles and then south to the Bay of Biscay; it also occurs along the east coast of America as far south as Cape Cod. It lives on gravelly or muddy bottoms from the laminarian zone to depths of about 200 m, deeper in the southern parts of its range. The animals are presumably carnivores, but their food is unknown. Their mode of breeding has not been described but is probably direct.

Shells of *Boreotrophon clathratus* (Linné) (Fig. 147C) are not uncommonly dredged on soft bottoms off British coasts, but these are subfossil, and living specimens have not been found nearer than Scandinavian waters. They resemble those of *B. truncatus* but have fewer, taller costae raised into conspicuous spurs at the edge of the subsutural flat area, and a longer and more obliquely set siphonal canal.

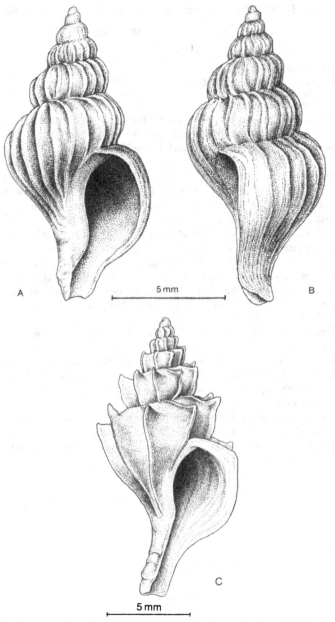

Fig. 147. *Boreotrophon truncatus*. A, shell in apertural view; B, in side view. C, *Boreotrophon clathratus*.

Genus TROPHONOPSIS Bucquoy, Dautzenberg & Dollfus, 1882

Trophonopsis muricatus (Montagu, 1803)
(Fig. 148)

Murex muricatus Montagu, 1803
Trophon muricatus (Montagu, 1803)

Diagnostic characters

Shell with a sharply pointed spire; whorls tumid, with both costae and spiral ridges and a distinct flat subsutural ramp. Aperture extended into a rather long, narrow, and open siphonal canal. Outer lip crenulated.

Other characters

The 7–8 whorls meet at deep sutures. There are 15–17 costae on the last whorl, where they fade towards the base, and 15–20 on the penult, which they cross. There are 5–6 spiral ridges on the penult whorl (none on the flat subsutural area) and about twelve on the last. Costae and ridges cross, forming a square network on the surface. The aperture is pear-shaped, broad apically, and the outer lip shows a shallow anal sinus near its origin. The shell is commonly white or cream with brown spiral bands, one below each suture and one at the base of the last whorl, but in some shells the bands are extremely pale or absent. Up to 19 mm high, 8 mm broad; last whorl occupies 70–80% of shell height, aperture 50–60%.

The body of the animal shows the same features and colouration as in the last species.

T. muricatus lives on sandy or gravelly bottoms from 20 to 300 m deep between the Mediterranean and the southern and western coasts of the British Isles. Like *Boreotrophon truncatus* (p. 360) it is a carnivore, but its precise prey is not known.

Eggs are laid between February and June (at Plymouth) (Lebour, 1936) in lens-shaped capsules attached by their flat base to the substratum; they have thick, wrinkled walls and measure 2–3 mm across. Development is direct.

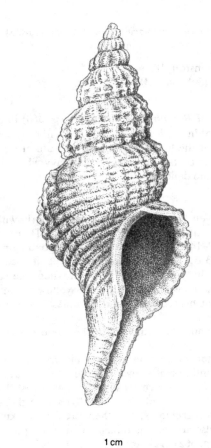

1 cm

Fig. 148. *Trophonopsis muricatus*.

Trophonopsis barvicensis (Johnston, 1825)
(Fig. 149)

Murex barvicensis Johnston, 1825
Trophon barvicensis (Johnston, 1825)

Diagnostic characters
Shell similar to that of *Boreotrophon truncatus* (p. 360) in having predominantly axial ornament in the form of well-spaced, narrow costae which show a marked elevation at the lower edge of the subsutural ramp. Spiral ridges also occur making the costae tuberculated or wavy where they cross them. Siphonal canal long and delicate.

Other characters
There are about seven tumid, semitransparent whorls with deep sutures. Spiral ridges are absent from the subsutural area. There are 11–13 costae on the last whorl, 12–14 on the penult; the corresponding numbers of spiral ridges are 6–8 and 3–5. The outer lip is thin and (if a costa lies along it) shows an out-turned hollow spine at the point where the costa is elevated. White or colourless, occasionally with some yellow along the costae. Up to 15 mm high, 7 mm broad; last whorl occupies 70–75% of shell height, aperture 60%.

The body of the animal is white, with the same organization as in the previous species.

The animals are not uncommon in dredgings from stony bottoms. They occur from Iceland and north Norway, where they live at a depth of a few metres, to northern and western localities in the British Isles, where they are more likely to occur 40–140 m deep. They live still further south at increasing depths. They are carnivores, but their prey is not known. Details of their breeding are also unknown, thought it may be presumed to be like that of *T. muricatus* (p. 362) in general.

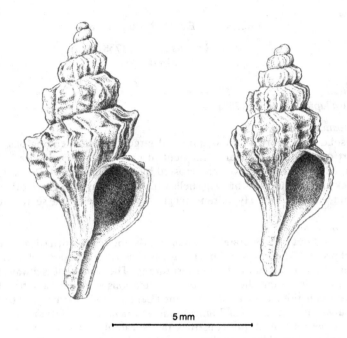

5 mm

Fig. 149. *Trophonopsis barvicensis*.

Genus NUCELLA Röding, 1798

Nucella lapillus (Linné, 1758)
(Fig. 150)

Buccinum lapillus Linné, 1758
Purpura lapillus (Linné, 1758)

Diagnostic characters

Shell solid with short, pointed spire and large last whorl ending basally in a short open siphonal canal. Ornament of low, strap-shaped spiral ridges, separated by narrow grooves and crossed by prosocline growth lines. Outer lip thick and toothed internally in shells which have ceased to grow, otherwise thin and without teeth. Hypobranchial gland produces a purple secretion.

Other characters

There are about six moderately swollen whorls, the spire with a slightly coeloconoid profile. The last whorl bears 11–14 spiral ridges, best developed near the periphery, reduced or absent subsuturally. The ornament is always most pronounced in young shells. In some populations – mainly sublittoral but also between tidemarks in North Kent (Largen, 1971) – the growth lines turn outwards to form small flounces (var. *imbricata*) (Fig. 150B). The aperture is large but the throat constricts rather rapidly. The siphonal canal is partly covered on the columellar side; the columella itself is broad with a slightly dished surface. A siphonal fasciole in the form of a blunt spiral keel lies along its abapertural side. The colour of *Nucella* shells is extremely varied: most are grey or white, but they may also be yellow, various brown hues or a pale purple. In addition, spiral bands of contrasting colour may occur, usually brown; these may be very broad or confined to the grooves between spiral ridges. The inner edge of the outer lip is often pigmented, uniformly or only between teeth. Moore (1936) explained the presence of dark pigment as a consequence of a diet of mussels, but this idea is too simple and does not always fit the facts. Up to 30 mm high, 20 mm broad; last whorl occupies about 85% of shell height, aperture about 70%.

The animal is white or cream with white speckles. The head is a flat transverse ridge with the opening of a proboscis pouch on its underside. Each tentacle has an eye about one third of its length up from the base.

The siphon hardly extends out of the canal. In males a recurved penis arises from the pallial floor on the right. The foot is shield-shaped and a short distance from the anterior end bears a pit opening to the sole in the mid-line in which is housed the accessory boring organ; in females there is also, a little behind this, another pit containing the ovipositor.

Fig. 150. *Nucella lapillus*. A, shell; B, *N. lapillus* var. *imbricata*, shell; C, egg capsules, the wall of one cut open to show the embryos feeding on food eggs.

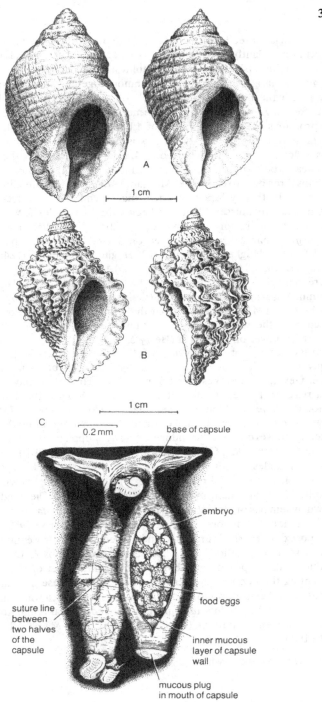

A

1 cm

B

1 cm

C

0.2 mm

base of capsule

embryo

food eggs

inner mucous
layer of capsule
wall

mucous plug
in mouth of capsule

suture line
between
two halves
of the
capsule

The species spreads throughout the littoral regions of the North Atlantic. It occurs abundantly on the middle and lower parts of all rocky shores throughout the British Isles and extends sublittorally to about 30 m. The animals are gregarious and are particularly common, together with their egg capsules, along with the barnacles and mussels on which they principally feed, the latter food slightly preferred to the former. Dog whelks feed either by pressing the proboscis between the valves of a barnacle or bivalve shell and rasping the flesh with the radula, or, if this is not possible, by drilling the shell. This is done by alternating periods of mechanical attack (by the radula) and chemical attack (by the accessory boring organ), the whole process taking perhaps three days to complete (Morgan, 1972). Periods of feeding and resting alternate. In this cycle a dog whelk may spend 1–3 tidal cycles on a bed of barnacles or mussels. If it has been eating barnacles it will then spend 2–4 tidal cycles resting in a crevice or other shelter before it emerges to attack more barnacles; if it has fed on mussels the resting phase may last for 7–9 tides (Hughes & Elner, 1979; Hughes & Dunkin, 1984a,b; Dunkin & Hughes, 1984; Hughes & Drewett, 1985).

Breeding may occur throughout the year but is maximal in spring and autumn. Vase-shaped egg capsules (their structure described by Ankel, 1937 and Fretter, 1941) are passed from the female duct into the ovipositor and attached to the walls of crevices or under stones where the animals live, or perhaps lower on the beach (Berry & Crothers, 1968). Each may contain up to 1000 eggs, more usually about 600; of these only 25–30 grow to emerge as juveniles with a shell about 1 mm high, the rest being used as food by these. Development takes about four months. The young may live alongside their parents (Feare, 1970) or at a lower level (Moore, 1938). They mature at about 2.5 years (Moore, 1938; Feare, 1970; Coombs, 1973), then stop growing and the outer lip thickens and develops internal teeth. Production of teeth, however, may be induced by any stoppage of growth, whatever its cause (Bryan, 1969; Cowell & Crothers, 1970; Feare, 1970).

Many females of this species from south-west England and the Channel and Atlantic coasts of France now exhibit a condition known as **imposex** (Smith, 1971), indicated by the possession of a penis. The condition is also shown by females of *Ocenebra erinacea* and *Hinia reticulata*, and it has become markedly more common since first observed by Blaber in 1970. The impact of imposex on *Nucella* is profound and has led to the nearly complete destruction of some populations. This has been shown (Gibbs & Bryan, 1986) to be due to the imposed penis blocking the genital duct of the female, thus preventing the deposition of egg capsules, though these are still produced. Imposex has been shown to be due to the presence in sea water of derivatives of the anti-fouling paints used in marinas and, in particular to their main toxic constituent, tributyltin (B. S. Smith, 1981; Féral & Le Gall, 1982; Bryan *et al.*, 1986).

Vernacular name: dog whelk.

Genus UROSALPINX Stimpson, 1865

Urosalpinx cinerea (Say, 1822)
(Fig. 151)

Fusus cinereus Say, 1822

Diagnostic characters
Shell with rather tall conical, sharply pointed spire of tumid whorls meeting at moderately deep sutures. Ornament of undulate costae and numerous spiral ridges. Siphonal canal rather short, open. Found only in Essex and North Kent in association with oysters.

Other characters
There are 7–8 whorls, the last with 10–12 costae and 16–18 prominent spiral ridges; the former die out towards the base, though they cross the whorls of the spire, and the latter are normally absent from the subsutural area. There is no, or only a weak, siphonal fasciole. Yellowish or grey, sometimes with irregular brown marks, and usually paler on the costae. Up to 40 mm high, 20 mm broad; last whorl occupies about 70% of shell height, aperture 45–50%.

The animal has external features similar to those of the dog whelk. It is cream with dark markings on the tentacles and mantle edge.

U. cinerea is a native of North America accidentally imported into Europe with oysters, probably last century, but first reported in Essex in 1927 (Orton & Winckworth, 1928; Orton, 1930). It may be found on the lower half of beaches but lives mainly sublittorally to a depth of about 12 m. It is a carnivore, taking oyster spat, the shell of which is bored in the same way as by *Nucella* (Carriker, 1981).

Eggs are laid in capsules attached to oyster shells, or stones, at a higher level than that at which adults live (Hancock, 1959, 1960); they are shaped like flattened vases. Each capsule has eleven or twelve eggs, most of which hatch as juveniles and there are no true food eggs. The newly hatched snails are strongly attracted by the odour from barnacles, presumably their normal food at this size, less strongly that from oysters (Rittschof, Williams, Brown & Carriker, 1983). Location of prey is achieved by sensing odours directionally (Brown & Rittschof, 1984) and avoiding those emanating from hungry and starved *Urosalpinx* (Pratt, 1976).

Vernacular name: oyster drill.

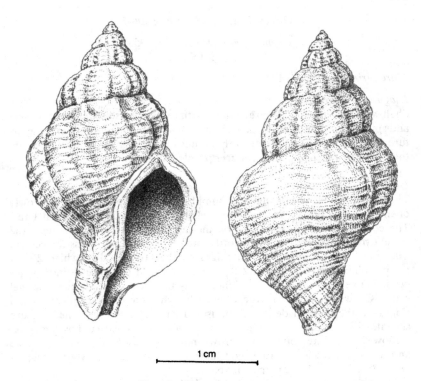

1 cm

Fig. 151. *Urosalpinx cinerea*.

Genus OCENEBRA Gray, 1847

Ocenebra erinacea (Linné, 1758)
(Fig. 152)

Murex erinaceus Linné, 1758

Diagnostic characters
Shell with tall, angulated spire, whorls with rather coarse sculpture of costae and spiral ridges, sutures deep. Siphonal canal closed in older animals; aperture with labial varix; no teeth within outer lip, though throat may be fluted there. Hypobranchial gland produces purple secretion.

Other characters
The shell has eight whorls which are tumid and a short siphonal canal narrowly open in young shells but closed by a flat growth of shelly material later. The costae are well spaced, 7–8 on the last whorl, 8–9 on the penult; the spiral ornament forms rounded cords, broader where they cross the costae. There are 8–9 prominent spiral ridges on the last whorl of which 2–3 at the periphery are the largest, 2–3 on each whorl of the spire. Minor ridges occur between all of these. A siphonal fasciole lies alongside the siphonal canal. Growth lines occur everywhere (though often eroded on costae and ridges) in the form of delicate, upraised arch-like structures. The aperture is oval, with a well-developed peristome, sometimes constricted by the varix. Yellowish or white, often with brown markings, especially on the costae and the spiral cords. Up to 40 mm high, 20 mm broad; last whorl occupies about 75% of shell height, aperture 60%.

The animal has a small flattened head, the mouth on its underside. Eyes lie about two thirds way up the tentacles. The siphon hardly projects, males have a penis and the foot is small. The flesh is cream, darker on the tentacles.

This is a southern species occurring from the Mediterranean to the southern half of the British Isles. Its southern range is evident in its susceptibility to cold: the animals were nearly killed off in the exceptionally cold winter of 1962–63 (Crisp, 1964), as they had been earlier in that of 1928–29, from which they took about 25 years to recover (Mistakidis & Hancock, 1955).

The animals are predominantly sublittoral in their occurrence, but may often be found in summer on the lower parts of stony beaches, especially where these are sheltered. They are predators with a rather wide range of prey organisms, which includes bivalves, barnacles and tubicolous worms (Hancock, 1960). Breeding takes place from April, when flattened vase-shaped capsules are laid in clumps on stones or shells. All the eggs in a capsule develop, hatching to juveniles in 12–13 weeks.

Vernacular names: sting winkle, tingle.

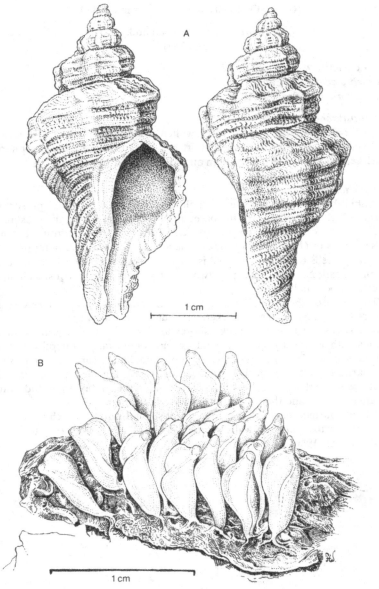

Fig. 152. *Ocenebra erinacea*. A, a young shell: in older ones the siphonal canal is closed to form a tube; B, egg capsules.

Genus OCINEBRINA Jousseaume, 1880

Ocinebrina aciculata (Lamarck, 1822)
(Fig. 153)

Murex aciculatus Lamarck, 1822
Ocenebra aciculata (Lamarck, 1822)
Murex corallinus Scacchi, 1836

Diagnostic characters
Shell broadly spindle-shaped, with swollen whorls ornamented with costae and spiral ridges. Siphonal canal short and closed, bordered by a siphonal fasciole. 6–7 elongated teeth within outer lip. Animal's flesh bright red.

Other characters
The shell has about seven whorls in all, the last with 8–10 broad prosocline costae which die out towards the base, and 18–20 spiral ridges, the subsutural 1–2 rather broader than the others. Growth lines are numerous, but not raised up as in *O. erinacea* (p. 372). The aperture is oval, the outer lip thin. In young shells the siphonal canal is open. Orange-buff. Up to 15 mm high, 10 mm broad; last whorl occupies two thirds to three quarters of shell height, aperture about half.

The body of the animal is like that of *O. erinacea* except in its colour. The red colour of the flesh is flecked with numerous yellow spots.

The snails are found on rocky shores at low water mark, usually under stones with *Balanus perforatus*, and extend sublittorally to depths of 15 m (Franc, 1952a). They are predators, their prey not known for sure, but possibly the barnacles with which they often occur. In Britain they are confined to south-west England and the Isles of Scilly. Elsewhere they are found in the Channel Islands and thence south to the Mediterranean.

About fourteen eggs are laid in a barrel-shaped capsule which is fastened with others to stones or walls of rock pools. The eggs develop to young snails (Franc, 1940).

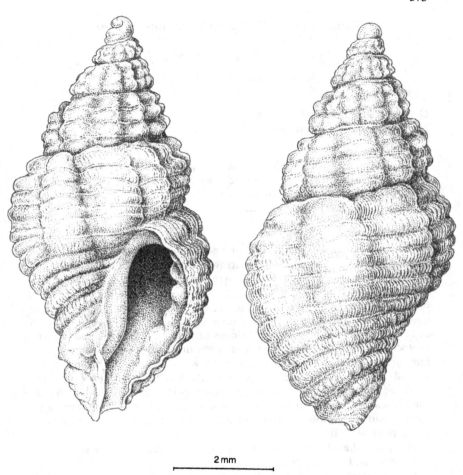

2 mm

Fig. 153. *Ocinebrina aciculata.* Young shell: in a fully mature one the siphonal canal is closed.

Family COLUMBELLIDAE Swainson, 1840

Genus AMPHISSA H. & A. Adams, 1853

Amphissa haliaeeti (Jeffreys, 1867)
(Fig. 154)

Columbella haliaeeti Jeffreys, 1867
Pyrene haliaeeti (Jeffreys, 1867)

Diagnostic characters

An approximately ovoid shell with short siphonal canal. Ornament of costae and less obvious spiral ridges. Columella with basal fold and (in some shells) up to two teeth; outer lip with internal teeth and varix when mature. Eyes at base of tentacles. From deep water.

Other characters

The shell is transparent and glossy, with 8–9 somewhat tumid whorls meeting at incised sutures. The costae are narrower than the intervening spaces and are flexuous, those on the last whorl dying away on the base. On this whorl, in mature shells, is a labial varix, often formed of a double costa, then, further away from the outer lip, an area on which costae are reduced in size or may be absent, then up to fourteen costae; usually fifteen or sixteen occur on other whorls. The spiral ridges are fine, numerous, and low. The protoconch has a complex pattern of spiral lines on the first whorl, zigzag lines and approximately axial ridges, often with a spiral cord below, on the second and third. The aperture is oval with a rather thick peristome and there are, in adult shells, up to six ridge-like teeth within the outer lip. White, sometimes with some brown marks; protoconch yellow. Up to 8 mm high, 4 mm broad; last whorl occupies two thirds to three quarters of shell height, aperture about half.

The animal has the usual neogastropod external features. The foot carries an operculum. Flesh whitish.

This species seems to occur rather widely in the North Atlantic, the animals living on soft bottoms 150–2000 m deep, though their way of life is unknown; Bouchet (1977) has suggested that the life history includes a free, possibly lengthy, larval stage. The animals have been recorded from only a few sites north and west of Scotland and Ireland.

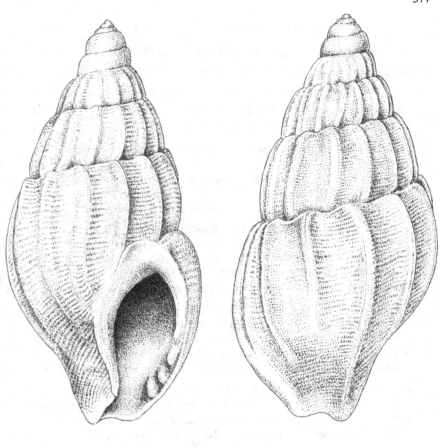

2 mm

Fig. 154. *Amphissa haliaeeti*.

Family BUCCINIDAE Rafinesque, 1815

Key to British genera and species of Buccinidae

1. Shell smooth to the naked eye or with only growth lines 2
   Shell with distinct ornament 7

2. Last whorl occupies 70% of more of shell height 3
   Last whorl occupies less than 70% of height 4

3. Shell bluish white, ovate, with blunt apex; operculum with terminal
   nucleus .... *Liomesus ovum* (p. 380)
   Shell pale brown, conical, with pointed apex; operculum with central nucleus; shell has delicate spirals if examined with lens ....
   .... *Buccinum humphreysianum* (p. 402)

4. Siphonal canal narrow, distinct from aperture and usually pointing
   to left; whorls slightly tumid and sutures shallow (these are worn
   shells: some spiral ridges and remains of periostracum may be
   visible in areas protected from erosion if examined with a lens)   5
   Siphonal canal broad, merging with aperture; whorls tumid and
   sutures deep   6

5. Axis of protoconch lying oblique to that of teleoconch; whorls
   nearly flat-sided .... *Colus gracilis* (p. 386)
   Axes of protoconch and teleoconch coincide; whorls a little turreted; periostracum often clearly hispid .... *Colus jeffreysianus* (p. 390)

6. Last whorl occupies two thirds of shell height; outer lip thin, not
   everted; opercular nucleus central ... *Buccinum hydrophanum* (p. 404)
   Last whorl occupies more than two thirds of shell height; outer
   lip everted; opercular nucleus terminal; throat often pinkish .....
   .... *Volutopsius norwegicus* (p. 384)

7. Ornament includes costae, with or without spiral ridges or lines   8
   Ornament of spiral elements only, with growth lines 11

8. Shell with subsutural shelf with prominent triangular spiral ridge
   at outer edge; spire turreted; costae thin, well-spaced, often
   forming varix-like projections where they cross spiral ridges; only
   found north of Shetland .... *Neptunea despecta* (p. 398)
   Shell not turreted, without a subsutural shelf   9

9. Shell breadth greater than half its height; aperture equals about
   half shell height; costae curved; siphonal fasciole prominent .....
   .... *Buccinum undatum* (p. 400)
   Shell breadth less than half its height; aperture less than half shell
   height   10

10. Costae straight, 9–10 on penult whorl; labial varix marked; up to 6 teeth in throat under varix in older shells; siphonal canal very short; shell not over 5 mm high .... *Chauvetia brunnea* (p. 406)

Costae curved, 12–17 on penult whorl; no varix, no teeth; canal moderately long; up to 45–50 mm high .. *Turrisipho fenestratus* (p. 392)

11. Shell without persistent periostracum; whorls tumid; siphonal fasciole marked; outer lip often thickened; shell sometimes reddish .... *Neptunea antiqua* (p. 396)

Shell with obvious periostracum except often for triangular patch near origin of outer lip; whorls flattish or tumid; siphonal fasciole inconspicuous or absent **12**

12. Whorls only slightly tumid, sutures shallow **13**

Whorls clearly swollen, sutures deep **15**

13. Apex of shell bulbous, its diameter greater than that of the next whorl, its axis oblique to that of the teleoconch; siphonal canal rather long and narrow .... *Colus islandicus* (p. 388)

Apex not bulbous, its diameter equal to or less than that of the next whorl **14**

14. Protoconch axis oblique to that of teleoconch; siphonal canal rather short and broad .... *Colus gracilis* (p. 386)

Protoconch axis coincident with teleoconch axis; whorls may be a little turreted and periostracum hispid .... *Colus jeffreysianus* (p. 390)

15. Apical whorls narrow, so spire has coeloconoid profile; other whorls tumid; spiral ridges often enlarged at periphery of last whorl; shell breadth less or at most equal to half its height; aperture large, confluent with siphonal canal; outer lip often thick and everted a little; shell large .... *Beringius turtoni* (p. 382)

Whorls very tumid and sutures very deep; shell breadth greater than half its height; outer lip thin .... *Turrisipho moebii* (p. 394)

Note on other buccinid species

A number of species belonging to the family Buccinidae have been recorded from deep waters in localities that are marginally British or Irish. The validity of such species is sometimes in considerable doubt and they may well prove to be merely ecological variants of the species described, especially as these are known to be variable, though the range of variability is uncertain. In other cases, where the specific validity is highly probable or certain, these are arctic forms which have extended their range a little southwards in deeper waters. All are rare here and are unlikely to be encountered unless access to deep water dredgings is possible. See Pain (1977, 1978, 1979).

Genus LIOMESUS Stimpson, 1865

Liomesus ovum (Turton, 1825)
(Fig. 155)

Buccinum ovum Turton, 1825
Buccinum dalei Sowerby, 1825
Buccinopsis dalei (Sowerby, 1825)

Diagnostic characters
Shell broad and conical, with blunt apex, whorls apparently smooth and glossy, the last one large; siphonal canal short and broad; aperture large.

Other characters
There are 5–6 tumid whorls which, though smooth to the naked eye, bear fine growth and spiral lines, and often retain a thin periostracum. The proto-conch is small and insunk, giving the blunt apex. White. Up to 35 mm high, 23 mm broad; last whorl occupies 80–85% of shell height, aperture 60%.

The animal has two rather short tentacles, each with a basal eye, arising from a transverse fold which represents the head and carries on its underside the opening of a proboscis pouch. The siphon is rather long. Males have a sickle-shaped penis attached behind the right tentacle, and the foot carries an oval operculum. The body is cream coloured.

The animals live on soft bottoms, 70–400 or more metres deep, over a range extending from the Arctic seas to the northern parts of the British Isles. Though common in the northern section of this range they are rare here. Their food is not known, but they presumably are carnivores, carrion eaters, or both. Development is direct, a young snail with a shell as much as 5 mm high emerging from a lens-shaped capsule fastened alongside the columellar region of the female's shell. When laid the capsule contains many small eggs all of which are eaten by the one successful embryo (Thorson, 1940).

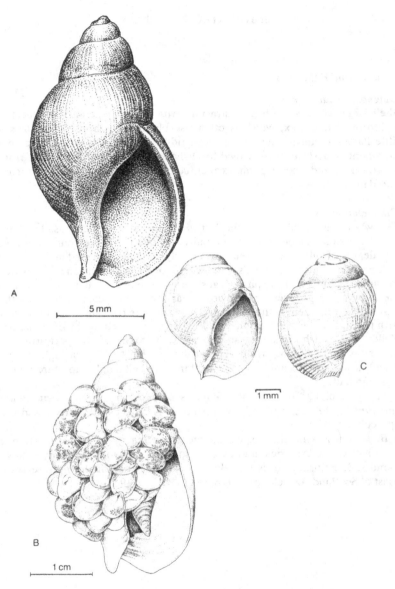

Fig. 155. *Liomesus ovum*. A, adult shell; B, shell with egg capsules; C, shells of newly hatched juveniles.

Genus BERINGIUS Dall, 1887

Beringius turtoni (Bean, 1834)
(Fig. 156)

Fusus turtoni Bean, 1834

Diagnostic characters

Shell large, covered with periostracum, with tall spire tending to become styliform at the apex, which is otherwise blunt; 7–8 whorls, sometimes a little flattened subsuturally, with spiral ridges and growth lines as the sole ornament. Aperture broadly oval to D-shaped, usually hardly distinct from short, wide, and open siphonal canal; columella nearly straight. Outer lip flared in old shells.

Other characters

The whorls are swollen and the last is somewhat extended basally; there may be a siphonal fasciole and umbilical groove but more commonly not. The degree of subsutural flattening is equally variable, reflected in the profile of the whorls which may be a smooth curve, or flat, or even concave towards the adapical suture. The spiral ridges are numerous, low, flat, broader than the intervening grooves, less obvious near the suture, coarser towards the base, most prominent at the periphery. The aperture typically connects widely with the siphonal canal but some shells show a constriction at their junction. Yellow or greenish, darker towards the apex; the colour is in the periostracum, the shelly matter being white. Throat darker. Up to 135 mm high, 64 mm broad; last whorl occupies about two thirds of shell height, aperture a little less than half.

The animal has the external features of a common whelk, *Buccinum undatum* (p. 400) save that the whitish flesh is flecked with purple, and the operculum is triangular.

B. turtoni is a northern species which ranges across the North Atlantic from Norway to America, and extends south to the North Sea and to Newfoundland. British records all relate to soft bottoms off the north-eastern coast of Scotland. Development is direct (Thorson, 1940).

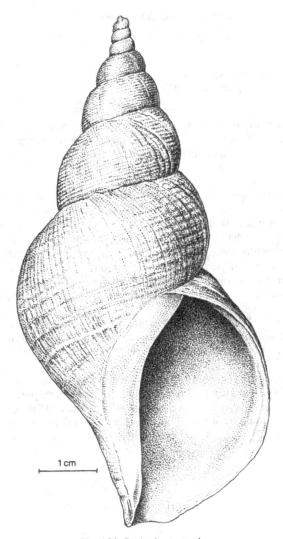

1 cm

Fig. 156. *Beringius turtoni*.

Genus VOLUTOPSIUS Mörch, 1857

Volutopsius norwegicus (Gmelin, 1791)
(Fig. 157)

Strombus norwegicus Gmelin, 1791
Fusus norwegicus (Gmelin, 1791)

Diagnostic characters
Shell like that of *Beringius* (p. 382) in shape (tall spire, blunt apex, swollen whorls) but with only 5–6 whorls, less obvious periostracum, and smooth. Columella distinctly flexuous; no siphonal fasciole.

Other characters
The surface, though smooth to the naked eye, always shows prosocline growth lines, whilst some slight spiral striae are occasionally visible. The subsutural area of each whorl is not flattened, and may show a tendency to thicken and wrinkle immediately below the suture. The aperture is widely oval, more pointed adapically than in *Beringius* and relatively higher and narrower; it is always widely confluent with a short siphonal canal. Cream or white, the apex darker; throat sometimes a little pink. Up to 100 mm high, 60 mm broad; last whorl occupies two thirds of shell height or a little more, aperture about 60%.

The animal shows the same external features as *Buccinum undatum* (p. 400) though the tentacles are stubbier, the foot larger, and the operculum a little more pointed. The flesh is cream or yellowish with purple-brown markings.

V. norwegicus is a northern species with a range stretching across the entire breadth of the Atlantic and reaching into the northern parts of the North Sea, though the animals are not commonly found there. It lives on soft bottoms to depths of about 600 m and preys on echinoderms (Kantor, 1985). Development is direct (Thorson, 1940), but there are no food eggs (Kantor, 1986).

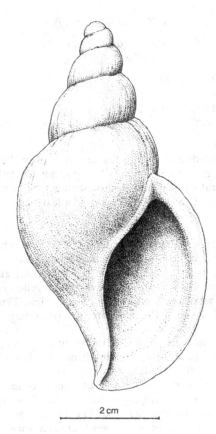

2 cm

Fig. 157. *Volutopsius norwegicus*.

Genus COLUS Röding, 1798

Colus gracilis (da Costa, 1778)
(Fig. 158)

Buccinum gracile da Costa, 1778
Fusus gracilis (da Costa, 1778)
Colus marshalli (Iredale, 1918)
Colus glaber (Kobelt, 1876)

Diagnostic characters
Shell a tall, straight-sided cone, protoconch only slightly swollen, set obliquely on the spire. Ornament of many low spiral ridges and growth lines. Aperture oval, siphonal canal rather short and broad. Shell covered with periostracum but this is absent from a triangular area along adapical end of inner lip.

Other characters
The shell is of rather variable proportions. It has 8–10 whorls, solid and opaque, only slightly tumid, with sutures a little channelled. The spiral ridges are numerous, there being about sixty on the last whorl and eighteen on the penult. The shell itself is white but it is covered by a yellow or rust-coloured periostracum, often worn in places. Up to 70 mm high, 28 mm broad; last whorl occupies 65–70% of shell height, aperture (including canal) 50–55%, siphonal canal by itself, 15%.

The head forms a transverse fold with the mouth (opening of a proboscis sac) on its underside, and a tentacle arising from each lateral margin; these are flat, rather short, with an eye half way up. The siphon projects a little from the canal as the animal crawls. The foot is more or less shield-shaped, straight-edged anteriorly with slightly recurved lateral points. Males have a large, flattened, curved penis on the right. The flesh is white or cream.

C. gracilis occurs on bottoms of sand or mud between Norway and Portugal, occasionally intertidally but commonly between 30 and about 600 m deep, the greater depths in the southern parts of its range. It has been recorded from all round the British Isles but it is scarce in the south. It is a scavenger, perhaps also a carnivore. A smooth-shelled form, sometimes separated as a distinct species, *C. glaber*, is probably a variety of this species; it has been found from Shetland northwards.

Lens-shaped capsules are laid singly on hard substrata, probably containing food eggs, since only one escaping juvenile has been found (Jeffreys, 1862–69).

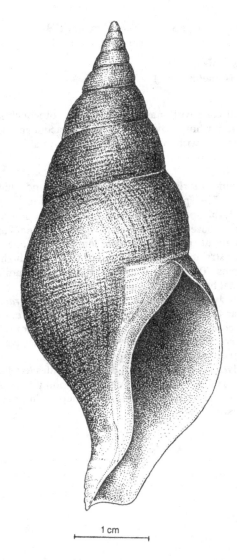

1 cm

Fig. 158. *Colus gracilis*.

Colus islandicus (Gmelin, 1791)
(Fig. 159)

Murex islandicus Gmelin, 1791
Fusus islandicus (Gmelin, 1791)

Diagnostic characters
Shell a rather tall cone with markedly bulbous protoconch set obliquely on the rest of the shell; whorls covered with periostracum and ornamented with spiral ridges. Aperture with long, narrow siphonal canal.

Other characters
There are up to nine whorls which are slightly tumid and which form a gently coeloconoid spire. The spiral ridges are coarser than in *C. gracilis* (p. 386), especially towards the apex, though the protoconch is smooth. The columella is less curved at its base than in *C. gracilis*, the siphonal canal appreciably longer (equal to about three quarters the length of the aperture). The shell is white, the periostracum a pale yellow. Up to 150 mm high, 50 mm broad; last whorl occupies 70% of shell height, aperture (including canal) about half, siphonal canal by itself, nearly 25%.

The animal's body is indistinguishable from that of *C. gracilis*. The species has a circumpolar distribution but extends south in the eastern Atlantic as far as northern Scotland, though the animals are rarely found there. Further south some subfossil shells may be collected. The animals live on soft bottoms from 10 to 3000 m deep. They are carnivores.

Breeding involves the attachment of domed capsules to a hard substratum; these contain a very large number of eggs (commonly over 6500). Only 2–3 embryos develp to hatch as juveniles, the others being used as food eggs (Dall, 1918; Thorson, 1935).

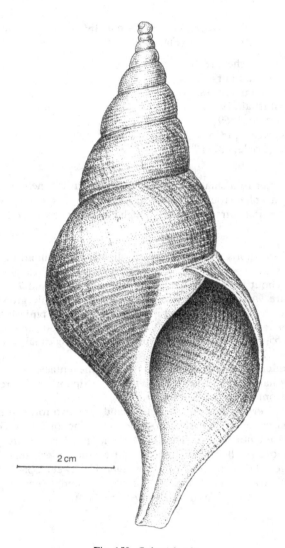

2 cm

Fig. 159. *Colus islandicus*.

Colus jeffreysianus (Fischer, 1868)
(Fig. 160)

Sipho jeffreysianus Fischer, 1868
Fusus propinquus of authors
Fusus buccinatus (Lamarck, 1822)
Colus howsei (Marshall, 1911)
Colus togatus (Mörch, 1869)
Colus tortuosus (Reeve, 1855)
Colus turgidulus (Friele, 1877) ?

Diagnostic characters
Superficially similar to a small *C. gracilis* (p. 386). Protoconch insunk, not dilated, coiling in a plane round the same axis as the rest of the shell. Siphonal canal rather short. Periostracum hairy, though hairs may be worn off.

Other characters
The shell has 7–8 whorls which are usually slightly more tumid than those of the two previous species. The ornament is like that of *C. gracilis* but there are only about thirty spiral ridges on the last whorl and 7–10 on the penult. They are well-spaced and sometimes only crescentic growth lines are visible. The shell is white, sometimes with a bluish or pinkish tint, the overlying periostracum a pale yellow. Up to 60 mm high, 30 mm broad; last whorl occupies 65–70% of shell height, aperture (including canal) about half, canal by itself, 15–20%.

The animal differs from *C. gracilis* only in details: the tentacles are relatively longer and less flat, the eyes nearer their tip; the siphon is also relatively longer and the front of the foot is rounder.

The species *jeffreysianus* is now held to include the two forms previously called *C. propinquus* and *C. howsei*, the former occurring south of St George's Channel, the latter north of it. The former is larger, has a more pointed apex, slightly more swollen whorls and a shorter and wider canal, but the two forms merge. The distribution of *jeffreysianus* is between Spain and Norway, the animals living on soft bottoms between 30 and 350 m deep. They are not common. Breeding is as in *C. islandicus* (p. 388).

Fig. 160. *Colus jeffreysianus*. A, shell of the type once called *C. howsei*, with a northern distribution; B, shell of the type once called *C. propinquus*, southern in its distribution; C, empty egg capsule; D and E, two views of a newly hatched juvenile.

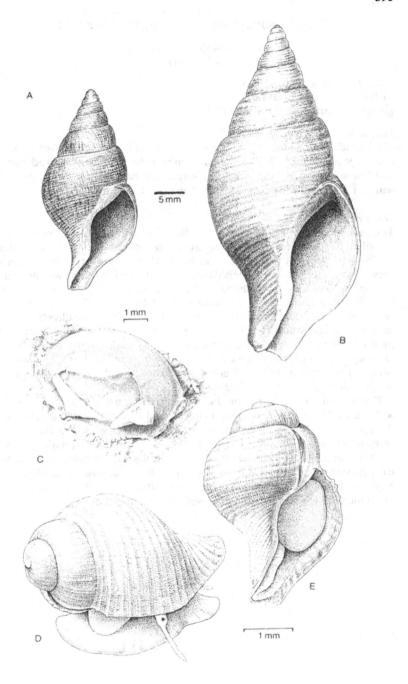

Genus TURRISIPHO Dautzenberg & Fischer, 1912

Turrisipho fenestratus (Turton, 1834)
(Fig. 161)

Fusus fenestratus Turton, 1834
Colus fenestratus (Turton, 1834)
Buccinum fusiforme Broderip, 1830

Diagnostic characters
Shell a rather tall cone with blunt apex, drawn out basally into a short, widely open siphonal canal. Whorls tumid, ornamented with many curved costae and low spiral ridges, the latter coarsest at the base. Surface dull, covered with periostracum which is raised into hairs along the spiral ridges.

Other characters
The shell has eight or nine whorls forming a straight-sided cone and meeting at rather deep sutures. The costae cross those of the spire but fade below the periphery of the last whorl. On most whorls there are about sixteen; they lie so as to form spiral keels down the shell when that is viewed apically. The spiral ridges are strap-shaped, often alternately larger and smaller; the last whorl has about thirty, the others about ten each. They run over the costae. The surface of the second and third whorls is reticulated. At the base of the last whorl is a rather blunt siphonal fasciole. The aperture is pyriform, the outer lip with a thin edge, the columella rather straight. The inner lip forms a triangualr area over the last whorl and is glossy. Shell white, periostracum cream. Up to 48 mm high, 20 mm broad; last whorl occupies 60–65% of shell height, aperture 40–45%.

The animal has no snout, the head taking the form of a transverse ridge with the proboscis pouch opening below it, and a short tentacle at each side. An eye lies a little way up each tentacle. The foot is large and shield-shaped, carrying an oval operculum. The flesh is white, the siphon with dark streaks.

These animals are rare, recorded sporadically off the west coast of the British Isles (but not in the Irish Sea), and off Scandinavia, Greenland and Newfoundland. They are found on gravelly bottoms 30 to about 1300 m deep. Their mode of life and reproductive methods are unknown.

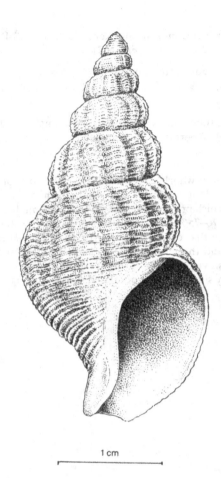

1 cm

Fig. 161. *Turrisipho fenestratus*.

Turrisipho moebii (Dunker & Metzger, 1874)
(Fig. 162)

Tritonofusus moebii Dunker & Metzger, 1874
Fusus sarsi Jeffreys, 1869

Diagnostic characters
Shell broad, whorls tumid and sutures deep, with numerous low spiral ridges
and shallow grooves. Aperture oval, confluent with wide siphonal canal.

Other characters
There are 7–8 swollen whorls covered with a rather obvious periostracum,
often slightly hairy along the spiral ridges. Shell white, but the periostracal
cover gives it a greenish brown appearance. Up to 40 mm high, 25 mm broad;
last whorl occupies 70–75% of total shell height, aperture about 55%.

The animal is like that of *T. fenestratus* (p. 392). The species is northern
in its distribution and the most northern Scottish waters mark its southern
limits. The animals live on soft bottoms at depths of 200–1000 m. Their way
of life is presumably similar to that of related species.

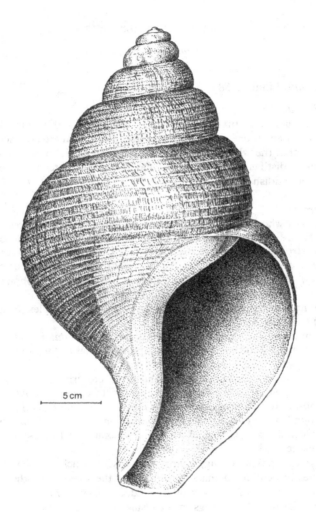

Fig. 162. *Turrisipho moebii*.

Genus NEPTUNEA Röding, 1798

Neptunea antiqua (Linné, 1758)
(Fig. 163)

Murex antiquus Linné, 1758

Diagnostic characters

Shell superficially like that of *Buccinum undatum* (p. 400) but devoid of costae, heavier and more solid for a given size. Subsutural area of each whorl often flat so that the whorl appears a little angulated. Siphonal canal short, widely open, distinct from aperture. Operculum with terminal nucleus. Animal often reddish.

Other characters

There are 7–8 whorls ornamented only with spiral lines and growth lines. The former are often finer than in *B. undatum* but in some shells a few may be so strong, especially in the more apical whorls, as to suggest small spiral keels. The growth lines on the last whorl are less flexuous than in *B. undatum*. The outer lip curves to meet the base of the siphonal canal rather than a point near its apex, as in *Buccinum*. Cream, sometimes reddish. Up to about 100 mm high, 55 mm across; last whorl occupies 70–80% of shell height, aperture about 60%.

The animal has the same external features as *B. undatum*. The siphon does not usually extend so far from the siphonal canal as the animal creeps as it does in that species. The operculum is dark.

N. antiqua occurs through most of the north-eastern Atlantic from the Bay of Biscay to Scandinavia, penetrating the most western part of the Baltic. It is common in the northern parts of its range, becoming rare in the south. It is not found between tidemarks but occurs on all kinds of bottom from 15 m to 1200 m deep. It is a general carnivore taking bivalves, annelids, and carrion (Pearce & Thorson, 1967; Taylor, 1978).

N. antiqua breeds in spring, laying egg capsules stuck together to form a tall, somewhat columnar mass, on stones or the shell of another whelk. Each capsule is squarish, convex on one side, concave on the other, about 20 × 15 mm. Of up to 5000 eggs in a capsule only one or two hatch about six months later as juveniles with shells 6–12 mm high (Dall, 1918; Thorson, 1946; Pearce & Thorson, 1967). Males mature at a shell height of 50–60 mm, females not until growth is complete when they may be ten years old.

Vernacular name: red whelk or buckie.

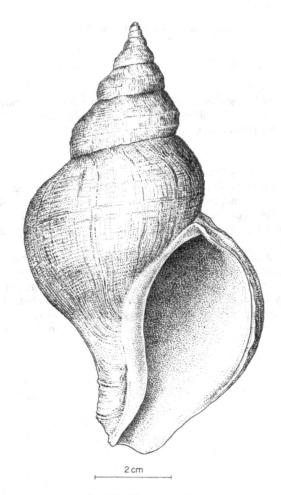

Fig. 163. *Neptunea antiqua*.

Neptunea despecta (Linné, 1758)
(Fig. 164)

Fusus despectus Linné, 1758

Diagnostic characters
Shell like that of *N. antiqua* (p. 396) but ornament coarser and more marked. Whorls with subsutural shelf bordered by a spiral ridge, triangular in section. Siphonal fasciole poorly developed.

Other characters
There are 7–8 whorls the spiral ridge on which gives the spire a turreted and angulated profile. There may be some well-spaced incipient costae on the subsutural area but they do not cross the whorls except in shells from Arctic areas. The aperture is like that of *N. antiqua*. White, yellow or brown. Up to 160 mm high, 80 mm broad; last whorl occupies 65–70% of shell height, aperture 45%.

The animal is like that of *N. antiqua*. *N. despecta* is a northern, probably circumpolar, species which extends southwards to waters north of Shetland. The animals live on soft bottoms 6–1400 m deep, probably feeding on polychaetes. They lay eggs in capsules which the female attaches to her shell. Although a capsule may contain 5000 eggs initially only one or two hatch as juveniles, the rest being eaten by these successful ones (Thorson, 1944, 1946).

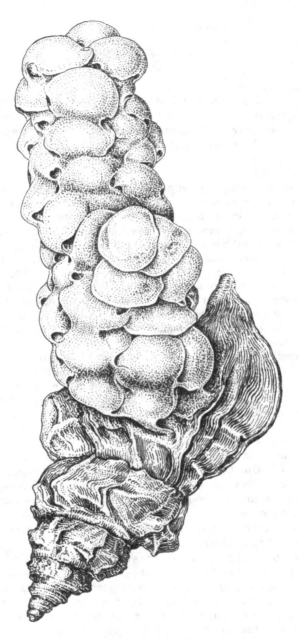

Fig. 164. *Neptunea despecta*, with egg capsules.

Genus BUCCINUM Linné, 1758

Buccinum undatum Linné, 1758
(Fig. 165)

Diagnostic characters

Shell large, last whorl tall and broad with short, oblique siphonal canal. Ornament of spiral ridges and costae; periostracum present but often inconspicuous. Animal (white with black marks) with long siphon, male with extremely large penis; foot bears operculum with central nucleus.

Other characters

The shell is solid, not glossy, with 7–8 tumid whorls. The costae are undulate in section and curved like a reversed-C, usually crossing the whorls of the spire but dying out on the base of the last whorl, which carries about twelve. The spiral ridges vary in size, the larger ones well spaced on the adapical half of the last whorl but becoming more crowded on its base. A prominent spiral keel (siphonal fasciole), sometimes double, runs round the base of the shell. The aperture is a broad oval, the outer lip a little turned out, smooth within, with a broad, shallow anal sinus. The inner lip forms an extensive shiny glaze over the last whorl. Yellowish brown with irregular spiral areas darker and lighter, the latter commonly near the periphery; lip and columella white. Up to 100 mm or more high, 60 mm broad; last whorl occupies about 70% of shell height, aperture about half.

The animal is fleshy with a large shield-shaped foot, broad anteriorly. The head is a flattened projection carrying the mouth on its underside and tentacles laterally, each with an eye about one third of its length above the base. The siphon projects some way from the canal as the snail creeps. Whelks eat worms and bivalves, opening the shell of the latter by driving the valves apart with the edge of their own shell, since they cannot bore (Nielsen, 1975; Taylor, 1978). They also take carrion.

B. undatum is abundant on soft bottoms from just below low water mark to depths of 1200 m throughout the North Atlantic, and it extends into some brackish areas. They are sometimes found intertidally. Their egg cases are often washed ashore as globular or irregularly hemispherical masses of capsules attached to one another. The species is extremely variable in the proportions of its shell and many varieties have been named (Pain, 1979), the taxonomic status of which is uncertain.

Breeding occurs predominantly in winter (Kristensen, 1959; Hancock, 1960). Eggs are enclosed in somewhat pyramidal capsules with one surface convex and the other concave, fastened together to form the masses referred to above. These may measure up to 50 cm in length and 25 cm in height, and contain many thousands of capsules. Several females contribute to their formation since the firm base needed for their attachment may be rare in the areas of soft bottoms on which the whelk lives. Most of the eggs in a capsule are food eggs (Portmann, 1925) and only 3–10 juveniles emerge

Fig. 165. *Buccinum undatum.*

after a developmental period that may be as long as nine months (Schäfer, 1955; Kristensen, 1959).

Vernacular name: whelk or buckie.

Buccinum humphreysianum Bennett, 1824
(Fig. 166)

Diagnostic characters

Shell a rather broad oval, thin, without obvious periostracum, a little transparent, apparently smooth. Spire rather regularly conical. Aperture oval, siphonal canal short and wide. Siphonal fasciole slight; expansion of inner lip over last whorl extremely thin. Animal with small, slightly angulated, oval operculum.

Other characters

The shell has eight whorls which are swollen and meet at pronounced sutures. When closely examined their surface is seen to be divided into square areas by a large number of fine spiral ridges crossing an equal number of prosocline growth lines of similar size. The aperture is narrower than in *B. undatum* (p. 400), the outer lip thinner, the siphonal canal more slender and distinct, and the siphonal fasciole is, at most, a slight fold. Yellowish white, with reddish brown streaks or spots, sometimes faint. Up to 45 mm high, 25 mm broad; last whorl occupies about 70% of shell height, aperture just over half.

The animal resembles *B. undatum* except in the shape and size of the operculum.

B. humphreysianum is apparently rare, though animals have been found on soft bottoms, in rather deep water, between the Mediterranean and Norway. Round the British Isles they have been met with off the Shetlands, the Hebrides, and west and south Ireland, 100–150 m deep. Their mode of life is not known but is perhaps not very different from that of *B. undatum*.

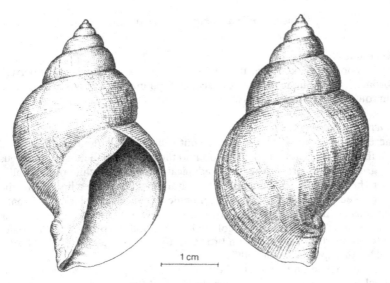

Fig. 166. *Buccinum humphreysianum*.

Buccinum hydrophanum Hancock, 1846
(Fig. 167)

Diagnostic characters
Shell smooth to naked eye, thin, with tall spire of tumid whorls. Aperture a broad oval confluent with siphonal canal. Siphonal fasciole absent. Operculum round with central nucleus.

Other characters
There are 6–7 whorls, tumid but without subsutural flattening and covered by a thin periostracum. Though smooth to the naked eye there are numerous fine growth lines, prosocline at the adapical suture, orthocline peripherally; older shells usually show equally fine spiral lines. The outer lip is thin, the inner reflected outwards, without an umbilical groove alongside. Pale bronze with darker areas where covered by the out-turned parietal lip and along some sutures; the outer lip, the columellar area and the protoconch are pale. Up to about 70 mm high, 40 mm broad; last whorl occupies 60–70% of total shell height, aperture about 50%.

The animal resembles that of other *Buccinum* species. This is an arctic animal with a range across the whole North Atlantic and extending south as far as Shetland. It lives on soft bottoms from 100 to 1200 m deep. Its mode of life is similar to that of other buccinids.

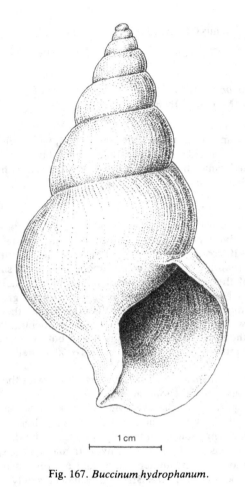

Fig. 167. *Buccinum hydrophanum*.

Genus CHAUVETIA Monterosato, 1884

Chauvetia brunnea (Donovan, 1804)
(Fig. 168)

Buccinum brunneum Donovan, 1804
Lachesis minima (Montagu, 1803)

Diagnostic characters
Shell small, thick and solid, but a little translucent, with a blunt tip; whorls with costae and spiral ridges; a varix along outer lip and about six teeth within it. Siphonal canal short, broad, open. Protoconch with minute spiral lines. Uniform chestnut brown.

Other characters
The shell is rather narrow (apical angle 25–30°), glossy, the spire cyrtoconoid. There are about five whorls, each a little tumid, the sutures distinct. The last whorl has 9–10 costae in addition to the varix, the penult ten; they are about equal in breadth to the intervening spaces, run from suture to suture on the whorls of the spire but disappear below the periphery on the last. The spiral ridges are broad, low, and broader than the grooves between them. They are more variable in number than the costae, there being 10–16 (usually about twelve) on the last whorl, 4–5 on the penult. They cross the costae but not the varix. The aperture is broad but not very high. Up to 5 mm high, 2–2.5 mm broad; last whorl occupies 60% of shell height, aperture 40%.

The animal's head forms a narrow transverse ridge with the mouth under it; two tentacles arise from it, one at each lateral corner, each with an eye half way up, the part below the eye twice as thick as the part above. The siphon extends some way from the canal as the animal crawls. The foot is narrow and rather long, square anteriorly, pointed behind, and carries an operculum. The flesh is pale brown or tawny, with many white spots.

C. brunnea is to be found under stones and in crevices near L.W.S.T. and to depths of about 60 m. In the British Isles it occurs only in the southern parts of England; it has also been recorded from the Channel Islands and south to the Mediterranean.

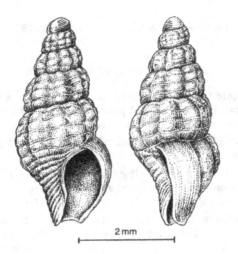

Fig. 168. *Chauvetia brunnea*.

Family NASSARIIDAE Iredale, 1916

Key to British species of Hinia

1. Shell large, spire tall, whorls nearly flat-sided; teeth may lie within outer lip, on columella and on parietal region; costae and spiral ridges about equal in height and breadth .... *Hinia reticulata* (p. 408)

   Shell not over 14 mm in height; whorls tumid; costae higher than spiral ridges **2**

2. Only 1 varix (at outer lip); 8–10 spiral ridges on penult whorl; up to 12 ridges on boss below spiral gutter near base of last whorl; black mark on wall of siphonal canal .... *Hinia incrassata* (p. 412)

   Varices on other whorls as well as at outer lip; 4–5 spiral ridges on penult whorl; 5 ridges on boss below spiral gutter; no black mark .... *Hinia pygmaea* (p. 414)

Genus HINIA Leach, 1852

Hinia reticulata (Linné, 1758)
(Figs 169, 170)

Buccinum reticulatum Linné, 1758
Nassa reticulata (Linné, 1758)
Nassarius reticulatus (Linné, 1758)
Nassa nitida Jeffreys, 1867
Nassarius nitidus (Jeffreys, 1867)

Diagnostic characters
Shell a rather tall, sharply pointed cone, with nearly straight sides, covered with a distinct periostracum. Ornament of numerous costae and spiral ridges crossing to give a reticulated surface. Aperture rather small, oval, with short, oblique siphonal canal which has an external spiral gutter round its base. Outer lip with internal teeth. Foot notched posteriorly, with two tentacles.

Other characters
The ten whorls are nearly flat-sided and meet at sutures emphasized by a rather thick subsutural spiral ridge. The costae are slightly flexuous, undulate in section, cross the whorls of the spire, and die out at the spiral gutter on the last whorl, on which (in addition to a labial varix) there are 20–22, with up to 25 on the penult. Spiral ridges are smaller and more numerous: 4–6 on the boss below the spiral gutter on the last whorl, 12–14 above it, usually five on each whorl of the spire. The aperture is oval, its outer lip thickened by a varix and, internally, by a ridge bearing 6–9 teeth. The columella and inner lip also bear teeth (or ridges), three usually at the base

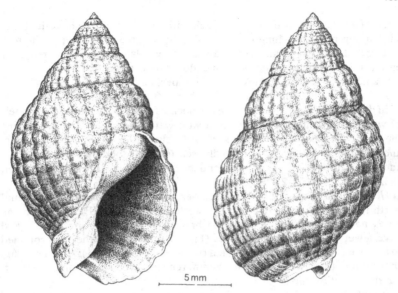

Fig. 169. *Hinia reticulata*.

of the columella, and up to three, less regularly and less prominently, on the semicircular expansion of the inner lip over the surface of the last whorl. The periostracum gives the shell a tan colour, but if it is lost it is pale, perhaps with a pinkish tinge. A slate-coloured band often lies below each suture and there may be brown ones near the periphery and base of the last whorl, but these are often obscure. Up to 30 mm high, 14 mm broad; last whorl occupies 60–70% of shell height, aperture 40–50%.

The head carries two tentacles each with an eye a little above its base. The siphon projects a long way from the canal as the snail creeps. The foot is long and narrow; its anterior end has recurved lateral points, its posterior end has a V-shaped notch, each lobe alongside it terminating in a tentacle. The animal is brown and black with white speckles.

H. reticulata is common on rocky shores from near L.W.S.T. to about 15 m deep, particularly where there are pockets of soft material under runnels of water at low tide. When covered with water the snails burrow, leaving the tip of the siphon above the surface (Ankel, 1929); they are scavengers and carrion feeders and crawl rapidly, sensing their food by the osphradium and, when close to it, by the siphon and the anterior end of the foot (Henschel, 1932; Crisp, 1971, 1972).

H. reticulata occurs from the Black Sea, through the Mediterranean, and north to Norway, extending into the western parts of the Baltic. The name *nitida* has been given to forms from brackish water which are smaller and have fewer costae and spiral ridges (Fig. 170). In Britain they are confined to the estuaries of some East Anglian rivers (Mistakidis, 1951); Collyer (1961) showed that there were biochemical difference between *nitida* and *reticulata*, but their relationship is still uncertain.

Breeding occurs in spring and summer, when egg capsules are attached to weeds, shells or stones. Each is a flattened vase shape but, unlike those of muricids and buccinids, is glassily transparent so that the pinkish eggs, of which there are 50–350 per capsule, are easily seen (Ankel, 1929). All the eggs develop and hatch as free veliger larvae, the shell of which has a prominent peripheral beak on the outer lip. The velum is at first 2-lobed with a marginal red-brown line; later it is 4-lobed. The larva is long-lived (Fretter & Shale, 1973).

Vernacular name: (netted) dog whelk.

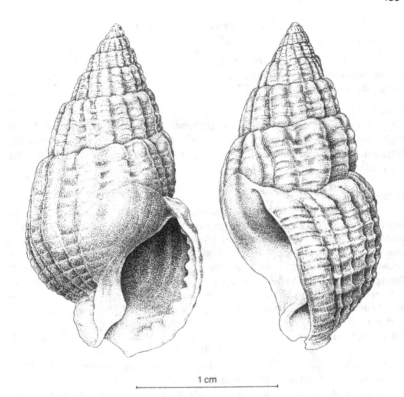

1 cm

Fig. 170. *Hinia reticulata* var. *nitida*.

Hinia incrassata (Ström, 1768)
(Fig. 171)

Buccinum incrassatum Ström, 1768
Nassa incrassata (Ström, 1768)
Nassarius incrassatus (Ström, 1768)

Diagnostic characters
Shell more or less like that of the previous species but rarely over 12 mm high, with a more prominent label varix which is white, and always distinguished by a dark spot within the siphonal canal and a white columella. Animal with two metapodial tentacles.

Other characters
Though like a small shell of *H. reticulata* (p. 408) in its general appearance this shows differences in detail: there are eight whorls, all a little more tumid; costae and spiral ridges are fewer (last whorl with labial varix and 10–12 costae, usually twelve on other whorls; last whorl with 10–11 spiral ridges above the spiral gutter, 4–5 on other whorls); the varix is more distinct; the spiral gutter is deeper; the aperture is relatively broader, the outer lip with 6–7 internal teeth and a similar number on the columella and inner lip. Reddish buff, often with a subsutural dark band, a peripheral white one, and a basal dark one on the last whorl; only the first may appear in the spire. The bands are clearer than those of *reticulata*. Up to 12 mm high, 6 mm broad; last whorl occupies 70% of shell height, aperture 40%.

The body of the animal is like that of *reticulata*, yellowish with black speckling, mainly on the siphon and foot.

H. incrassata is common in the Mediterranean and all western European coasts, living gregariously in rather silty places on rocky shores, under stones and in tufts of weed, on the lower part of the shore and sublittorally. In the most northern parts of its range it goes to depths of about 200 m. Breeding follows the same pattern as in *H. reticulata* though the egg capsules are smaller. The late veliger larva may be distinguished by the great length of the four lobes of the velum, equal to twice the height of the shell (Lebour, 1931a).

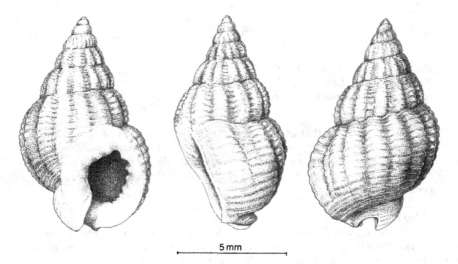

5 mm

Fig. 171. *Hinia incrassata*.

Hinia pygmaea (Lamarck, 1822)
(Fig. 172)

Ranella pygmaea Lamarck, 1822
Nassa pygmaea (Lamarck, 1822)
Nassarius pygmaeus (Lamarck, 1822)

Diagnostic characters
Shell in general like that of *H. incrassata* (p. 412) but with narrower spire, more marked ornament, wider siphonal canal with a wider and shallower spiral gutter at its base, a less well developed labial varix which is rarely all white. Other varices usually present on spire. Columella not white. Metapodial tentacles long.

Other characters
The costae are usually more numerous than in *incrassata* (varix plus thirteen on last whorl) and are not flexuous; where costae and spiral ridges cross more prominent tubercles occur than in that species. There are nine or ten teeth within the outer lip. The colour pattern consists of spiral bands (one subsutural, one peripheral, one basal on the last whorl, only the first usually visible in the spire); these are usually more obvious than in *incrassata* and cross the varices. The siphonal canal and columella are brown, the former lacking the intensely dark spot of *incrassata*. Up to 14 mm high, 8 mm broad; last whorl occupies about two thirds of shell height, aperture 40–45%.

In animals of this species the siphon is longer than in *incrassata* and the anterior end of the foot is more markedly angulated. The flesh is pale yellow or buff with black markings.

H. pygmaea has been recorded from the western parts of the Channel and north along western coasts to the Shetlands, absent, however, from the Irish Sea. It lives on sandy bottoms from just below L.W.S.T. to about 200 m deep, but is rather infrequent. Its mode of life is similar to that of the other *Hinia* species, and the animals have the same mode of reproduction (Vestergaard, 1935).

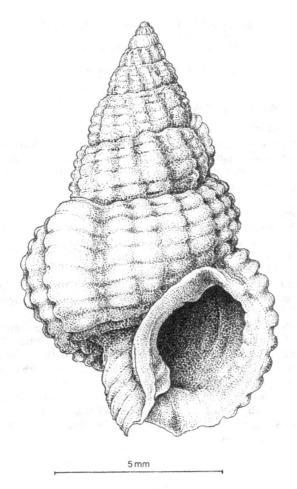

5 mm

Fig. 172. *Hinia pygmaea.*

Family FASCIOLARIIDAE Gray, 1853

Genus TROSCHELIA Mörch, 1876

Troschelia berniciensis (King, 1846)
(Fig. 173)

Fusus berniciensis King, 1846

Diagnostic characters
Shell with moderately high spire with blunt tip; whorls swollen and rather angulated in profile, marked with spiral ridges and less obviously with growth lines. Aperture with open siphonal canal bent to left. Periostracum bears hairs along crests of ridges. Shell often with pink tinge.

Other characters
The shell has about eight whorls. Of the spiral ridges about four, peripherally placed, are more marked than the others. The aperture is oval or pear-shaped, tapering below. The outer lip has a distinct anal sinus and becomes a little everted in older shells. The periostracum is yellowish, but a pink colour often appears in the underlying shell, most evident in the throat. Up to 120 mm high, 50 mm broad; last whorl occupies about 70% of shell height, aperture about half.

The tentacles arise close together from the transverse elevation which is the head. The mouth lies under the fold. The siphon extends clearly from the siphonal canal as the animal creeps. The foot is large, with a curved anterior edge and a rounded posterior end. The flesh is pinkish cream.

T. berniciensis has a wide range throughout the north-eastern Atlantic, commoner in the northern parts, but always rather rare. It has been recorded from the northern half of the North Sea and from off the west coasts of Ireland and Scotland, living on soft bottoms between 140 and 2000 m deep. Its habits are not known.

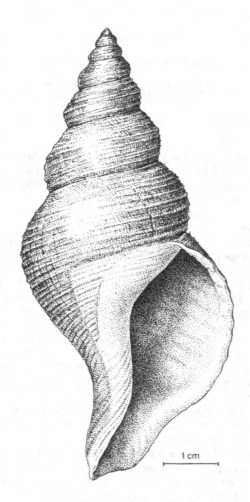

Fig. 173. *Troschelia berniciensis*.

Family CANCELLARIIDAE Forbes & Hanley, 1853

Genus ADMETE Kröyer, 1842

Admete viridula (Fabricius, 1780)
(Fig. 174)

Tritonium viridulum Fabricius, 1780

Diagnostic characters

Shell relatively broad for its height, whorls tumid with subsutural ramp, costae and spiral ridges, the former often limited to adapical parts of whorls, the latter to peripheral parts. Aperture large, broadly confluent with siphonal canal; outer lip thin. No operculum on foot.

Other characters

This is a variable shell, especially in its ornament; the spire is moderately tall with a sharp apex and the last whorl forms its bulk. There are 5–6 whorls. Costae are prosocline, die out at the periphery of the last whorl or even on the apical side of that, and also near the outer lip. There are 12–18 on the last whorl, 15–20 on the penult. The spiral ridges are absent from the subsutural ramp except sometimes on some older whorls. The number of ridges is even more liable to vary than the number of costae – 15–22 on the last whorl, 5–8 on the penult, those at the periphery always the most prominent. The outer lip never develops internal teeth. There is neither umbilical groove nor umbilicus. White or yellowish, sometimes greenish. Up to 15 mm high, 8 mm broad; last whorl occupies 60–75% of shell height, aperture 40–50%.

The tentacles are digitiform, an eye placed about a third of the length up from the base. The siphon is short. Males have a tentaculiform penis. The body is white or cream.

A. viridula is an arctic species, probably circumpolar, reaching south to Massachusetts and to the extreme northern borders of the North Sea. The animals live on soft bottoms at depths varying from a few metres in the Arctic to about 1000 m at the species' southern limits. They are said to eat ophiuroids. They lay egg capsules (Thorson, 1935, then erroneously ascribed to *Velutina undata*) which are egg-shaped and attached to the substratum by a short stalk. Development is direct.

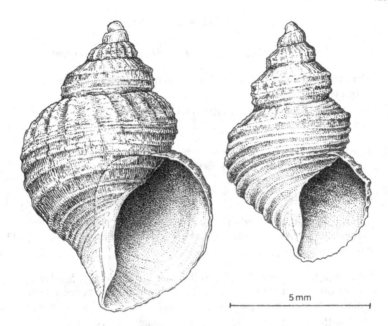

Fig. 174. *Admete viridula*.

Family TURRIDAE Swainson, 1840

Key to British genera and species of Turridae

1. Ornament of spiral ridges and grooves only; growth lines present but without costae **2**

   Ornament includes costae (size variable) as well as spiral ridges and grooves **5**

2. 1 peripheral spiral keel on each whorl; spire tall; anal sinus deep and narrow; growth lines flexuous; outer lip with peripheral bulge; protoconch smooth .... *Spirotropis monterosatoi* (p. 424)

   Shell not like this **3**

3. 2 spiral ridges on each whorl of spire; whorls with clear subsutural shelf devoid of spiral lines; apex blunt; rare *Oenopota violacea* (p. 432)

   Shell not like this **4**

4. Shell with 8–9 whorls; 20–25 spiral ridges on last whorl, 5–6 large ones on penult; anal sinus deep and narrow; protoconch with elaborate criss-cross pattern .... *Teretia teres* (p. 462)

   Shell with 5–6 whorls; about 30 spiral ridges on last whorl; anal sinus wide and shallow; protoconch with irregular, fine, spiral lines .... *Thesbia nana* (p. 436)

5. Breadth of shell equals half or more of total height **6**

   Breadth of shell less than half the total height **7**

6. Spiral ridges prominent, few on base of shell; protoconch brown .... *Taranis borealis* (p. 464)

   Spiral ridges weak, frequent on base of shell; protoconch white .... *Taranis moerchi* (p. 464)

7. 12 costae or more (size variable) on penult whorl **8**

   Fewer than 12 costae on penult whorl **17**

8. Siphonal canal long (= aperture length) and narrow **9**

   Siphonal canal not like this **11**

9. 10 whorls, each with concave subsutural area; 14–15 costae on penult whorl, reduced or absent on subsutural area; anal sinus deep and narrow; up to 18 teeth within outer lip (mature shells); protoconch with spiral keel; yellow with brown spiral bands .... *Comarmondia gracilis* (p. 452)

   Shell not like this **10**

10. 6–7 spiral ridges and 16–17 costae on penult whorl; costae cross
whorls; tubercles at intersection of ridges and costae; outer lip
with varix and internal teeth; anal sinus shallow; protoconch
with criss-cross pattern .... *Raphitoma asperrima* (p. 458)

Many fine spiral ridges on penult whorl; costae on this and other
young whorls reduced to peripheral tubercles; outer lip thin,
without internal teeth; anal sinus deep; protoconch smooth ......
.... *Typhlomangelia nivalis* (p. 434)

11. Spire turreted, each whorl with distinct subsutural shelf **12**

Spire not turreted, shelf absent or nearly so **14**

12. Shell with two prominent spiral ridges on whorls of spire; costae
many and fine; protoconch without spiral keels; rare ....
.... *Oenopota violacea* (p. 432)

Costae and spirals about equally prominent; 2–3 spiral keels on
protoconch **13**

13. Subsutural shelf broad; not more than 16 costae on last whorl;
second whorl of protoconch with three spiral keels ....
.... *Oenopota turricula* (p. 428)

Subsutural shelf narrow; 20 or more costae on last whorl; second
whorl of protoconch with more than three keels ....
.... *Oenopota trevelliana* (p. 430)

14. Shell with 6–7 whorls; 13–15 flexuous costae on penult whorl; outer
lip without internal teeth; second whorl of protoconch with 3
spiral keels; often reddish; operculum on foot . *Oenopota rufa* (p. 432)

Shell with 8–9 whorls; outer lip thick with internal teeth in mature
animals, protoconch with criss-cross pattern; no operculum on
foot **15**

15. 18–24 costae and 8–12 spiral ridges on penult whorl; costae slightly
opisthocline; anal sinus narrow and deep; protoconch yellow,
rest dark reddish purple with white-cream blotches, outer lip
white .... *Raphitoma purpurea* (p. 456)

14–17 costae and 6–9 spiral ridges on penult whorl; costae orthocline; shell not purple-red **16**

16. 8–9 spiral ridges and 14–17 costae on penult whorl; aperture long
(about half shell height); embryonic shell white ....
.... *Raphitoma leufroyi & Cenodagreutes* (p. 460)

6–7 spiral ridges and 16–17 costae on penult whorl; aperture short
(less than half shell height); marked tubercles where ridges and
costae cross .... *Raphitoma asperrima* (p. 458)

17. 4–7 narrow spiral ridges which broaden over 10–11 costae on penult whorl; sutures deep; anal sinus narrow, at origin of outer lip; up to 9 internal teeth in mature animals; protoconch with criss-cross pattern; shell pale with brown marks on spiral ridges .... *Raphitoma linearis* (p. 454)

Shell not like this **18**

18. Spiral ridges numerous (more than 10 on penult whorl), all approximately similar; protoconch smooth or with tuberculated third whorl **19**

Spiral ridges few (less than 10 *prominent* ones on penult whorl) and of different width; protoconch with tuberculated third whorl or with criss-cross pattern **24**

19. Whorls nearly flat-sided, sutures shallow and wavy; 7–9 costae per whorl, commonly aligned to form keels along spire; spiral ridges small, strap-shaped; outer lip with varix and internal thickening but without teeth; protoconch smooth or punctate; operculum on foot .... *Haedropleura septangularis* (p. 426)

Shell not like this **20**

20. Shell tall, slender; sutures deep with swollen band along lower edge; spiral ridges narrow, beaded; 9–11 costae on each whorl of spire; outer lip without internal teeth; chestnut brown with darker spiral bands, paler on costae .... *Mangelia powisiana* (p. 442)

Shell not like this **21**

21. Spiral ridges narrow, low, square in section with sharp clefts between; anal sinus narrow and deep with out-turned margin **22**

Spiral ridges low and flat; anal sinus shallow **23**

22. Spiral ridges rather coarse, about 45 on penult whorl; shell breadth about one third of shell height; 9 whorls; 7–11 costae on penult whorl; last whorl with several brown spiral bands on pale background .... *Cytharella smithi* (p. 446)

Spiral ridges fine, many more than 45 on penult whorl; shell breadth more than one third of its height; 7–8 whorls; 8–9 costae on penult whorl; shell either chocolate on spire and pale with brown spiral lines on last whorl, or tawny throughout with brown spiral lines .... *Cytharella coarctata* (p. 448)

23. Breadth of shell about 40% of height; penult whorl with 7–9 broad costae and 11–30 close-set spiral ridges which are moderately nodose or scaly; sometimes a dark peripheral spiral band .... *Mangelia nebula* (p. 438)

Breadth of shell about 30% of height; penult whorl with 8–10 tall, narrow, flexuous costae and many low, strap-shaped spiral ridges; base of shell extended; sometimes a thickening, but no teeth, within outer lip; often many narrow brown spiral lines on pale background .... *Mangelia attenuata* (p. 444)

24. Shell with 6–7 whorls; nine costae on penult whorl and 5 *prominent* spiral ridges, plus perhaps minor ones, all squarish in section; last whorl occupies about two thirds of shell height, aperture about half; rare, only from south-west Britain and Channel Islands .... *Cytharella rugulosa* (p. 450)

Shell with 7–9 whorls; 8–10 costae on penult whorl and 5–6 large spiral ridges, plus perhaps minor ones; ridges on lower border of suture prominent; all ridges wide apart, nodose or scaly; last whorl occupies about 60% of shell height, aperture about 40%; sometimes a dark band on outer lip ..... *Mangelia brachystoma* (p. 440)

Genus SPIROTROPIS Sars, 1878

Spirotropis monterosatoi (Locard, 1897)
(Fig. 175)

Pleurotoma monterosatoi Locard, 1897
Spirotropis carinata of authors

Diagnostic characters
Shell with rather tall spire, each whorl with 1 prominent spiral keel and fine growth lines, prosocline above and opisthocline below the keel. Aperture with pronounced anal sinus, a bulge on outer lip and moderately long, open siphonal canal.

Other characters
The shell is a little translucent, has a blunt apex, and 9–10 tumid whorls. The aperture is narrowly pear-shaped, the outer lip with a thin edge. White. Up to 26 mm high, about ten across, northern specimens smaller; last whorl occupies about 60% of shell height, aperture 40%.

The animal shows the same external features as other turrids, has an operculum on the foot, and is white.

S. monterosatoi is normally found in deeper waters offshore, on soft bottoms. There are few British records (off the Shetlands), but the animal is common on the continental shelf off western Europe (Bouchet & Warén, 1980).

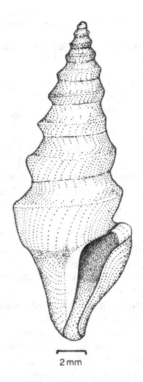

Fig. 175. *Spirotropis monterosatoi.*

Genus HAEDROPLEURA Bucquoy, Dautzenberg & Dollfus, 1883

Haedropleura septangularis (Montagu, 1803)
(Fig. 176)

Murex septangularis Montagu, 1803
Pleurotoma septangularis (Montagu, 1803)
Mangelia septangularis (Montagu, 1803)

Diagnostic characters

Shell solid, nearly fusiform, opaque and glossy; spire cyrtoconoid with rather flat whorls. Ornament (visible to naked eye) of costae, usually seven per whorl. Aperture rather long and narrow, confluent with short siphonal canal; outer lip with shallow anal sinus and usually with thick edge.

Other characters

The apex is rather blunt, the 7–8 whorls only a little tumid, flatter adapically and more rounded abapically. The costae are slightly opisthocline, though on the last whorl their apical ends curve to become prosocline. They cross the whorls of the spire but fade towards the base of the last whorl, and are often aligned to form keels running the length of the spire. In addition to costae minute spiral ridges cover the whole surface but are usually eroded from the costal summits. The inner lip is often thickened near its junction with the outer, which is thick except in young shells. Without internal teeth. Pale brown; costal crests, columella, much of the canal wall, and sometimes the subsutural areas are pale. Up to 14 mm high, 6 mm broad; last whorl occupies 55–60% of shell height, aperture about 40%.

The tentacles are short, thick basally, and arise from the head rather close together; each has an eye about two thirds of its length from the base. The siphon projects from the canal as the snail crawls. Males have a large sickle-shaped penis. The flesh is white.

H. septangularis occurs from the Mediterranean to Norway. It has been collected off most parts of the British Isles, except in the North Sea, from gravelly bottoms 10–50 m deep. It is rare. The life history probably includes a free larval stage (Lebour, 1936; Thiriot-Quiévreux & Babio, 1975).

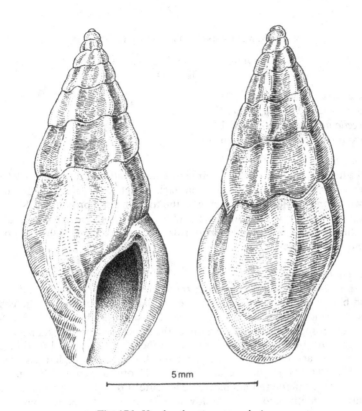

Fig. 176. *Haedropleura septangularis*.

Genus OENOPOTA Mörch, 1852

Oenopota turricula (Montagu, 1803)
(Fig. 177)

Murex turricula Montagu, 1803
Lora turricula (Montagu, 1803)
Mangelia turricula (Montagu, 1803)
Pleurotoma turricula (Montagu, 1803)

Diagnostic characters

Shell a little translucent (alive), spire moderately tall with its profile markedly turreted by a flat subsutural area on each whorl. Ornament: 12–16 costae per whorl and numerous spiral ridges, the former more prominent. Siphonal canal rather short, widely open. Second whorl of protoconch (often eroded) with three spiral keels and numerous ridges crossing the intervening grooves. Colour pattern absent.

Other characters

The spire is cyrtoconoid in profile; there are 6–7 whorls, each moderately tumid below the flat area and often nearly flat at the periphery. The costae cross the whorls of the spire but disappear towards the base of the last. They run prosoclinally as low ridges over the subsutural area but then increase in height and become orthocline or opisthocline, varying in breadth. The spiral ridges are delicate on the flat area, broader and higher elsewhere; there are about thirty on the last whorl, 9–11 on the penult. The aperture is a long oval, angulated adapically. The outer lip is thin, has neither varix nor internal teeth, and has a shallow anal sinus with its deepest part at the outer edge of the subsutural shelf. Biscuit-coloured. Up to about 12 mm high, 5 mm broad; last whorl occupies about two thirds of shell height but is variable, sometimes only half, sometimes nearly three quarters, aperture occupies about half.

The head is a transverse fold carrying the mouth ventrally. The tentacles have an eye about one third of their length below the tip. The siphon extends far beyond the tip of the canal in an active animal. The foot is shield-shaped, angulated anterolaterally, and bears a small oval operculum posteriorly. The flesh is grey or brownish, with opaque white flecks.

O. turricula ranges north from the British Isles to the Arctic and is probably circumpolar, being found as far south as Massachusetts and Washington on the east and west coasts of North America. Off the British Isles the animals are locally not uncommon on sandy bottoms 20–200 m deep. Their food is small polychaete worms. Eggs are laid in lens-shaped capsules fastened to shells or stones. Of the only three capsules containing embryos which have been found (Vestergaard, 1935) one had well-developed veliger larvae which were probably going to be liberated.

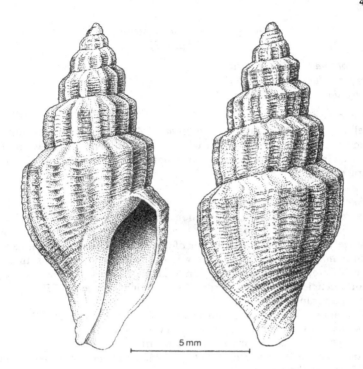

Fig. 177. *Oenopota turricula*.

Oenopota trevelliana (Turton, 1834)
(Fig. 178)

Pleurotoma trevelliana Turton, 1834
Lora trevelliana (Turton, 1834)
Mangelia trevelliana (Turton, 1834)

Diagnostic characters
Shell in general like that of *O. turricula* (p. 428) but smaller and with more numerous costae (20–27 on last whorl, 17–23 on penult); spire less clearly turreted in profile. If present, second whorl of protoconch shows 12–15 small spiral ridges.

Other characters
The relatively short spire and siphonal canal of this species give the shell a more oval profile than that of *O. turricula*, and this is smoothed by the relatively slight subsutural flattening of the whorls. The anal sinus is narrower than in *turricula*, lies closer to the wall of the last whorl, and its lip is often slightly flared. Biscuit-coloured. Up to about 12 mm high, 5 mm broad; last whorl occupies about two thirds of shell height (range 58–75%), aperture about half (range 40–56%).

The animal resembles *O. turricula* in all respects save that the eyes lie about half way along the tentacles. Flesh white or grey.

O. trevelliana is circumpolar in its distribution; it extends south to the British Isles in the eastern Atlantic, to Maine in the western Atlantic, to California in the eastern Pacific. It has been recorded from sandy bottoms in the northern parts of the North Sea, 25–30 m deep, locally frequent, feeding on small annelid worms (Pearce, 1966). Egg capsules are lens-shaped and have been found (Thorson, 1946; Pearce, 1966) on stones and shells from April to August. Veliger larvae escape from them.

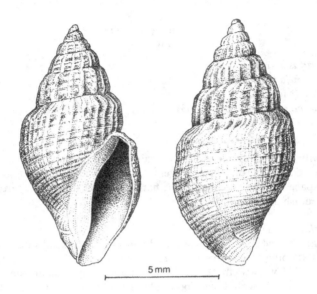

Fig. 178. *Oenopota trevelliana*.

Oenopota rufa (Montagu, 1803)
(Fig. 179)

Murex rufus Montagu, 1803
Mangelia rufa (Montagu, 1803)
Lora rufa (Montagu, 1803)
Pleurotoma rufa (Montagu, 1803)
Pleurotoma ulideana Thompson, 1845

Diagnostic characters

Shell with moderately tall spire, not turreted in profile; apex often a little bulbous. Ornament of costae and less obvious spiral ridges confined to inter-costal spaces. Siphonal canal short, hardly distinct from the aperture. Shell often reddish in whole or in part.

Other characters

There are 6–7 whorls which are tumid and usually lack subsutural and peripheral flattening. There are 13–15 rather narrow and flexuous costae on the youngest 3–4 whorls, those on the last whorl dying out on the base. The aperture is an elongated oval, pointed adapically. The outer lip is thin, with a small anal sinus near its adapical end. The columella and siphonal canal are paler than the rest of the shell; dead shells are often more uniformly pale, perhaps due to fading. Up to 15 mm high, 6 mm broad; last whorl occupies about two thirds of shell height (range 56–75%), aperture usually less than half.

The animal resembles that of *O. turricula* (p. 428), its flesh white with some grey; a bluish colour may be present on the siphon and the anterior end of the foot.

O. rufa has been found between Norway in the north and the Atlantic coast of France in the south. It has been collected from soft bottoms from about L.W.S.T. to 70 m deep on most coasts of the British Isles, but it is not common.

Oenopota violacea (Mighels & Adams, 1842)

Pleurotoma violacea Mighels & Adams, 1842
Oenopota bicarinata (Couthoy, 1838)

Diagnostic characters

Spire tall, turreted, apex blunt. Whorls with many fine and two prominent spiral ridges together with opisthocline growth lines. Aperture lanceolate.

Other characters

There are 5–6 whorls bearing many spiral ridges, of which two, one at the margin of the subsutural shoulder and one at the periphery, enlarge to form rounded keels. The finer spirals become coarser on the base of the last whorl.

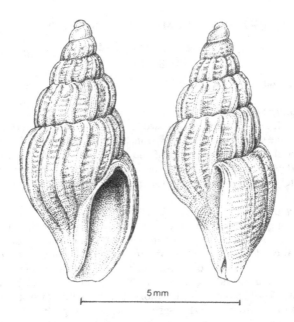

5 mm

Fig. 179. *Oenopota rufa.*

Numerous fine costellae (or thickened growth lines), opisthocline in direction, cross the whorls. The aperture is broadly connected to a short, widely open siphonal canal. The outer lip has a rather wide and shallow anal sinus. Protoconch smooth. Reddish white. Up to 11 mm high, 5 mm broad; last whorl occupies about 70% of shell height, aperture about 50%.

This is strictly a circumpolar species, but it extends south in deep water at least as far as the British Isles, though recorded there only once, off the west of Ireland at a depth of 990 m.

Genus TYPHLOMANGELIA Sars, 1878

Typhlomangelia nivalis (Lovén, 1846)
(Fig. 180)

Pleurotoma nivalis Lovén, 1846

Diagnostic characters
Shell rather tall and slender, whorls swollen and angulated at the periphery.
Ornament of fine spiral cords (barely visible to the naked eye) and opisthocline
costae on basal half of each whorl of the spire and peripheral region only
of last whorl. Outer lip with wide and deep anal sinus, peripheral bulge,
and moderately long siphonal canal. Operculum present on foot.

Other characters
The shell is not glossy, and the spire is very slightly cyrtoconoid (apical angle
of large shells about 20°). The protoconch is a little swollen, smooth or nearly
so. The spiral cords are numerous, about fifty on the last whorl and twenty
on the penult. The costae tend to disappear on the last two whorls, especially
towards the aperture, and are very variable in number, but 12 may occur
on the penult, 15–19 on the previous whorl. There is neither varix nor internal
thickening on the outer lip. White-grey, sometimes slightly yellow near the
apex and outer lip. Up to 17 mm high, 5 mm broad; last whorl occupies about
half shell height, aperture about 40%.

The head is a transverse fold with the mouth under it; it also bears a
short tentacle on each side, but the animals are blind and without eyes. Males
have a sickle-shaped penis on the right of the head. The foot is large, both
long and broad. The body is white or cream.

T. nivalis is widespread throughout the north-east Atlantic and extends
into the Mediterranean and along the arctic coast of Asia. The animals have
been found only east of Shetland in the area of the British Isles. They live
on soft bottoms, 40–400 m deep in the northern part of their range, 600–3000 m
in the southern. Their mode of life is probably comparable to that of other
turrid species.

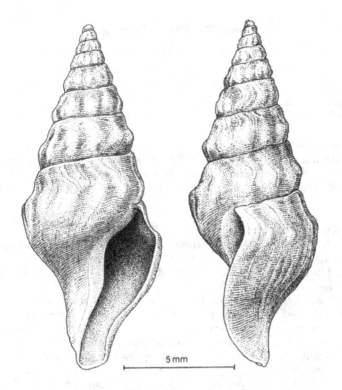

Fig. 180. *Typhlomangelia nivalis*.

Genus THESBIA Jeffreys, 1867

Thesbia nana (Lovén, 1846)
(Fig. 181)

Tritonium nanum Lovén, 1846
Columbella nana (Lovén, 1846)
Mangelia nana (Lovén, 1846)

Diagnostic characters
Shell like a small *Colus* species (p. 386–390), with blunt apex, tumid whorls and ornament of many small spiral ridges and grooves. Aperture elongated, siphonal canal short and wide; anal sinus at origin of outer lip. No operculum on foot.

Other characters
The shell is thin and semitransparent with 5–6 whorls. The spiral ridges are strap-shaped and number about 30 on the last whorl, 10–12 on the others. White. Up to 6 mm high, 2.5 mm broad; last whorl occupies about 70% of shell height, aperture about half.

The head, a transverse fold, bears short and slender tentacles, each with a basal eye, and has the mouth on its underside. The siphon is short, the foot narrow, straight-edged anteriorly, pointed behind. The flesh is white.

T. nana has only a limited distribution between the northern North Sea on the south, the Norwegian coast on the east, Spitzbergen in the north and Rockall in the west. British records are confined to the area between the Northern Isles and Rockall. The animals live on sandy bottoms 75–1000 m deep, are not common, and their way of life is not known, though presumably not essentially different from that of other turrids.

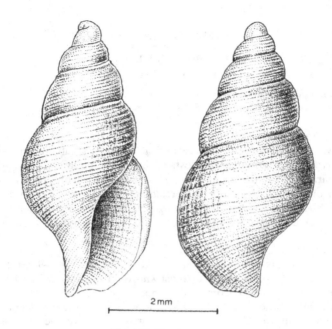

Fig. 181. *Thesbia nana*.

Genus MANGELIA Risso, 1826

Mangelia nebula (Montagu, 1803)
(Fig. 182)

Murex nebula Montagu, 1803
Pleurotoma nebula (Montagu, 1803)

Diagnostic characters

Shell with dull and rough surface and a moderately tall, sharply pointed and narrow spire. Ornament of costae (nine or ten on last whorl) and many weak spiral lines (25–30 bigger ones on last whorl) which often appear slightly beaded or scaly. Third whorl tuberculated. Sutures wavy, with a slightly thicker spiral ridge along their lower edge. Neither teeth nor thickening within outer lip. Operculum absent from foot.

Other characters

The shell, alive, is not quite opaque, and has commonly 8–9 gently tumid whorls. The costae number 7–9 on most whorls, arise a little below the suture, are prosocline there but become orthocline; on the last whorl they are flexuous and fade on its base. The first and second whorls are smooth, the third has a reticulate pattern. The aperture is a narrow oval extended into a short, rather wide, open siphonal canal. The outer lip is thin and not inturned, its adapical half with a broad anal sinus; there may or may not be a costa close to it. White, yellow brown, or nearly orange, the costae paler. Up to 14 mm high, about 5 mm broad; last whorl (including canal) occupies about 60% of shell height, aperture about 45%.

The head is a transverse shelf, the mouth on its underside, which carries the cephalic tentacles laterally, each with an eye two thirds of its length above the base. The foot is shield-shaped, broad anteriorly. The animal is white with yellow spots.

M. nebula occurs from the Mediterranean to Norway. The animals have been found all round the British Isles except in the eastern Channel and the southern North Sea. They live on sandy bottoms 10–30 m deep and may be locally common, more so in the south than in the north. They probably feed on small polychaetes.

Breeding takes place in spring and early summer. The egg capsules are lens-shaped (Lebour, 1936). Veliger larvae emerge which have at first a 2-lobed velum marked with orange spots; later the velum is 4-lobed and its margins become red-purple. Characteristic of this (and of the larvae of other *Mangelia* species) is the way in which the velar lobes are folded over the shell as the animal swims (Thiriot-Quiévreux, 1969; Richter & Thorson, 1975).

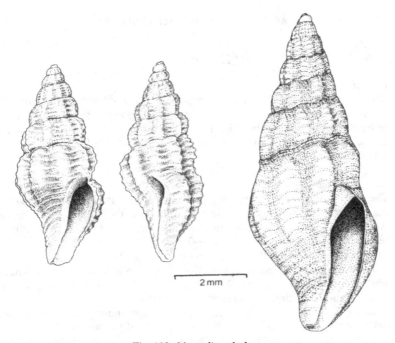

Fig. 182. *Mangelia nebula*.

Mangelia brachystoma (Philippi, 1844)
(Fig. 183)

Pleurotoma brachystoma Philippi, 1844

Diagnostic characters

Shell much like that of a small *M. nebula* (p. 438) but with more prominent spiral ridges and growth lines which interact to give the spiral ridges a scaly or beaded appearance; the ridge below the suture is thicker than the others. Commonly with seven or eight whorls, the last with 7–9 costae and 16–19 major spiral ridges.

Other characters

The apical angle of shells belonging to this species is usually rather greater than in *nebula*, the whorls are more tumid, and the sutures more incised. As in that species the third whorl is tuberculated. Pale cream to brown, reddish brown, or tawny. Up to 6–7 mm high, 2–3 mm broad; last whorl occupies 55–60% of shell height, aperture about 40%.

The animal has the same organization as in the previous species. It is nearly white, with opaque white marks, especially on the back end of the foot and on the siphon, which may also show a brown tint.

M. brachystoma lives on bottoms of sand and muddy sand 4–60 m deep between the Mediterranean and Norway. It is locally common off most shores of the British Isles, except those facing the southern North Sea. It almost certainly feeds on small polychaetes and reproduces in the same manner as *M. nebula*, a species to which it is very close.

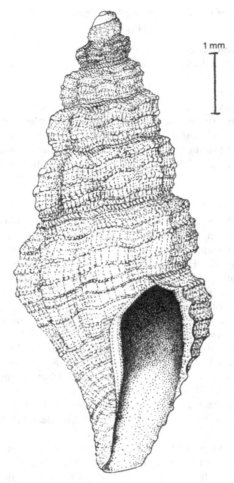

1 mm

Fig. 183. *Mangelia brachystoma*.

Mangelia powisiana (Dautzenberg, 1887)
(Fig. 184)

Bela powisiana Dautzenberg, 1887
Pleurotoma laevigata Philippi, 1836

Diagnostic characters

Shell slightly glossy, with moderately high and rather narrow and straight-sided spire; sutures often nearly straight, bounded below by a broad, thickened, spiral band. Costae often absent near outer lip, which is thin, not inturned, has only a shallow anal sinus, and neither varix nor internal thickening. Third whorl tuberculated. Dark spiral bands usually visible. No operculum on foot.

Other characters

There are 9–10 whorls, a little tumid. The costae number 9–11 on each whorl except the last, where they die out towards the aperture and the base. The surface also bears small spiral ridges, about 30 on the last whorl, half that number on the penult. They may be affected by growth lines to give pimple-like elevations, but not scales as in the two previous species. The aperture is long and narrow, the siphonal canal short and wide. The general colour of the shell is chocolate brown, sometimes with a purplish cast, and usually darker towards the apex. The costae and the swollen subsutural band are pale. Spiral bands of darker colour usually lie (a) below the subsutural thickening; (b) at the periphery of the last whorl and above the suture of those in the spire; (c) at the base of the last whorl. The throat, canal and columella are dark. Up to 15 mm high, 6 mm broad; last whorl occupies 60–65% of shell height, aperture 40–45%.

The animal has the same structure as in the previous species except that the eyes lie close to the tips of the tentacles. The flesh is white with whiter markings.

M. powisiana ranges from the Bay of Biscay to Norway. There are few records from British localities and apparently none from Irish: off west Scotland, the Isles of Scilly, Plymouth and the Channel Islands. The animals have been collected (rarely) at L.W.S.T. and (more commonly) on sandy bottoms 5–100 m deep.

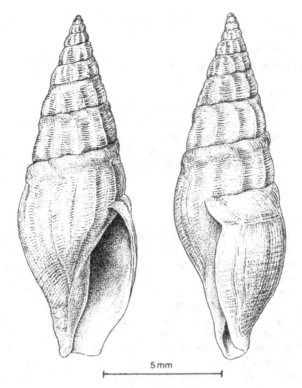

5 mm

Fig. 184. *Mangelia powisiana*.

Mangelia attenuata (Montagu, 1803)
(Fig. 185)

Murex attenuatus Montagu, 1803
Pleurotoma attenuata (Montagu, 1803)

Diagnostic characters
Shell glossy, tall, rather slender, the base elongated, with open siphonal canal. Sutures wavy. Costae usually extend to base of last whorl; they are narrow and flexuous, the spaces between filled with small, strap-shaped spiral ridges. Outer lip commonly with varix, a thin edge often inturned, and sometimes a slight internal thickening under the varix. Third whorl tuberculated. Shell often with many narrow spiral brown lines, sometimes fewer and broader ones. Operculum absent.

Other characters
There are up to nine whorls, tumid, but often rather flat at the periphery. The costae are prosocline near the suture but rapidly become a little opisthocline; they have steep sides. The last whorl has, except in young shells, a varix, plus 8–9 other costae; whorls in the spire have 8–9 in all. The aperture is long and narrow, pointed above, merging with the canal below. The outer lip has a rather narrow and shallow sinus near its adapical end. Yellowish brown, with darker brown spirals: these may be numerous and narrow (the common pattern) or there may be only a broad subsutural one together with a subperipheral one on the last whorl. Up to 13 mm high, 5 mm broad; last whorl occupies nearly two thirds of shell height, aperture nearly half.

The animal differs from other *Mangelia* species only in having the eyes near the base of the tentacles. It is white with opaque white spots.

M. attenuata has been recorded from the Mediterranean to northern Norway, except in the southern half of the North Sea. The animals occur sporadically on sandy or clayey bottoms 5–150 m deep off most parts of the British Isles. Little or nothing is known for sure about their mode of life.

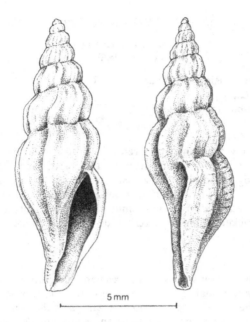

Fig. 185. *Mangelia attenuata*.

Genus CYTHARELLA Monterosato, 1875

Cytharella smithi (Forbes, 1844)
(Fig. 186)

Mangelia costulata Risso, 1826 var. *smithi* Forbes, 1844
Mangelia striolata Risso, 1826
Pleurotoma striolata (Risso, 1826)

Diagnostic characters

Spire moderately tall, slightly turreted in profile and gently cyrtoconoid. Sutures wavy. Costae high and narrow, the intervening spaces filled with small spiral ridges and grooves, which are almost square in section, and all alike. Aperture long, narrow, hardly distinct from the canal; outer lip with thin, inturned edge, narrow anal sinus close to wall of last whorl, varix, and internal thickening under varix. Third whorl tuberculated. No operculum on foot.

Other characters

The shell is slightly translucent when fresh and has eight tumid whorls which usually curve rather sharply away from the suture and may be rather flattened at their periphery. The costae cross the whorls of the spire and run to the base of the last whorl; there are 7–10 on most whorls. The appearance of the spiral ridges is characteristic of *Cytharella* species and separates them clearly from species of *Mangelia*. There are about 80 on the last whorl and they are a little beaded where crossed by growth lines. The anal sinus is semicircular and its edge slightly flared. The siphonal canal is wide and straight. The apex is yellow or brown, the rest of the shell yellowish, brown or whitish, with some darker spiral bands. There may be about six narrow ones on the last whorl, or only one subsutural and two peripheral ones. The base is nearly always pale. Up to 15 mm high, 5 mm broad; last whorl occupies 60–65% of shell height, aperture 40–45%.

The animal is like *Mangelia* species, the eyes close to the tips of the tentacles, the flesh white with opaque white markings and with some pink on the siphon.

C. smithi ranges from the Mediterranean to the British Isles where it has been found off western coasts as far north as the Hebrides on soft bottoms 20–50 m deep. They are common locally.

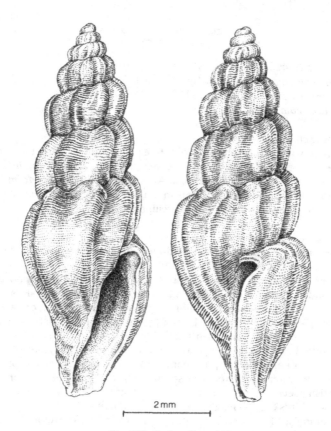

Fig. 186. *Cytharella smithi.*

Cytharella coarctata (Forbes, 1840)
(Fig. 187)

Pleurotoma coarctata Forbes, 1840
Mangelia coarctata (Forbes, 1840)
Pleurotoma costata (Donovan, 1803)
Mangelia costata (Donovan, 1803)

Diagnostic characters
Shell squat, with a clearly cyrtoconoid spire and a blunt tip. Ornament: costae, and the same type of spiral ridge (square in section) as in *C. smith* (p. 446), but here finer and more numerous. Anal sinus a little away from wall of last whorl, more pronounced than in *smithi*, and with internal thickening of outer lip better developed. No operculum on foot.

Other characters
The shell has 7–8 whorls, tumid, and meeting at deep and sinuous sutures. There are 7–8 costae on the last whorl, where they reach the base, 8–9 on the others, which they cross. Spiral ridges number about 100 on the last whorl, and, in some shells, occasional ridges may be larger than others, especially on the more adapical whorls. The shells exhibit one of two colour patterns: (*i*) spire and adapical half of the last whorl chocolate brown, base cream or tawny, sometimes with spiral chestnut band(s) on it, a brown blotch on the middle of the outer lip, a second in the throat; (*ii*) the whole shell tawny with orange-brown spiral bands, the most obvious at the periphery; they may be interrupted to form streaks. Up to 11 mm high, 5 mm broad, but British specimens are commonly about half that size; last whorl occupies about two thirds of shell height, aperture about half.

The animal is like *C. smithi* in colour and form, with the eyes about half-way up the tentacles.

C. coarctata ranges from the Mediterranean to Norway. It has been recorded as occurring in many localities in the British Isles, mainly in the south and west, dredged on sandy bottoms from just below L.W.S.T. to 250 m deep. Though not known in detail its way of life is presumably like that of related species.

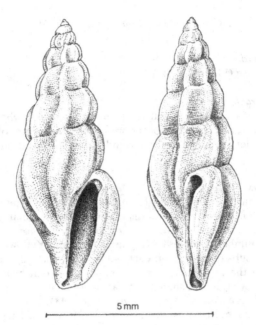

Fig. 187. *Cytharella coarctata*.

Cytharella rugulosa (Philippi, 1844)
(Fig. 188)

Pleurotoma rugulosa Philippi, 1844
Mangelia rugulosa (Philippi, 1844)

Diagnostic characters
Shell like that of *C. coarctata*, but relatively broader; 6–7 whorls with eight costae on the last, nine on those of the spire; spiral ridges are often regularly larger and smaller, becoming broader and flatter near and on the siphonal canal.

Other characters
The costae are narrow subsuturally but broaden towards the periphery. There are 7 thick spiral ridges on the walls of the siphonal canal and 11 major ones on the rest of the last whorl; over the penult whorl run 4–5 major ridges and the number becomes less higher in the spire. Yellowish with darker spiral bands: a subsutural one on each whorl, a peripheral and a basal one additionally on the last whorl. Up to 6 mm high, 3 mm broad; last whorl occupies about two thirds of shell height, aperture about half.

The animal is presumably like *C. coarctata* (p. 448). The snails have been found off the Channel Islands, the extreme south west of England, and the west of Ireland. They are rare and there are no recent records of their occurrence. They live on soft bottoms 10–75 m deep. The life history is known to include a free veliger larval stage (Richter & Thorson, 1975) but nothing else is known about their way of life.

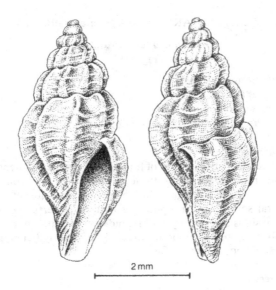

2 mm

Fig. 188. *Cytharella rugulosa*.

Genus COMARMONDIA Monterosato, 1884

Comarmondia gracilis (Montagu, 1803)
(Fig. 189)

Murex gracilis Montagu, 1803
Defrancia gracilis (Montagu, 1803)
Mangelia gracilis (Montagu, 1803)
Philbertia gracilis (Montagu, 1803)

Diagnostic characters
Spire tall and slender, turreted in profile; siphonal canal prominent. Ornament of costae and spiral ridges. Subsutural area on each whorl marked, nearly flat, with only fine spiral ridges, the costae arising below it. Third whorl with peripheral spiral keel and punctate pattern. Aperture oval, outer lip with deep anal sinus and usually a varix; older shells with internal thickening under varix bearing teeth. Siphonal canal open, usually bending to right of apertural axis. No operculum on foot.

Other characters
There are 10–11 tumid whorls meeting at deep sutures. The costae are broad and slightly opisthocline, more obviously so on the last whorl on which they die out at the periphery; there are 12–15 on each of the last two whorls. The spiral ridges are strap-like, up to 50 on the last whorl, 15–20 on the penult. In the grooves between ridges, and prominent on the subsutural area, minute papillae lie in short, irregular axial rows. The outer lip has a thin edge. The varix and teeth are absent in young shells. Buff, usually a pale band near the periphery of the last whorl with a darker one below it; a darker subsutural one on each whorl; costae and varix paler. Up to 25 mm high, 10 mm broad; last whorl occupies 60% of shell height, aperture 40–45%.

The head is a transverse shelf with the mouth underneath, giving rise to tentacles each with an eye near the tip. The siphon is short, hardly extending out of the canal as the animal moves. Males have a penis on the right. The foot is rounded anteriorly, with short lateral projections, and is pointed behind. The animal is white with red-brown markings.

C. gracilis is found between the Mediterranean, the Azores and the British Isles, where it occurs in the western parts of the Channel, off western coasts, off the Northern Isles and in the Moray Firth. The animals live on soft or gravelly bottoms 7–150 m deep, probably eating polychaetes.

The animals breed in spring (Lebour, 1933d, 1934) laying lens-shaped capsules with a rough and radially ridged surface, about 3.5 mm in diameter, and containing 40–80 eggs. All hatch as veliger larvae with a bilobed velum, later changing to one with four long and narrow arms each with a reddish margin and orange spots (Thiriot-Quiévreux, 1969; Richter & Thorson, 1975).

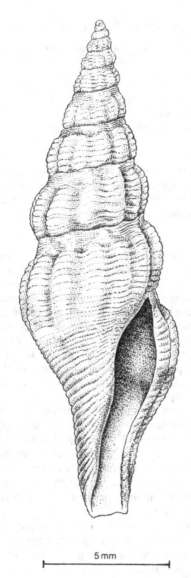

Fig. 189. *Comarmondia gracilis*.

Genus RAPHITOMA Ballardi, 1848

Raphitoma linearis (Montagu, 1803)
(Fig. 190)

Murex linearis Montagu, 1803
Philbertia linearis (Montagu, 1803)
Defrancia linearis (Montagu, 1803)
Mangelia linearis (Montagu, 1803)

Diagnostic characters
Protoconch with criss-cross linear pattern on basal parts of whorls. Whorls tumid, with costae and spiral ridges, sutures deep; 10–11 costae on penult whorl with 4–7 spiral ridges. Outer lip with internal teeth. Brown streaks on spiral ridges. Operculum absent.

Other characters
The shell has 7–9 whorls the whole surface of which is minutely papillated. The costae, which are more conspicuous than the spiral ridges, are broad, often a little prosocline, and cross the whorls of the spire though the 11–12 on the last whorl disappear about the base of the siphonal canal. The spiral ridges are thin and cross the costae, broadening as they do so. The aperture is a rather long oval, narrow adapically. The outer lip has a slight anal sinus near its origin and its edge is made wavy by the ends of the spirals. The canal is short. White with brown lines along the spiral ridges, sometimes also on the subsutural parts of whorls and the base of the last; protoconch often purplish brown. Up to 12 mm high, 7 mm broad; last whorl occupies 65–70% of shell height, aperture 45–50%.

The head forms a transverse ridge with the opening of the proboscis sac under it, and carries the tentacles each with an eye half way along it, its distal half narrower than the basal. The siphon of an active animal projects far from the siphonal canal. The penis in males arises behind the right tentacle. The foot has a deep embayment anteriorly with recurving anterolateral points, and is narrow and pointed behind. Body yellowish with white markings.

R. linearis has been found in most parts of the British Isles living on sandy or stony bottoms 10–200 m deep, and occasionally encountered at L.W.S.T. in southern regions. Its broader range is from the Mediterranean to northern Norway. The animals probably eat small polychaetes. In spring and summer clear, hemispherical capsules about 2 mm across are laid, each with 200–500 eggs which hatch as veliger larvae.

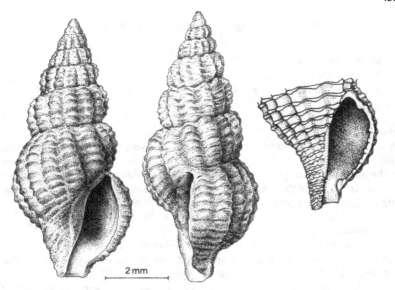

Fig. 190. *Raphitoma linearis*.

Raphitoma purpurea (Montagu, 1803)
(Fig. 191)

Murex purpureus Montagu, 1803
Philbertia purpurea (Montagu, 1803)
Defrancia purpurea (Montagu, 1803)
Mangelia purpurea (Montagu, 1803)

Diagnostic characters
Shell tall and rather narrow, whorls tumid and sutures deep; penult whorl
with about nineteen costae and 8–12 spiral ridges with prominent tubercles
at the intersections; protoconch with reticulate pattern of lines; aperture
narrow, outer lip with internal teeth; siphonal canal short. Usually purplish
with irregular pale blotches, outer lip pale.

Other characters
The shell apex is narrow. There may be twelve whorls though 8–9 are commoner. Costae and spiral ridges are numerous: about twenty costae on the
last whorl which fade a little below the periphery, and 25 spiral ridges, both
sets about equal in size. The aperture is a somewhat elongated oval, narrow
adapically, and slightly separated from the siphonal canal, which is widely
open. Though usually purple the shell may be brownish or reddish. Up to
20 mm high, 8 mm broad; last whorl occupies about 60% of shell height,
aperture 45%.

   The external features of the animal are like those of other *Raphitoma*
species (p. 454–460), the body white with a purplish cast. The species ranges
from the Mediterranean to northern Norway, the animals living on sandy
or gravelly bottoms 10–100 m deep. They have been recorded mainly off
the southern and western coasts of the British Isles, in the most southern
parts occasionally being found in crevices at L.W.S.T. They are predators,
their food probably polychaetes (Franc, 1952b).

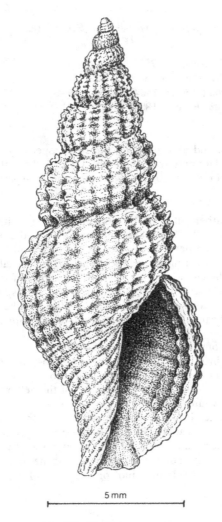

5 mm

Fig. 191. *Raphitoma purpurea*.

Raphitoma asperrima (Brown, 1827)
(Fig. 192)

Fusus asperrimus Brown, 1827
Philbertia asperrima (Brown, 1827)
Philbertia reticulata (Renier, 1804)
Defrancia reticulata (Renier, 1804)

Diagnostic characters

Spire sharply pointed and distinctly turreted, especially adapically; proto-conch with reticulate pattern; penult whorl with 6–7 spiral ridges and 16–17 costae, usually prominent tubercles where they cross. Aperture small.

Other characters

The shell has a rather tall spire, its most apical whorls with a flat subsutural area lying at right angles to the axis, the lower ones without this. The sutures are deep. The costae have steep sides, run orthoclinally, those on the last whorl (17–19) disappearing at the base of the siphonal canal. There are 16–19 spiral ridges on the last whorl. On the subsutural part of each whorl a series of C-shaped lines marks former positions of the anal sinus. The outer lip has a rather small anal sinus and there are 9–10 ridge-like teeth on its inner surface. Siphonal canal not very long. Yellow-brown with some darker areas arranged without apparent pattern. Up to 12 mm high, 7 mm broad; last whorl occupies 65–70% of shell height, aperture 40–45%.

The animal is in general like *R. linearis* (p. 454) but the ridge bearing the tentacles shows a division in the mid-line so that the tentacles appear longer. The body is pale yellow-white, most of the yellow on the head and foot, with opaque white points.

The species ranges from the Mediterranean to northern Norway but has not been found in the southern North Sea and eastern Channel. It is apparently uncommon and recent finds of living animals round the British Isles are rare. Neither the way of life nor the reproduction of the animals is known, but they are presumably predators, and, on the basis of the protoconch, have a free larval stage.

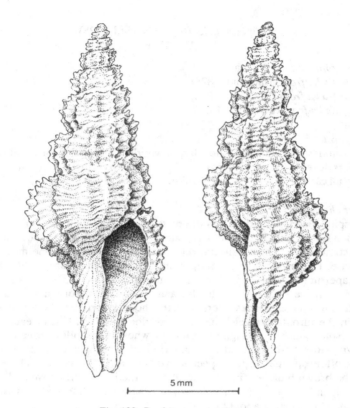

5 mm

Fig. 192. *Raphitoma asperrima*.

Raphitoma leufroyi (Michaud, 1828)
(Fig. 193)

Pleurotoma leufroyi Michaud, 1828
Philbertia leufroyi (Michaud, 1828)
Defrancia leufroyi (Michaud, 1828)
Mangelia leufroyi (Michaud, 1828)

Diagnostic characters

Spire moderately tall, last whorl large; whorls tumid with prominent costae and less developed spiral ridges (14–17 costae and 6–7 ridges on penult whorl); siphonal canal short; protoconch white.

Other characters

The spire is cyrtoconoid and a little turreted in its most apical region. There are 8–9 whorls with deep and wavy sutures. The costae are broad, undulate in section, low in the subsutural region; on the last whorl some may reach the base. The spiral ridges are narrow and often alternately larger and smaller. The aperture is rather broad and is confluent with the siphonal canal. The outer lip has a shallow semicircular anal sinus near its origin. It is usually a little inturned and lacks internal teeth, though there may be a thickening within the throat. Brownish, usually paler on the basal half of each whorl in the spire and at the periphery of the last whorl; spiral ridges often chestnut brown, those at the edge of paler areas often especially dark, those within them often opaque white, especially where they cross costae. There may also be brown blotches, most noticeably on the last whorl above and below the pale area. Up to 15 mm high, 6 mm broad; last whorl occupies 70% of shell height, aperture about half.

The body of the animal is like that of other *Raphitoma* species (p. 454–460), yellowish in colour. The animals have been widely and not uncommonly found round the British Isles except in the eastern Channel and southern North Sea. They live on sandy or stony bottoms from L.W.S.T. to depths of 150 m. There is a veliger stage in the life cycle.

Genus CENODAGREUTES Smith, 1967

Cenodagreutes aethus Smith, 1967
Cenodagreutes coccyginus Smith, 1967

These two species, which have been found only once, in the Firth of Clyde, and are all but indistinguishable, so far as the shell is concerned, from *Raphitoma leufroyi*, were placed by Smith (1967) in a separate genus because they lacked the radula and poison gland which are present in *R. leufroyi*.

C. aethus was distinguished by having the shell surface between costae more pustular than in *R. leufroyi*, by having a broader siphonal canal and a V-shaped rather than a semicircular anal sinus.

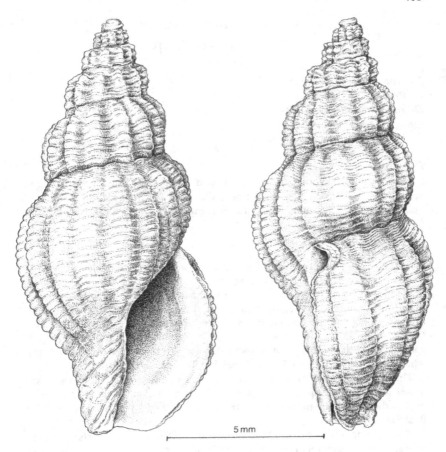

Fig. 193. *Raphitoma leufroyi*. The same drawings also illustrate the two species of the genus *Cenodagreutes*, since these differ significantly only in their internal anatomy, lacking radula and poison gland.

C. coccyginus could be separated by its slightly more prominent ornament, less pustular surface, less tumid whorls, thicker outer lip, narrower anal sinus, and the frequently purplish colour of the body. All these points are well-nigh unrecognizable unless a series of shells of *R. leufroyi* is available for comparative purposes.

The status of this genus and the two species assigned to it must be regarded as still doubtful since the degree of variation normally found in *Raphitoma leufroyi* is not known.

Genus TERETIA Norman, 1888

Teretia teres (Reeve, 1844)
(Fig. 194)

Pleurotoma teres Reeve, 1844
Philbertia teres (Reeve, 1844)
Defrancia teres (Reeve, 1844)
Mangelia teres (Reeve, 1844)
Philbertia anceps (Eichwald, 1830)

Diagnostic characters

Spire rather tall, whorls ornamented with spiral ridges and grooves only, apart from opisthocline growth lines. Outer lip with deep anal sinus and marked peripheral bulge; aperture merging with canal. No operculum on foot.

Other characters

The shell has a pointed apex formed from a protoconch with a reticulate pattern. There are up to seven postlarval whorls which are tumid and meet at deep sutures. The spiral ridges are broader than the grooves between them, are undulate in section, tend to be alternately larger and smaller and are absent from the band below the suture running back from the anal sinus; there are 20–25 on the last whorl and 5–6 large ones on the penult. White, sometimes with reddish brown spots. Up to 10 mm high, 4 mm broad; last whorl occupies about two thirds of shell height, aperture about 45%.

The tentacles arise from a ridge with the opening of the proboscis sac under it; they are long, blunt-tipped and have each a basal eye. The anterior end of the foot has a median embayment and anterolateral projections. The flesh is white.

T. teres ranges between the Mediterranean and the southern parts of Scandinavia. It has been recorded from nearly all parts of the British and Irish coasts, but it is not common. The animals live on sandy bottoms 30–900 m deep. They have no radula (Bouchet & Warén, 1980) and their food and method of feeding are not known. The life history includes a free veliger stage which may grow to have six whorls in the shell, suggesting a prolonged larval life.

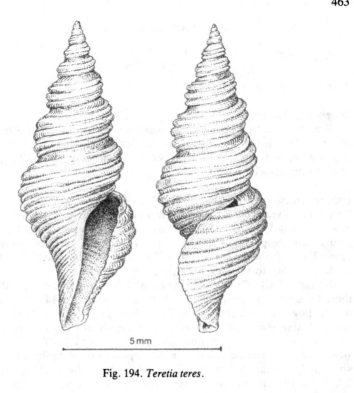

5 mm

Fig. 194. *Teretia teres*.

Genus TARANIS Jeffreys, 1870

Taranis moerchi (Malm, 1861)
(Fig. 195A)

Trophon moerchii Malm, 1861

Diagnostic characters
Shell with blunt tip and markedly keeled whorls often with a periostracal covering. Ornament of many fine costae and spiral ridges, the latter confined to the basal half of each whorl. Protoconch white, marked with spiral rows of diamond-shaped points.

Other characters
This shell is variable in proportions and ornament but the spire is usually moderately tall; there are 5–6 whorls each with a peripheral keel and the sutures are deep and slightly incised. The costae are numerous, running prosoclinally from the suture to the keel and then opisthoclinally below the keel. They run to the lower suture in the spire but fade towards the base of the last whorl. The spiral ridges are about the same size as the costae; there are about three on each whorl of the spire and 10–14 on the last whorl, where they become fainter towards the base and disappear on the siphonal canal. Costae and spiral ridges produce a rhomboidal reticulation where they cross. The aperture is approximately oval, pointed adapically, merging below with the siphonal canal which is broadly open and bent to the left. The periostracum is orange yellow, the underlying shell white. Up to 4 mm high, 2–2.5 mm broad; last whorl occupies 60–75% of shell height, aperture 40–50%.

The animal has short tentacles, each with a basal eye. The siphon is short, the foot without an operculum.

T. moerchi is a widely ranging species in the Mediterranean and North Atlantic (southern Scandinavia to Florida), but its claim to inclusion in the British fauna rests on a single capture east of Shetland. The animals have been taken on soft bottoms from 80–2000 m deep. Their way of life is unknown though they appear to have direct development (Bouchet, 1977).

A second species of this genus, *T. borealis* Bouchet & Warén, 1980 (Fig. 195B), may prove to occur locally though so far only found off western Norway and in the Skagerrak. This is distinguished by the smaller number of spiral ridges, lesser angulation of the whorls, shallower sutures and by the brown colour of the protoconch.

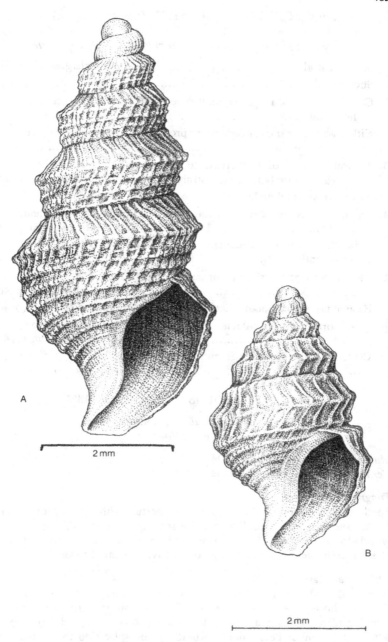

Fig. 195. A, *Taranis moerchi*; B, *Taranis borealis*.

Family CERITHIOPSIDAE H. & A. Adams, 1853

Key to British genera and species of Cerithiopsidae

1. Shell sinistral .... *Laeocochlis granosa* (p. 478)
   Shell dextral **2**

2. Costae and spiral ridges equally developed, surface tuberculated;
   shell brown **3**
   Either costae or spiral ridges more prominent, surface not tubercu-
   lated; shell white **6**

3. 4 spiral ridges (3 or 4 tuberculated) on penult whorl; 15 costae on
   last whorl; spire tall, slender, straight .... *Cerithiopsis metaxa* (p. 470)
   3 spiral rows of tubercles on penult whorl; spire cyrtoconoid **4**

4. Tubercles well spaced with obvious rectangular areas between them;
   apex blunt .... *Cerithiopsis pulchella* (p. 472)
   Tubercles practically touching, only narrow grooves between them;
   apex pointed **5**

5. Base of last whorl with spiral ornament; shell rather stout ....
   .... *Cerithiopsis tubercularis* (p. 466)
   Base of last whorl smooth; shell slender .... *Cerithiopsis barleei* (p. 468)

6. Spiral ornament prominent, costae slight; basal canal narrow,
   oblique .... *Cerithiella metula* (p. 476)
   Costae prominent, spiral ornament slight; basal canal wide, straight
   .... *Eumetula costulata* (p. 474)

Genus CERITHIOPSIS Forbes & Hanley, 1851

Cerithiopsis tubercularis (Montagu, 1803)
(Fig. 196)

Murex tubercularis Montagu, 1803
Cerithiopsis clarki Forbes & Hanley, 1851

Diagnostic characters
Shell with tall, cyrtococoid spire and small aperture with basal siphonal notch.
Up to fourteen whorls each with three rows of squarish tubercles, the last
whorl also with 2–3 spiral ridges, sometimes nodose, below the rows of tubercles; tubercles near outer lip elongated axially. Chestnut brown.

Other characters
The shell is glossy and often loses the protoconch, which has about four
smooth whorls. The whorls are nearly flat-sided but the sutures lie in rather
deep furrows. There may be up to 24 axial rows of tubercles on the last
whorl, the number decreasing to 16–18 up the spire. Up to 6.5 mm high,
2.25 mm broad; last whorl occupies 35–40% of shell height, aperture 20–25%.

1 mm

Fig. 196. *Cerithiopsis tubercularis.*

The head has no snout, the mouth (the opening of an introvert) being a small opening under a transverse fold from which two tentacles arise close together, each long and slender, with a lateral basal eye. The mantle edge extends to the left to form a small siphon. Within the mantle cavity the genital duct (in both sexes) is an open groove, and males have no penis. The animals are nearly white but are grey on the head and the sides of the foot; opercular lobes yellow.

C. tubercularis is spread between the Black Sea, the Azores and Norway, but is confined to southern and western coasts of the British Isles. The animals are not uncommon and are mainly found at L.W.S.T. and sublittorally to about 100 m on sponges (*Halichondria panicea, Hymeniacidon perleve*), especially those growing at the base of tufts of red weeds (usually *Lomentaria, Corallina*). They eat the sponges and also any epiphytic or epizooic detritus (Fretter, 1951b; Fretter & Manly, 1977b).

Breeding occurs spring and summer (Lebour, 1933b, 1936) when males and females come together; the males emit spermatozeugmata which enter the mantle cavity of the female in the inhalant water stream. Egg capsules, each with about 200 pinkish eggs, are inserted into holes which have been rasped out of the tissues of the sponges on which the adults live. The eggs develop to veliger larvae, described by Lebour (1933b), Thiriot-Quiévreux (1969), Fretter & Pilkington (1970) and Richter & Thorson (1975). The veligers have a long larval life and do not settle until autumn.

Cerithiopsis barleei Jeffreys, 1867
(Fig. 197)

Diagnostic characters

This shell differs from that of *C. tubercularis* (p. 466) only in being thinner and more slender, with deeper sutures, in having the base below the prominent spiral keel of the last whorl smooth, and in being paler. Flesh yellowish.

Other characters

The shell has up to twelve whorls, meeting at deep sutures. Up to 7 mm high, 2 mm broad; last whorl occupies 35–40% of shell height, aperture 20–25%.

Animals of this species are associated with the sponge *Suberites domunculus* and are not found between tidemarks, but to depths of about 90 m. The species is more southern in distribution than the last and is restricted to the south-western parts of Britain, where it is not uncommon. Breeding in this species is much as in *C. tubercularis* (Lebour, 1933b; Thiriot-Quiévreux, 1969; Fretter & Pilkington, 1970; Richter & Thorson, 1975).

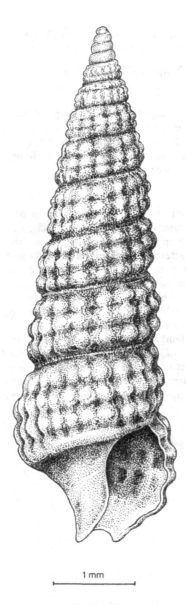

1 mm

Fig. 197. *Cerithiopsis barleei*.

Cerithiopsis metaxa (delle Chiaje, 1828)
(Fig. 198)

Murex metaxa delle Chiaje, 1828

Diagnostic characters
A tall, slender, nearly straight-sided shell of up to fifteen tumid whorls; sutures deep; five spiral ridges on last whorl, four on those of the spire, the most adapical spiral smooth or nodose, the others markedly tuberculate except the lowest on the last whorl which is smooth or nodose. Aperture small, with basal notch. Yellow or whitish.

Other characters
The shell is both taller and more slender than in other *Cerithiopsis* species (p. 466–472). It has usually lost the protoconch, but, if this is present, it has 4–5 whorls slightly more swollen than those of the rest of the shell, each showing a complex pattern of axial lines adapically and of axial lines crossing spiral lines abapically; this pattern is usually eroded and appears only as some pitting of the surface. Up to 8 mm high (protoconch present), 2 mm broad; last whorl occupies 25–30% of shell height, aperture only 10–15%

The animals presumably resemble *C. tubercularis* (p. 466) but are not properly known. They are sublittoral and probably live on sponges. The species ranges between Shetland and the Mediterranean, but the animals have been found only off Shetland, south-west England and the Channel Islands, and none, apparently, this century. It is doubtful whether any of the local finds was of more than empty shells. Veliger larvae occur in the life history and have been described by Thiriot-Quiévreux (1969) and by Richter & Thorson (1975).

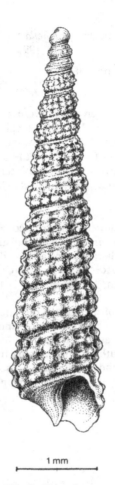

1 mm

Fig. 198. *Cerithiopsis metaxa*.

Cerithiopsis pulchella Jeffreys, 1858
(Fig. 199)

Cerithiopsis jeffreysi Watson, 1885

Diagnostic characters
Shell tall with up to fourteen tumid whorls, spire cyrtoconoid. There are three spiral ridges on each whorl of the spire, four on the last, the most basal ridge less tuberculated than the others. There are 20–22 costae per whorl and blunt tubercles form where these cross the spiral ridges; both ridges and costae are narrower than the intervening spaces. Aperture oval with basal notch but without an extended canal. Yellow-brown.

Other characters
The shell apex is rather blunt and the protoconch often lost so that the number of whorls is usually about ten. The shell is distinct from that of other species of *Cerithiopsis* in which the tubercles are the dominant element of the ornament, whereas here the spaces between tubercles are the feature which catches the eye. Up to 6 mm high, 1.5 mm broad; last whorl occupies 40% of shell height, aperture 20%.

The external features of animals of this species have not been described but presumably resemble those of others, and this is probably also true of the animals' mode of life as they have been found among sponges 50–500 m deep. The species is a southern one, ranging from the Mediterranean northwards to the Channel Islands, Scilly, and off some south-western British and Irish coasts. The animals are rare and have probably not been taken alive in these areas.

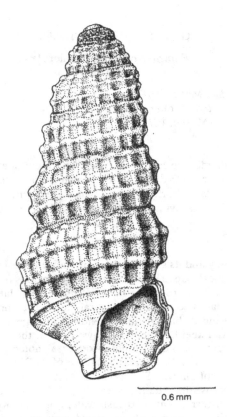

0.6 mm

Fig. 199. *Cerithiopsis pulchella*.

Genus EUMETULA Thiele, 1912

Eumetula costulata (Möller, 1842)
(Fig. 200)

Turritella costulata Möller, 1842
Cerithiopsis costulata (Möller, 1842)
Eumetula arctica (Mörch, 1875)

Diagnostic characters
Shell tall and slender, spire straight-sided with blunt apex. There are 12–14 slightly tumid whorls with deep sutures. Ornament of costae and spiral ridges; sixteen costae on last whorl, fifteen on penult, two spiral ridges more prominent on older whorls, with, in addition, two keels on base of last whorl. White.

Other characters
The shell is glossy and its sutures are commonly emphasized by a spiral cord along their adapical side. The costae are more obvious than the spiral ridges and interact with them to give a marked square reticulation of the surface, though on the last two whorls the spiral ridges (but not the keels at the base) are so reduced that this effect tends to disappear. On the last whorl the adapical basal keel lies along the lower end of the costae and is usually nodose. The aperture has a rather wide basal notch. Up to 7 mm high, 2–3 mm broad; last whorl occupies about 35% of shell height, aperture 20%.

The animal is not known. The species is Arctic in distribution and spreads from Norway across the Atlantic to the Bay of Fundy. In Europe it extends south to the Skagerrak but in British waters only empty shells have been found, off Shetland. There is also a record from off western Ireland. There are no recent records and it is likely that the animals have been retreating northwards since the end of the Ice Age.

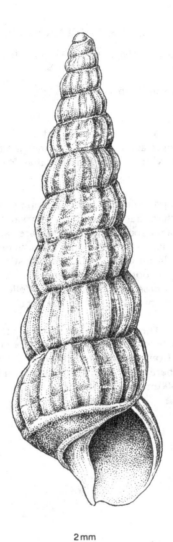

2 mm

Fig. 200. *Eumetula costulata*.

Genus CERITHIELLA Verrill, 1882

Cerithiella metula (Lovén, 1846)
(Fig. 201)

Cerithium metula Lovén, 1846

Diagnostic characters
Shell tall, slender, glossy, spire nearly flat-sided. Whorls (up to eighteen in all) with spiral ridges and costae, the former the more obvious. Aperture small, quadrangular, with oblique siphonal canal. Protoconch rather bulbous.

Other characters
The spire is slightly cyrtoconoid and the diameter of the protoconch is usually marginally greater than that of the first postlarval whorls. The spiral ridges (3–4 on each whorl of the spire, 4–5 on the last) are narrower than the intervening spaces and are marked by small tubercles where they meet the costae, of which there are 24 on each whorl. The most basal spiral ridge on the last whorl is usually smooth. The first whorl of the protoconch is smooth, the second has many crescentic axial ridges. Yellowish white or white. Up to 12 mm high, 4 mm broad; last whorl occupies 30–40% of shell height, aperture about 20%.

The flesh is white. There are two slender tentacles which appear curved and ringed, with an eye at the base of each. The foot is long and narrow.

C. metula occurs from the Mediterranean to Iceland, but British records are very few and there are no Irish ones. The animals have been found on soft bottoms 40–400 m deep; neither their way of life nor their reproduction is known.

2 mm

Fig. 201. *Cerithiella metula*.

Genus LAEOCOCHLIS Dunker & Metzger, 1874

Laeocochlis granosa (S. Wood, 1848)

Cerithium granosum S. Wood, 1848
Laeocochlis macandreae (H. Adams, 1858)

Diagnostic characters
This shell is easily recognized as it is sinistral, has a tall spire of 11–12 slightly tumid whorls ornamented with narrow spiral ridges. There are seven ridges on the last whorl, eight on each of those in the spire. The two apical whorls, which belong to the protoconch, are more tumid than the others and also bear small costae and so have a reticulated surface. The base of the last whorl is smooth. The aperture is small and is drawn out basally into a short siphonal canal twisted markedly to the right. White. Up to 20 mm high, 6 mm broad; last whorl occupies about 40% of shell height, aperture 25%.

L. granosa is strictly Arctic in distribution but extends south to latitudes close to those of the British Isles, occurring on sandy or muddy bottoms at considerable depths.

Family TRIPHORIDAE Gray, 1847

Key to British species of Triphora

1. Shell with 5 spiral rows of tubercles on last whorl near outer lip,
   lying above 2–3 spiral keels (Fig. 202c); flesh yellow ....
   .... *Triphora pallescens* (p. 482)

   Shell with 3 spiral rows of tubercles lying above 2 spiral keels which
   may be nodose, on last whorl near outer lip **2**

2. 4 spiral rows of tubercles on whorl above aperture; flesh reddish
   .... *Triphora erythrosoma* (p. 482)

   3 spiral rows of tubercles above aperture **3**

3. Tubercles on last whorl near outer lip elongated axially; flesh white;
   the common species .... *Triphora adversa* (p. 480)

   Tubercles there not elongated; flesh dark .... *Triphora similior* (p. 483)

Note on Triphora *species*

Until the work of Bouchet & Guillemot appeared (1978) all specimens of
Triphora found in the British Isles were assigned to the species *perversa*
Linné, 1758, which these workers believe to be strictly Mediterranean. The
accurate distribution of the four species now described is not known and
any specimens collected should therefore be checked for their specific identity.

Genus TRIPHORA Blainville, 1828

Triphora adversa (Montagu, 1803)
(Fig. 202A,B)

Murex adversus Montagu, 1803
Cerithium perversum (Linné, 1758)
Triphora perversa of authors

Diagnostic characters
Shell sinistral, superficially like that of *Cerithiopsis tubercularis* (p. 466). Only three tuberculated spiral ridges on last whorl near outer lip, with tubercles there axially elongated. Body of animal, including tentacles white.

Other characters
The shell is glossy and solid, with about fifteen whorls of which four belong to the protoconch. The spire is markedly cyrtoconoid with nearly flat-sided whorls and shallow sutures, the upper edge of which is marked by a spiral ridge. The last whorl has three tuberculated ridges plus 2–3 basal ones which are smooth or slightly nodose; in the spire the more basal whorls have three tuberculated ridges, but the number falls to two, then one, towards the apex. Tubercles arise where costae cross ridges. There are 20–23 costae on the last whorl and about two less on each successive whorl up the spire. The whorls of the protoconch have fine transverse lines which cross 1–2 spiral ones. Yellow-brown, the edge of the aperture dark. Up to 7 mm high, about 2 mm broad; last whorl occupies about one t' ird of the shell height, the aperture a fifth to a quarter.

The head has a short snout with the opening of the introvert on its underside. The tentacles are long and slender, each with a basal eye. A short siphon arises from the mantle edge on the right. Pallial organs are reversed in position because of the animal's sinistrality, with the gill on the right and anus and genital duct on the left. The male has no penis.

T. adversa occurs from Spain to Norway and penetrates the Baltic as far as Kiel. The animals live on sponges, especially *Hymeniacidon* and *Halichondria*, are often partly embedded in them, and feed off them (Fretter, 1951b); they may also be found under stones or on algae near L.W.S.T. and to about 100 m deep. This is the commonest species of the genus in the British Isles and possibly the only one in their northern parts. Absent from the southern North Sea.

Lebour (1933b) found veliger larvae at Plymouth July-October. They appear to have a long planktonic life since at metamorphosis the shell has four whorls.

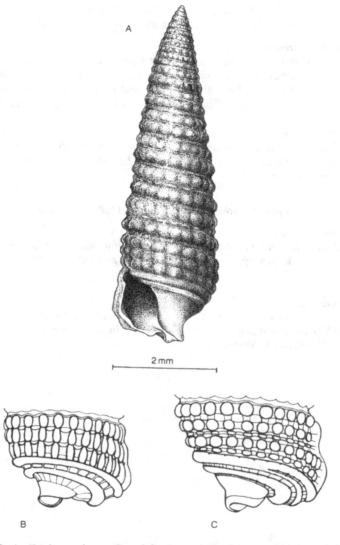

Fig. 202. A, *Triphora adversa*; B and C, views of the abapertural surface of the last whorl of (B) *T. adversa*, (C) *T. pallescens*.

Triphora pallescens Jeffreys, 1867
(Fig. 202C)

Diagnostic characters
Differs from the previous species in that there are five tuberculated spiral ridges on last whorl near the outer lip, in that at the same level the adapical untuberculated spiral ridge splits into two, and in that the tubercles are not markedly elongated axially. Body white, but tentacles have yellow marks.

Other characters
So far as is known this animal has been found only in the extreme south-western parts of England and in the Channel Islands. Abroad it ranges south to the north of Spain.

Triphora erythrosoma Bouchet & Guillemot, 1978

Diagnostic characters
Like *T. adversa* but with four tuberculated spiral ridges and one plain one on last whorl above aperture. Body with red and yellow marks, tentacles yellow.

Other characters
This is a southern species recorded in the western Mediterranean and the Bay of Biscay. It appears to occur occasionally in the extreme south-western parts of England and Wales.

Triphora similior Bouchet & Guillemot, 1978

Diagnostic characters
Very similar to *T. adversa* but the body has many black speckles, except on the tentacles.

Other characters
A southern species found in the western Mediterranean and the Bay of Biscay. The sole record from the British Isles so far is from western Ireland.

Family EPITONIIDAE Berry, 1910

Key to British genera and species of Epitoniidae

1. Ornament of well spaced costae only; spiral ridges absent though fine lines may occur; aperture circular or nearly so **2**

   Ornament as above but with 1 prominent basal spiral on last whorl; spiral brown band below suture; rare   *Cirsotrema commutatum* (p. 486)

2. Each costa with spur near adapical end (Fig. 203, *3*); no brown spiral bands on most shells .... *Epitonium trevelyanum* (p. 492)

   Costae without such spur, though they may enlarge adapically; with or without brown spiral bands **3**

3. 18–22 costae per whorl, erect, not bent up spiral (Fig. 203, *4*); no brown spiral bands, but brown on columella; shell not over 12 mm high .... *Epitonium clathratulum* (p. 494)

   Less than 18 costae per whorl, all bent up spiral (Fig. 203, *1&2*) brown spiral bands or streaks present; shell over 12 mm high **4**

4. Costae flat, closely applied to whorl surface, their adapical ends running along sutural area to meet next costa (Fig. 203, *2*) ....
   .... *Epitonium turtonis* (p. 490)

   Costae elevated, their adapical ends not meeting (Fig. 203, *1*) ....
   .... *Epitonium clathrus* (p. 488)

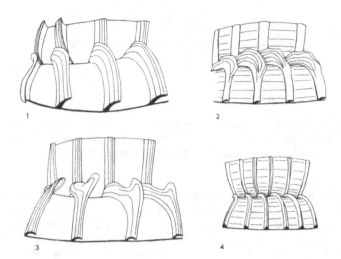

Fig. 203. Details of costal structure in species of *Epitonium*. Each figure shows the adapical ends of a few costae and their relations to the suture and to neighbouring costae. At their abapical ends the costae are seen in section. *1, Epitonium clathrus; 2, E. turtonis; 3, E. trevelyanum; 4, E. clathratulum.*

Genus CIRSOTREMA Mörch, 1852

Cirsotrema commutatum (Monterosato, 1877)
(Fig. 204)

Scalaria commutata Monterosato, 1877

Diagnostic characters

Shell tall, conical, glossy; whorls very tumid and sutures very deep; ornament of upstanding lamellar costae; on the base of the last whorl these cross a single spiral cord. Aperture oval, with costa along outer lip.

Other characters

There may be up to ten whorls including the protoconch, but this is often lost. The costae are gently prosocline and lie so as to form spiral keels down the length of the shell (examine in apical view), each costa becoming taller where it runs over a deep sutural space. The long axis of the aperture lies at about 30° to that of the spire, and the outer lip arises level with the spiral cord. The shell is yellowish and has a dark brown spiral band below the suture and, sometimes, on the adapical side of the spiral cord, a second, placed at the base of C-shaped streaks between the costae; the streaks are often present without the lower spiral. Up to 20 mm high, 10 mm broad; last whorl occupies about half the shell height, aperture about a third.

The animal of this species is not known, but presumably resembles an *Epitonium* species (p. 488–494) and like that probably feeds on some coelenterate. The species ranges from the Mediterranean to south-west Britain, though only a few empty shells have been found there. The animals seem to be rare everywhere but are said to live on sandy bottoms.

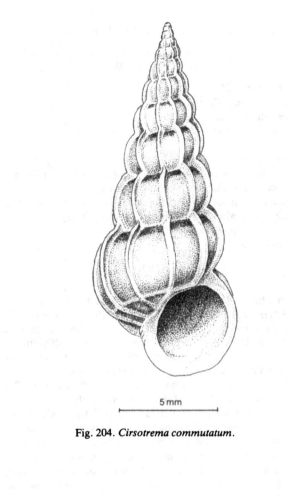

5 mm

Fig. 204. *Cirsotrema commutatum*.

Genus EPITONIUM Röding, 1798

Epitonium clathrus (Linné, 1758)
(Figs 203, *1*; 205)

Turbo clathrus Linné, 1758
Clathrus clathrus (Linné, 1758)
Scalaria communis (Lamarck, 1819)
Clathrus communis (Lamarck, 1819)

Diagnostic characters
Shell a tall cone, the swollen whorls just failing to meet but linked and braced
by prosocline costae which lie so as to form spiral keels down the shell.
At its adapical end each costa meets the corresponding one on the next higher
whorl, but not those alongside it on the same whorl; over the surface of
the whorl it forms a lamella curved so as to be convex on the side towards
the aperture (Fig. 203, *1*).

Other characters
There may be up to fifteen whorls but usually less since apical ones are
lost. The spaces between costae appear smooth. There is a varix along the
outer lip together with eight other costae on the last whorl; other whorls
have 8–9. The aperture is oval, and alongside the columella the ends of about
four costae unite to form a thickening. The shell is greyish yellow, the costae
lighter, and has two narrow peripheral spiral bands of brown, with a third
basal one on the last whorl; all are usually more distinct on the costae than
in between. Other brown marks may occur. Up to 40 mm high, 12 mm broad;
last whorl occupies about a third of shell height, aperture about a fifth.

The head lacks a snout and forms a thin fold from which the tentacles
arise, and which carries the opening of an introvert on its underside. The
tentacles are long and slender, each with a basal eye. The mantle cavity
is narrow and there is no penis in males. The foot is also narrow, truncated
anteriorly, rounded posteriorly, carrying there a black paucispiral operculum.
The body of the animal is white, with purple-black markings, the tentacles
black except near the eyes.

E. clathrus lives on sandy-muddy bottoms 5–70 m deep, but may be found,
usually in spring, at L.W.S.T. on sandy beaches where it has come to deposit
a string of somewhat pyramidal egg capsules, each covered in sand grains.
The animals feed on anemones, perhaps mainly *Anemonia sulcata*. The
species has a range from the Black Sea to Norway and animals have been
found off most shores of the British Isles, more frequently in the south and
west than in the north and east.

The animals are consecutive hermaphorodites (Ankel, 1936), changing
sex each season. Sperm are transferred from male to female by means of
spermatozeugmata (enlarged apyrene sperm to which large numbers of
eupyrene ones become attached) which swim into the mantle cavity of the

5 mm

Fig. 205. *Epitonium clathrus*.

female or are passively carried there in the inhalant respiratory stream. The egg capsules are attached to sand grains; veliger larvae emerge which are easily recognized by the bluish tint of their shell and the red central tissues of the larval body (Richter & Thorson, 1975).

Vernacular name: wentletrap.

Epitonium turtonis (Turton, 1819)
(Figs 203, 2; 206)

Turbo turtonis Turton, 1819
Clathrus turtonis (Turton, 1819)
Scalaria turtonis (Turton, 1819)
Scalaria turtonae (Turton, 1819)

Diagnostic characters
A tall conical shell similar in general appearance to that of *E. clathrus* but with a more nearly straight-sided spire, closer contact between successive whorls, more numerous costae (12–15 on last whorl) bent so as to lie almost flat on the whorl surface (Fig. 203, 2); costae alternate from whorl to whorl and bend at the suture so as to join with the next costa on the same whorl; intercostal spaces with clear spiral striae.

Other characters
There are about fifteen whorls, those of the protoconch often lost. The aperture is usually more angulated at the base of the columella than in *E. clathrus* (p. 488), the columellar lip more everted, and the labial varix less clear, though varices are often visible elsewhere. The costae, because of the way in which they bend to lie nearly parallel to the surface of the whorls, appear broader and less erect than those of *E. clathrus*. The colour is much the same as in that species, though often darker and with more obvious spiral bands. Up to 40 mm high, 12 mm broad; last whorl occupies about 40% of the shell height, aperture about 25%.

The body of the animal is like that of the previous species but is more heavily pigmented.

This species occurs between the Mediterranean and Norway, and has a similar distribution within the British Isles to that of *E. clathrus*. The animals live on sandy bottoms 5–20 m deep; their food is unknown, but it is probably a coelenterate as in other species (Perron, 1978). Breeding is also presumably similar, the egg capsules described by Vestergaard (1935) as those of *E. clathrus* having, according to Thorson (1946), been erroneously identified and really belonging to this species.

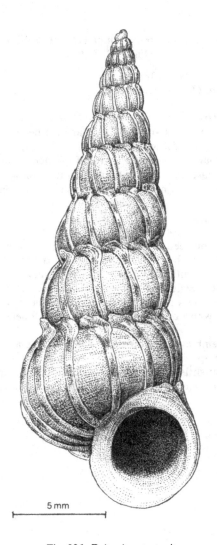

Fig. 206. *Epitonium turtonis*.

Epitonium trevelyanum (Johnston, 1841)
(Figs 203, *3*; 207)

Scalaria trevelyana Johnston, 1841
Clathrus trevelyanus (Johnston, 1841)

Diagnostic characters
In general appearance this shell is like that of other species of the genus but distinguished by the following: not over 20 mm high; spire cyrtoconoid; a flat triangular spur near the adapical end of each costa, giving the spire a turreted profile; base of aperture angulated and forming a spout; 8–9 costae converge on thickening alongside columella; shell commonly without brown spiral bands.

Other characters
There are usually fourteen costae on each whorl; each is bent towards the surface of the whorl rather more than in *clathrus* but less than in *turtonis* (Fig. 203, *3*). In the sutural region each costa curves down the spiral of the shell to join its neighbour. Between costae some spiral lines are often visible, as may also be a peripheral brown spiral band. Up to 20 mm high, 10 mm broad; last whorl occupies about 40% of shell height, aperture about a quarter.

The animal resembles *E. clathrus* (p. 488) but has a more delicate foot and is paler. The species ranges from the Mediterranean to Norway but records off the British Isles are few (off western Ireland and northern Scotland). The animal's way of life is not known, but is presumably similar to that of *E. clathrus*.

5 mm

Fig. 207. *Epitonium trevelyanum*.

Epitonium clathratulum (Kanmacher, 1798)
(Figs 203, *4*; 208)

Turbo clathratulum Kanmacher, 1798
Scalaria clathratula (Kanmacher, 1798)
Clathrus clathratulus (Kanmacher, 1798)

Diagnostic characters
Shell like that of the other *Epitonium* species but not over 12 mm high, thin
and glossy; up to 18–22 costae on each whorl, thin vertical lamellae not curving
up the spire, making only slight contact with those in the whorl above, and
none with their neighbours (Fig. 203, *4*); outer lip hardly or not at all turned
out basally.

Other characters
The shell has about 11 whorls which form a straight-sided spire. White, with
a yellow stain on the columella. Up to 12 mm high, 4 mm broad; last whorl
occupies about 40% of shell height, aperture about a quarter.

The animal is white and shows the same appearance as the other species
of this genus. The species ranges from the western Mediterranean to Norway
and the animals live 30–100 m deep, presumably in the same manner as
E. clathrus (p. 488). In the British Isles they have been found mainly in
the south west, but there are recent records from Galway Bay, Orkney and
the southern North Sea.

2 mm

Fig. 208. *Epitonium clathratulum*.

Family JANTHINIDAE Leach, 1823

Key to British species of Janthina

1. Shell with close-set V-shaped transverse ridges (apex of V pointing
   up spire); aperture with peripheral sinus .... *Janthina exigua* (p. 500)
   Shell without such ridges, with growth lines **2**

2. Meeting of outer lip and columella rounded; protoconch and teleoconch nearly coaxial; shell pale; animal oviparous ....
   .... *Janthina pallida* (p. 498)

   Meeting of outer lip and columella rectangular; protoconch axis
   oblique; shell dark; animal ovoviviparous .... *Janthina janthina* (p. 496)

Genus JANTHINA Röding, 1798

Janthina janthina (Linné, 1758)
(Fig. 209)

Helix janthina Linné, 1758
Janthina communis Lamarck, 1799
Janthina brittanica Forbes & Hanley, 1853
Janthina rotundata Dillwyn, 1840

Diagnostic characters
Shell thin, fragile, deep violet, the adapical half of the last whorl paler than
the base; surface more or less smooth; base of outer lip and columella meet
nearly at right angles; protoconch axis set at about 45° to axis of rest of
shell.

Other characters
The shell may be relatively high or low. There are up to seven whorls though
only three or four belong to the adult shell; they are tumid, meet at excavated
sutures, and the last is bluntly keeled at the periphery. Apart from fine growth
lines, which are prosocline on the whorls of the spire but become opisthocline
on the base of the last whorl, the surface is smooth, but there may be some
spiral grooves, especially near the periphery of the last whorl. The aperture
is large and its lip is thin, usually with a little spout where outer lip and
columella meet. Up to 40 mm high, 30 mm broad; last whorl occupies 75–80%
of shell height in high shells, about 90% in low ones, aperture correspondingly
occupies about 45% and 70%.

The head has a long cylindrical snout with a terminal mouth; tentacles,
which are bifid and without eyes, arise from its sides about half way along
it. Behind, but separate from the left tentacle a 3-lobed lappet marks the
beginning of a lobed epipodial fold which runs to the posterior end of the
foot where it joins another which begins at the right cephalic tentacle. Males

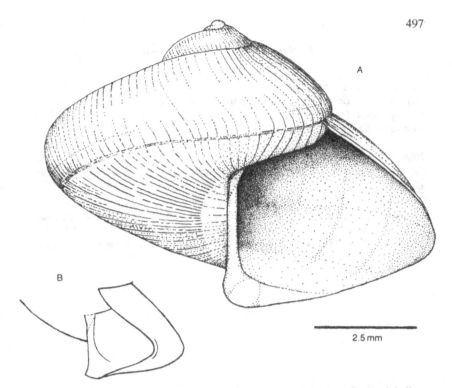

Fig. 209. *Janthina janthina*. A, shell, apertural view; B, side view of base of shell.

are aphallic. The foot is broad anteriorly and tapers to a posterior point; a broad depression ('funnel') occupies the anterior part of the sole, the posterior part bearing longitudinal grooves. An operculum is absent. The flesh is a deep violet or black.

J. janthina, like other janthinids, is holopelagic, drifting at the surface of the sea sustained by a float of mucus with entangled air bubbles, made in the funnel and attached to the foot (Fraenkel, 1927); if this is lost the animals are liable to sink and without contact with the air are incapable of forming another. The species is cosmopolitan in its distribution, the animals living primarily between 50° N and 40° S, but liable to be blown or drifted beyond these limits and stranded on south-western and western coasts of the British Isles. They prey on the pelagic coelenterates *Porpita* and *Velella* (Bayer, 1963). The animals seem to be protandrous consecutive hermaphrodites, males producing spermatozeugmata which swim to the female, enter the genital duct and effect fertilization of the eggs while they are still in the ovary. The animals are ovoviviparous, the eggs developing within the genital duct of the female. Later they are shed as veliger larvae from the left side of the mantle cavity (Wilson & Wilson, 1956).

Vernacular name: violet snail.

Janthina pallida Thompson, 1840
(Fig. 210)

Diagnostic characters
Shell globose, only pale violet in colour; basal lip of aperture out-turned, curving smoothly into columella; axis of protoconch not, or only very slightly, oblique to main shell axis.

Other characters
The shell is less glossy than that of *J. janthina* (p. 496) and the sutures are deeper. Up to 25 mm high, 23 mm broad; last whorl occupies 90–95% of shell height, aperture 80–90%. The animal is like that of *janthina* but paler in its colour, is oviparous, the eggs being laid in flattened, pear-shaped capsules which are attached to the underside of the float (Laursen, 1953). This species has the same way of life as the previous and the same general distribution, but its members reach the British Isles less frequently.

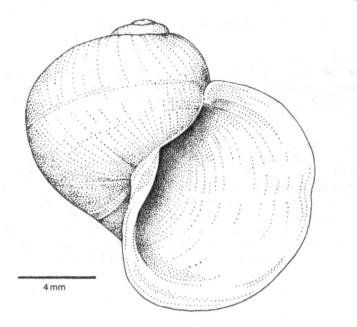

Fig. 210. *Janthina pallida*.

Janthina exigua Lamarck, 1816
(Fig. 211)

Diagnostic characters
Much like a small *J. janthina* (p. 496) in shape and colour but identifiable at once by the occurrence of prosocline ridges crossing the whorls of the spire and the upper half of the last whorl; on its base they become opisthocline. Protoconch tilted as in *janthina*.

Other characters
The spire is relatively taller than in the other *Janthina* species. There is often a shallow groove at the periphery of the last whorl and this ends at a deep V-shaped notch on the outer lip; the base is out-turned and meets the columella more or less at right angles. Up to 17 mm high, 14 mm broad; last whorl occupies 80% of shell height, aperture 55–60%.

The body of the animal is like that of other species, as is, too, the mode of life. The animals are oviparous. *J. exigua* has the same range as other janthinids but, like *J. pallida*, is infrequently cast ashore in the British Isles.

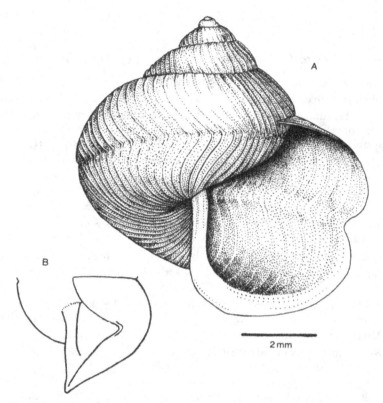

Fig. 211. *Janthina exigua*. A, shell, apertural view; B, side view of base of shell.

Family ACLIDIDAE G. O. Sars, 1878

Key to British genera and species of Aclididae

1. Shell with obvious ornament on basal whorls **2**

   Shell smooth or with growth lines only **4**

2. Shell breadth about 25% of its height; ornament of many fine flexuous costae and fainter spiral threads; umbilicus absent; not over 3 mm high .... *Graphis albida* (p. 512)

   Shell breadth 40–50% of height; ornament of spiral cords only, usually at peripheral part of whorls **3**

3. 7–8 whorls, each with 4–5 spiral cords; umbilicus small; 1–2 mm high .... *Aclis ascaris* (p. 505)

   10–12 whorls, each with 3–5 spiral cords; umbilicus wide; 3–4 mm high .... *Aclis minor* (p. 503)

4. Shell conical, apex pointed; 8–12 whorls; over 3 mm high **5**

   Shell rather columnar, apex blunt; 4–7 whorls; not over 3 mm high **7**

5. 10–12 whorls, the last occupying about 40% of shell height; umbilicus wide .... *Aclis minor* (p. 503)

   Up to 10 whorls; umbilicus small **6**

6. 8–10 whorls, very tumid with deep sutures; last whorl occupies about 35% of shell height; sutures lie far below the periphery of the upper whorl .... *Aclis walleri* (p. 506)

   6–9 whorls, slightly tumid; last whorl occupies about 50% of shell height; sutures lie just below the periphery of the upper whorl .... *Hemiaclis ventrosa* (p. 508)

7. 6–7 whorls, moderately tumid; outer lip with anal sinus and great basal flare; umbilicus present .... *Pherusina gulsonae* (p. 514)

   4–5 whorls, very tumid; outer lip with anal sinus but only slight basal eversion or none; umbilicus at most a chink *Cima minima* (p. 510)

Genus ACLIS Lovén, 1846

Aclis minor (Brown, 1827)

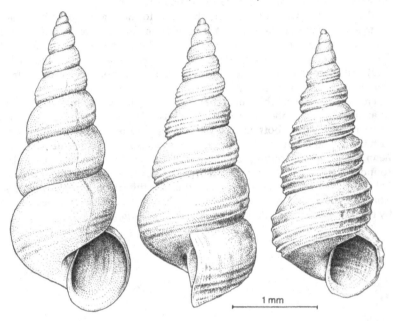

Fig. 212. *Aclis minor.* The umbilicus usually appears larger and more prominent than is shown here.

Turritella minor Brown, 1827
Aclis supranitida (Wood, 1842)

Diagnostic characters
Shell small, glossy, semitransparent when fresh, conical; whorls tumid and sutures deep; apart from growth lines ornament variable, but if present always in the form of spiral cords. Aperture small and oval, outer lip angulated if spiral ornament is present, and a little everted basally; umbilicus obvious.

Other characters
There are 10–11 whorls, about four of which belong to the protoconch and are always smooth; the others may also be smooth but more commonly bear some spiral ridges of varying degrees of clarity: they may be evenly spaced or concentrated on the basal half of each whorl, and may be either sharp or rounded in section. There may be 3–5 on the last whorl, 2–4 on those

of the spire. There is always a smooth area between suture and the first spiral ridge. White. Up to about 6 mm high, 2.5 mm broad; last whorl occupies 40–45% of shell height, aperture about a quarter.

The animal is without a snout, the head forming a ledge with tentacles at its anterior margin and the opening of an introvert below it. The tentacle bases are close together and an eye lies behind and lateral to each. The foot has a mentum (an enlarged lobe of the propodium, dorsal to the anterior end) and is broad in front, tapering behind. The right opercular lobe is larger than the left. The body is white.

The way of life of this and other aclidids is unknown, but the presence of an acrembolic proboscis strongly suggests that the animals are carnivores. They live on sandy bottoms 15–150 m deep and are either genuinely rare or hard to find because of their size. The species ranges from the Mediterranean to southern Scandinavia. There is a wide scatter of records from places round the British Isles but few recent catches of live animals. Their breeding habits are not known; larvae described under this name by Thorson (1946), Thiriot-Quiévreux (1969) and Fretter & Pilkington (1970) have been shown by Richter & Thorson (1975) to be those of *Epitonium*.

Aclis ascaris (Turton, 1819)

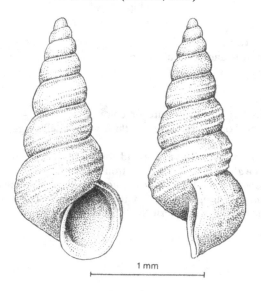

1 mm

Fig. 213. Aclis ascaris.

Turbo ascaris Turton, 1819

Diagnostic characters
Shell like that of *A. minor* but much smaller, with only 7–8 whorls, and prominent spiral ridges which may fade before reaching the outer lip; aperture relatively large and basal eversion of outer lip marked; umbilicus present but not conspicuous.

Other characters
The spiral ridges are of regular occurrence, the last whorl with 4–5. White. Up to 2 mm high, 0.5 mm broad; last whorl occupies just under half of the shell height, the aperture about a third.

The body of the animal is like that of the previous species. The snails live on sandy bottoms 10–50 m deep and are even less frequently found than *A. minor* (p. 503), nor is their way of life any better known. The distribution of the species is similar.

Aclis walleri Jeffreys, 1867
(Fig. 214)

Diagnostic characters

Shell tall, delicate, glossy, semitransparent when fresh; whorls very tumid with deep sutures and without any ornament. Aperture oval, outer lip with broad anal sinus; umbilicus small.

Other characters

The shell has ten whorls. White or colourless. Up to 4 mm high, 1.5 mm broad; last whorl occupies about one third of the shell height, aperture about a quarter.

The body of the animal is white and shows the same features as *A. minor* (p. 503). *A. walleri* lives on soft bottoms but at greater depths than the other *Aclis* species; its biology is unknown and the animals appear to be rare everywhere, though less so in the north. British records are limited to off Plymouth and the northern North Sea.

Fig. 214. *Aclis walleri*.

Genus HEMIACLIS G. O. Sars, 1878

Hemiaclis ventrosa (Friele, 1874)
(Fig. 215)

Aclis ventrosa Friele, 1874
Hemiaclis glabra Sars, 1878

Diagnostic characters
Shell small, glossy, semitransparent; apex blunt; whorls only slightly tumid and without ornament, the last more swollen than those in the spire. Aperture oval, pointed adapically, usually with a basal spout; umbilicus small.

Other characters
There are nine whorls which may show a few growth lines but no spiral ones. The breadth of the aperture is variable and the narrower it is the more marked is the basal out-turning. Colourless or white. Up to 5 mm high, 2 mm broad; aperture occupies about one third of shell height, last whorl about half.

The animal has the same organization and colouring as *Aclis* species (p. 503–506). *H. ventrosa* lives on soft bottoms 100–300 m deep between the Bay of Biscay and the Lofoten Islands. The animals are rare round the British Isles and have been found only in the most northerly parts of the North Sea. Their mode of life is not known.

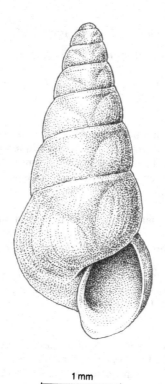

1 mm

Fig. 215. *Hemiaclis ventrosa*.

Genus CIMA Chaster, 1898

Cima minima (Jeffreys, 1858)
(Fig. 216)

Odostomia minima Jeffreys, 1858

Diagnostic characters
Shell minute, rather columnar, with blunt apex and globular protoconch; whorls rather tumid with deep and distinctly oblique sutures, devoid of ornament other than growth lines. Aperture oval, outer lip with deep anal sinus and peripheral bulge but only a slight basal eversion.

Other characters
The shell is glossy and nearly transparent, with about five whorls; growth lines are C-shaped. There is a narrow umbilicus. White or colourless. About 1.25 mm high, 0.5 mm broad; last whorl occupies about half the shell height, aperture nearly a third.

The head forms a tapering ledge with the opening of an introvert on its underside and the two cephalic tentacles springing from it anteriorly. The animal has no eyes and lacks a ctenidium. A penis arises behind the right tentacle in every animal, which therefore seems to be hermaphrodite. The flesh is white.

The distribution of this species is improperly known since it has almost certainly been overlooked on account of its minuteness; its habitat is equally little known as the animals are often found by chance. Most British finds are from sites on southern and western coasts and are mainly of dead shells.

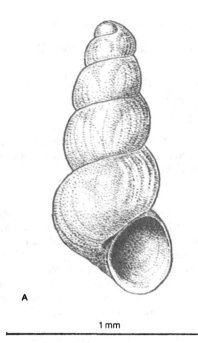

A

1 mm

B

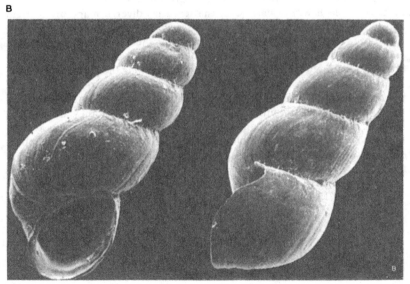

Fig. 216. *Cima minima*. A, shell; B, shell showing outer lip in side view. SEM ×
25.

Genus GRAPHIS Jeffreys, 1867

Graphis albida (Kanmacher, 1798)
(Fig. 217)

Turbo albidus Kanmacher, 1798
Aclis unica (Montagu, 1803)

Diagnostic characters
Shell small, needle-shaped, with blunt, bulbous apex; whorls tumid and sutures rather deep; ornament of many delicate costae with spiral ridges in the intervening spaces. Aperture small.

Other characters
There are 9–10 whorls of which two belong to the protoconch and are smooth; of the rest the basal ones have thirty or more costae each, the others about twenty. The costae are narrow and flexuous and extend to the base of the last whorl. Between costae lie the spiral ridges, 10–11 on the more basal whorls, fewer up the spire. They are usually absent from the subsutural part of each whorl and from the base of the last one. A groove lies along the columellar lip, but is without an umbilicus at its end. Cream. Up to about 2 mm high, about 0.5 mm broad; last whorl occupies 35–40% of shell height, aperture about 20%.

The head has a narrow snout, the mouth lying at its tip and two tentacles at its base, each with a basal eye; the tentacles diverge markedly to right and left as the snail creeps. The mantle edge has a tentacle on the right. The foot is slender, with its anterolateral corners tentaculiform, its posterior end pointed and carrying a thin operculum. The flesh is white, with dark brown on the snout and the sides of the head and along the gill and rectum.

G. albida has been found from the Mediterranean north to Norway. There are many records from all parts of the British Isles but most refer to dead shells. The animals are either rare or overlooked because they are so small or both. They live on muddy and sandy bottoms and on weeds from L.W.S.T. to about 30 m deep, but their biology is unknown, though they are hermaphrodite and, in the absence of jaws, radula and proboscis, probably suctorial feeders.

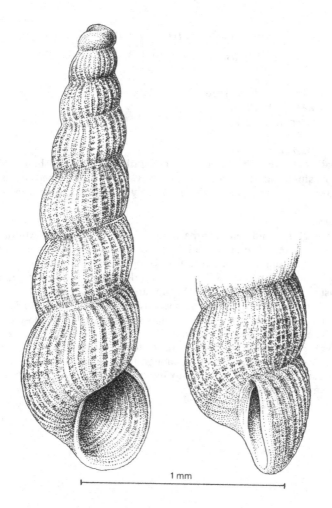

Fig. 217. *Graphis albida*.

Genus PHERUSINA Norman, 1888

Pherusina gulsonae (Clark, 1850)
(Fig. 218)

Chemnitzia gulsonae Clark, 1850
Aclis gulsonae (Clark, 1850)
Odostomia gulsonae (Clark, 1850)

Diagnostic characters
Shell small, delicate, with blunt apex and rather cylindrical shape; whorls without ornament apart from growth lines. Outer lip with deep anal sinus, marked peripheral bulge, and with base greatly everted to form a trough.

Other characters
The shell is glossy and semitransparent. There are 6–7 tumid whorls with rather deep sutures. The aperture is oval but distinct by reason of the basal flare of the lip, though this is not fully developed until the animal is mature. There is a small umbilicus. White or colourless. Up to 2 mm high, 1 mm broad; last whorl occupies half shell height, aperture one third.

The mouth lies under a ledge formed by the head. The tentacles arise from the head and have their bases united. An eye lies at the base of each tentacle. The foot is narrow. Flesh white.

P. gulsonae spreads from Madeira north to the British Isles where it has been recorded from many localities, though finds of living animals are rare and not recent. The animals seem to live sublittorally on sandy bottoms, but their way of life is unknown.

Fig. 218. *Pherusina gulsonae.* A, a young shell in which the trough-like eversion of the base of the aperture has not yet developed; B, adult shell. SEM × 40.

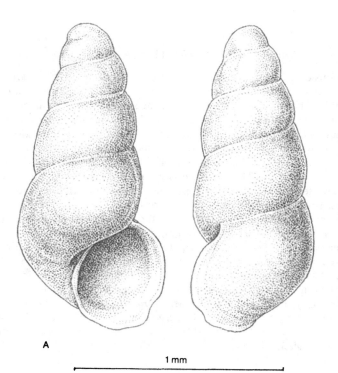

A

1 mm

B

Family EULIMIDAE H. & A. Adams, 1853

The shells of animals classified in this family are all nearly transparent when fresh so that much of the internal shell structure as well as the viscera may be visible. The whorls are flat-sided and the sutures correspondingly shallow. What at first examination of a shell seem to be sutures are parts of the internal partitions which separate neighbouring whorls and are known as **false sutures**. The true sutures may be seen only when the surface of the shell is examined with reflected light. The same is true for the extremely slight ornament which these shells possess. Their surface is highly polished, but by examining it with reflected light and using a stereomicroscope, growth lines and, in some species, spiral lines may be seen. Previous positions occupied by the outer lip (**varices**) appear, in some species, as the only distinct ornamental feature on the surface of the whorls. Old, empty shells lose their transparency and become opaque. Most shells are strictly colourless when fresh but may appear coloured because their transparency allows pigmented viscera to show.

Identification of eulimids may be difficult. Where such diagnostic characters as the colour bands of *Eulima* species, the regular growth lines of *Melanella alba* (p. 526), the spiral lines of *M. lubrica* (p. 528), the broad shells of *Vitreolina petitiana* (p. 538) and *V. curva* (p. 540), or the red flesh of species of *Polygireulima* are absent (p. 532), the most helpful features are the proportions of the shell (see Fig. 219) and the shape of the outer lip in side view (see Fig. 220).

Family EULIMIDAE H. & A. Adams, 1853

Key to British genera and species of Eulimidae

1. Shell with brown spiral bands **2**

   Shell without such bands **3**

2. Last whorl with 2 peripheral brown bands; shell breadth equals 30% of height; outer lip with marked peripheral bulge (Fig. 220, 2); not over 7 mm high (Fig. 219, 2) .... *Eulima bilineata* (p. 522)

   Last whorl with 2 peripheral brown bands and 1 subsutural, plus others on base in some shells; shell breadth equals 25% of height; outer lip without peripheral bulge (Figs 219, *1*; 220, *1*) .... *Eulima glabra* (p. 520)

3. Shell slender, straight, apex blunt; last whorl occupies less than half shell height; aperture narrow, about 35% of shell height, its long axis curved; sutures with white band along them (Figs 219, *3*; 220, *3*) .... *Haliella stenostoma* (p. 524)

   Shell not like this **4**

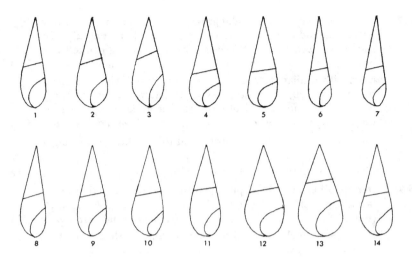

Fig. 219. Diagrammatic profiles of the shells of eulimid species all drawn to the same absolute height to bring out differences in relative proportions. The line crossing the spire obliquely marks the adapical limit of the last whorl and the angle at which it cuts the shell axis. The basal oval shows the height and breadth of the aperture. *1, Eulima glabra*; *2, E. bilineata*; *3, Haliella stenostoma*; *4, Melanella alba*; *5, M. lubrica*; *6, M. frielei*; *7, Eulitoma compactilis*; *8, Polygireulima sinuosa*; *9, P. monterosatoi*; *10, Vitreolina philippii*; *11, V. collinsi*; *12, V. petitiana*; *13, V. curva*; *14, V. incurva* (? = *V. philippii*).

4. Shell solid, nearly opaque; spire with straight profile; last whorl less than 45% of shell height; aperture occupies about 25% of shell height; flesh not red 5

 Shell not like this, always transparent if fresh; flesh red or not 8

5. Whorls with microscopic growth lines (examine with reflected light), regular and close-set; last whorl angulated at periphery; outer lip with anal sinus and peripheral bulge (Fig. 220, 6); up to 20 mm high (Fig. 219, 4) .... *Melanella alba* (p. 526)

 Shell not like this 6

6. Whorls with growth lines as in *M. alba* but also with similar spiral lines; outer lip without anal sinus but with peripheral bulge (Fig. 220, 7); not over 10 mm high (Fig. 219, 5) .... *Melanella lubrica* (p. 528)

 Shell not like this 7

7. Whorls smooth or with irregular growth lines; outer lip with narrow anal sinus and turning forward to join last whorl (Fig. 220, *8*); suture between penult and last whorls not curving markedly to origin of outer lip; northern (Fig. 219, *6*) ...... *Melanella frielei* (p. 530)

   Whorls with similar surface; outer lip similar but with peripheral bulge (Fig. 220, *9*); penult–last whorl suture curves markedly to origin of outer lip; southern, not recorded in British Isles this century (Fig. 219, *7*) .... *Eulitoma compactilis* (p. 530)

8. Shell slender, its breadth at most equal to one third of its height **9**

   Shell stout, breadth greater than one third of its height; rare **12**

9. Outer lip rather straight adapically, without anal sinus and reaching far along last whorl (Fig. 220, *5*); flesh red; 5–6 mm high (Fig. 219, *9*) .... *Polygireulima monterosatoi* (p. 532)

   Outer lip with anal sinus and peripheral bulge; flesh red or white **10**

10. Flesh red; shell usually nearly straight; last whorl occupies about 45% of height; aperture long and narrow; sutures very oblique; 7–8 mm high (Figs 219, *8*; 220, *4*) .... *Polygireulima sinuosa* (p. 532)

    Flesh white; shell bent or straight; last whorl occupies 45% of shell height or more **11**

11. Last whorl just under half shell height; aperture 30% of shell height or just less; columella equals half aperture height; sutures dip markedly to adapical end of varices; the commonest local eulimid (Figs 219, *10*; 220, *10*) .... *Vitreolina philippii* (p. 534)

    Last whorl over half shell height; aperture occupies more than 30% of shell height; columella less than half aperture height; sutures dip only a little to adapical end of varices ....
    .... *Vitreolina collinsi* (p. 536)

12. Aperture a broad oval, its breadth about 70% of its height; last whorl occupies about half the shell height (Figs 219, *12*; 220, *13*) .... *Vitreolina petitiana* (p. 538)

    Aperture breadth about 60% of its height; last whorl occupies more than half shell height (Figs 219, *13*; 220, *12*) .. *Vitreolina curva* (p. 540)

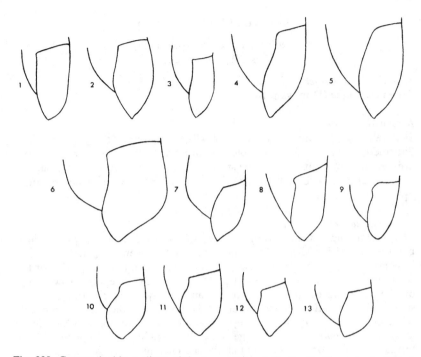

Fig. 220. Camera lucida outlines of the base of the shell of eulimid species in lateral view to show the shape of the outer lip. *1, Eulima glabra; 2, E. bilineata; 3, Haliella stenostoma; 4, Polygireulima sinuosa; 5, P. monterosatoi; 6, Melanella alba; 7, M. lubrica; 8, M. frielei; 9, Eulitoma compactilis; 10, Vitreolina philippii; 11, V. collinsi; 12, V. curva; 13, V. petitiana.*

Genus EULIMA Risso, 1826

Eulima glabra (da Costa, 1778)
(Figs 221; 219, *1*; 220, *1*)

Strombiformis glaber da Costa, 1778
Eulima subulata (Donovan, 1802)

Diagnostic characters
Shell tall, sharply pointed, slender, very glossy and nearly transparent; whorls flat-sided, sutures nearly invisible; no ornament. Aperture long, narrow, pointed apically; outer lip nearly straight in side view and forming shallow, open canal basally. Usually three spiral brown bands on each whorl of spire, up to six on last whorl.

Other characters
The shell has up to twelve whorls, three or four belonging to the protoconch, but these, which are a little tumid, are often broken off. The last whorl is slender and, in apertural view, the outer lip continues the profile of the spire. Its edge is rounded, not sharp. There are a few irregular growth lines on the shell and occasional prosocline and nearly straight markings which show former positions of the outer lip. Some brown spiral lines, especially those at the periphery, may be represented by separate streaks which curve axially. Yellow-white, with orange-brown bands. Up to 10 mm high, 2.5 mm broad; last whorl occupies about half the shell height, aperture a third.

The head is a thin ledge carrying the opening of an introvert on its underside and a pair of tentacles anteriorly. These are long and tapering, each with an eye behind and medial to its base. A pallial tentacle arises from the mantle edge on the right. In males (small, young animals) a penis with an open seminal groove on its dorsal side arises behind the right tentacle; in females (larger, older) a vestige of this persists. The foot is large but narrow, broad anteriorly, with conspicuous opercular lobes behind. The animal is white.

E. glabra ranges from the Mediterranean to the British Isles where it has been found from south-west England, the south and west of Ireland, as far north as Shetland; not in the southern North Sea. The animals are not common. They are ectoparasitic on echinoderms, probably ophiuroids, and live sublittorally to depths of about 200 m. The animals are consecutive hermaphrodites and the life history probably includes a free veliger stage (Warén, 1983a).

Fig. 221. *Eulima glabra*.

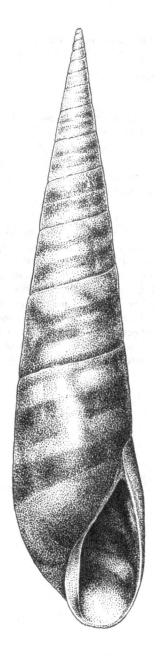

2mm

Eulima bilineata Alder, 1848
(Figs 222; 219, 2; 220, 2)

Eulima trifasciata of authors

Diagnostic characters
Shell much like that of *E. glabra* (p. 520) but half the size, the whorls just recognizably less flat-sided, the last whorl more swollen. Outer lip, in apertural view, curves outwards beyond the continuation of the profile of the spire and, in side view, shows a marked anal sinus and peripheral bulge. Only one spiral brown band on each whorl of spire (this may sometimes appear double), two on last whorl.

Other characters
The shell has 10–11 whorls, their surface showing only some faint, irregular growth lines. Colourless except for the brown spirals, one of which lies near the suture, the second near the periphery on the last whorl; both tend to fade as they approach the outer lip. Up to 6 mm high, 2 mm broad; last whorl occupies about half the shell height, aperture about a third.

The animal is like *E. glabra*. The species extends from the Mediterranean to the north of Norway, the animals being found on soft bottoms 20–250 m deep. They have been collected from many sites in the British Isles, where they are much commoner than *E. glabra*. They parasitize ophiuroids (Warén, 1983b) but are usually found free since they fall off the body of their host when disturbed.

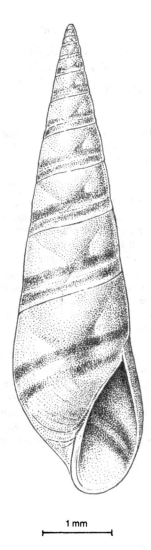

1 mm

Fig. 222. *Eulima bilineata*.

Genus HALIELLA Monterosato, 1878

Haliella stenostoma (Jeffreys, 1858)
(Figs 223; 219, *3*; 220, *3*)

Eulima stenostoma Jeffreys, 1858

Diagnostic characters
Shell tall and slender, with blunt apex; whorls slightly but distinctly tumid; aperture long, very narrow, with long axis apparently curved. Colourless (fresh), white (old and empty).

Other characters
The shell has about ten whorls; though they are slightly tumid the sutures (which lie rather more obliquely than is usual) are extremely inconspicuous. Their position is sometimes made clearer by a white band of shelly material on their abapical side, with the false suture appearing as a dark line below that. The surface shows only a few faint growth lines. In side view the outer lip is a little flexuous with a shallow anal sinus in its adapical half, a slight bulge in its abapical half. Up to 10 mm high, 2.5 mm broad; last whorl occupies rather more than half the shell height, aperture just over a third. The shell has the narrowest aperture of all the local eulimids, its breadth being only about 40% of its height, well justifying the specific name (derived from Greek words meaning 'narrowed-mouthed').

The animal is like that of *Eulima* species (p. 520–2), but is blind and has no pallial tentacle. It is white with some brown pigment on the mantle edge.

H. stenostoma occurs between the Mediterranean and Norway. The animals live, probably with ophiuroids, from about 70–3000 m deep, on soft bottoms. They are rarely found; local records are few, mainly northern, and, apart from one off Shetland, all of empty shells. Like *Eulima* species the animals are probably consecutive hermaphrodites (Warén, 1983a).

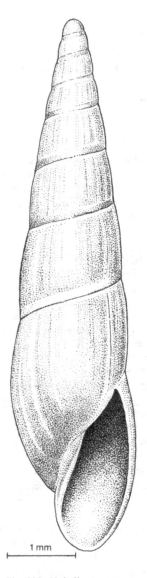

Fig. 223. *Haliella stenostoma*.

Genus MELANELLA Bowdich, 1822

Melanella alba (da Costa, 1778)
(Figs 224; 219, *4*; 220, *6*)

Strombiformis albus da Costa, 1778
Balcis alba (da Costa, 1778)
Eulima polita of authors

Diagnostic characters
Shell tall, conical, sometimes slightly twisted, the broadest part close to the base; it is generally opaque and its surface very highly polished; whorls just detectably tumid, sutures extremely slight and not very oblique; the surface marked with fine, regular growth lines. Aperture small, outer lip, in side view, with marked anal sinus and pronounced peripheral bulge.

Other characters
The shell tends to be slightly transparent at its apical end which is sharply pointed. There are 16–17 whorls, 3–4 less if the protoconch has been lost. The spire may look straight, or may show a change of direction about a third of its length below the apex. The last whorl is roundly angulated at its periphery. Though smooth and glossy to the naked eye the surface of each whorl bears many regularly arranged and closely set microscopic growth lines; these cross some equally minute spiral grooves, variably spaced. The spire usually shows varices, often lying one below the other on successive whorls, and appearing as shallow crescentic and opisthocline marks. Milky white, bluish white adapically, sometimes creamy white below. Up to 20 mm high, 6 mm broad; last whorl occupies about 40% of shell height, aperture about a quarter; the breadth of the aperture is about two thirds of its height.

The head forms a flat projection bearing two tentacles anteriorly and the opening of an introvert ventrally. The tentacles are narrow, their bases approximated, with an eye behind each. There is no pallial tentacle. Males have a curved penis attached beside the right tentacle, with an open seminal groove along its length. The foot is rather small. The body is mainly white, but the tentacles, the front of the foot, its sides and the opercular lobes all bear orange markings.

M. alba is one of the two eulimids of the British Isles which may be described as reasonably common, and it has been widely recorded there, mainly from western parts. The animals parasitize the holothurian *Neopentadactyla mixta* (Cabioch, Grainger, Keegan & Könnecker, 1978) but must, presumably, feed by sucking fluid since they are without a radula (Warén, 1983a). They live on soft bottoms 16–135 m deep. The species ranges from the Mediterranean to Norway.

Lebour (1935b) found animals at Plymouth breeding in spring and summer, when thick-walled capsules were laid containing some hundreds of pink eggs. These gave rise to free veliger larvae.

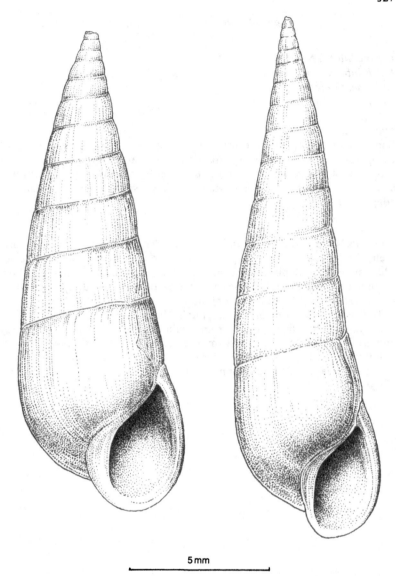

5 mm

Fig. 224. *Melanella alba*.

Melanella lubrica (Monterosato, 1891)
(Figs 225; 219, *5*; 220, 7)

Eulima lubrica Monterosato, 1891
Balcis lubrica (Monterosato, 1891)
Eulima intermedia of authors
Balcis intermedia of authors

Diagnostic characters
Shell smaller than, but otherwise very similar to that of *M. alba* (p. 526); recognizable by the following: spire usually straight; ornament of whorls formed by fine but distinct spiral lines about 8 μm apart, with growth lines as in *M. alba*; outer lip in side view showing a smooth curve, opisthocline adapically.

Other characters
The whorls are slightly more tumid than in *alba* and the sutures more easily seen; the last whorl is less angulated at the periphery than in that species. The varices are simple opisthocline lines. Milky white. Up to 9 mm high, 3 mm broad; last whorl occupies about 40% of shell height, aperture about a quarter; the breadth of the latter is about two thirds of its height.

The animal is like that of *M. alba*, but is white.

M. lubrica has been found from Spain to Norway on muddy, sandy, and gravelly bottoms between 14 m and 100 m deep. The animals parasitize holothurians. They are common locally but most records from the British Isles (west coasts only) are of empty shells.

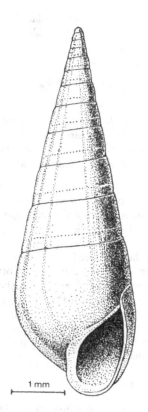

Fig. 225. *Melanella lubrica*.

Melanella frielei (Jordan, 1895)
(Figs 226; 219, *6*; 220, *8*)

Eulima frielei Jordan, 1895
Balcis frielei (Jordan, 1895)
Balcis anceps (Marshall, 1901) (=young *frielei*)

Diagnostic characters

Shell very similar to that of *M. lubrica* (p. 528) but distinctly narrower and without spiral lines on the whorls; it also lacks such regularly arranged growth lines as occur in *M. alba* (p. 526), though scattered ones are visible. Outer lip in side view turns forwards above anal sinus to join last whorl.

Other characters

The shell is a narrow cone of about twelve whorls, often with a blunt tip, and usually straight. Varices are common and appear as shallow crescentic lines to which the adapical suture dips. White. Up to 10 mm high, 2.5 mm broad; last whorl occupies about 40% of shell height, aperture a little over a quarter; the breadth of the aperture is about half its height.

The animal has the same external features as in other species; it is white on head and foot, the digestive gland orange-yellow, showing through the shell.

M. frielei is a northern species with the British Isles at the southern limit of its range. There it has been found (shells only) off the Scillies, but live specimens have been dredged off the west coast of Ireland and in the northern North Sea. The animals live on muddy and sandy bottoms 20–100 m deep as parasites of holothurians.

Genus EULITOMA Laseron, 1955

Eulitoma compactilis (Sykes, 1903)
(Figs 219, *7*; 220, *9*)

Eulima compactilis Sykes, 1903

Diagnostic characters

Shell like that of *M. frielei* (above) and without ornament. Outer lip in side view turns forwards to join last whorl but less clearly than in *frielei*; the adapical suture, however, dips markedly to meet it.

Other characters

Previous apertural positions (not frequent) are marked by curved varices with the suture above dipping markedly to their adapical ends. White. Up to 7 mm high, 1.75 mm broad; last whorl occupies about 40% of shell height, aperture about a quarter; it is appreciably broader than in *M. frielei*, reaching nearly two thirds of its height.

E. compactilis is a southern species reaching from the British Isles to the Mediterranean. Local records are few (western Irish and Scottish coasts), are of shells only, and are old. The animals live on soft bottoms to a depth of about 120 m, but their way of life is not known.

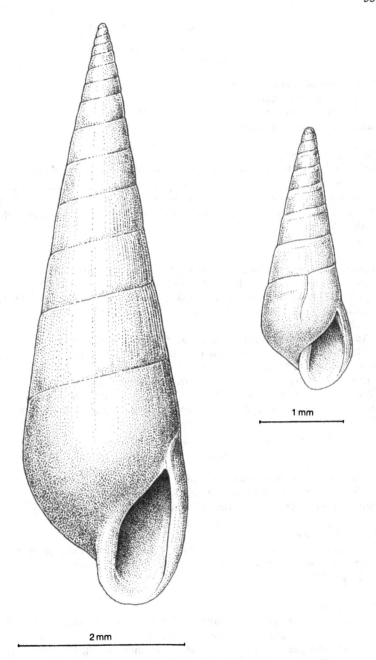

Fig. 226. *Melanella frielei*. Left, adult shell; right, young shell.

Genus POLYGIREULIMA Sacco, 1892

Polygireulima sinuosa (Scacchi, 1836)
(Figs 227; 219, *8*; 220, *4*)

Rissoa sinuosa Scacchi, 1836
Balcis sinuosa (Scacchi, 1836)
Eulima intermedia var. *rubrotincta* Jeffreys, 1867
Balcis intermedia of authors

Diagnostic characters
Shell similar to that of *Melanella frielei* (p. 530) in its proportions, but the following features identify it: spire usually gently curved; indistinct and irregular growth lines the only ornament; aperture long and narrow; outer lip with anal sinus and peripheral bulge but not turning forwards to join last whorl. Flesh reddish.

Other characters
There are 13–14 whorls which are flat-sided and do not dip to the fine sutures. Varices are present in the form of simple curved lines. Whitish. Up to 7 mm high, 2 mm broad; last whorl occupies slightly less than half of the shell height, aperture about a third; the breadth of the aperture equals about 60% of its height.

Apart from their reddish colour the animals show the same external features as other eulimids, but a penis is only briefly present during an early male phase (Warén, 1983a).

P. sinuosa ranges from the Mediterranean to Scandinavia. There are only a few records from western British and Irish coasts. The animals occur rarely, on soft bottoms 30–150 m deep, presumably associated with some echinoderm.

Polygireulima monterosatoi (Monterosato, 1890)
(Figs 219, *9*; 220, *5*)

Eulima monterosatoi Monterosato, 1890
Balcis monterosatoi (Monterosato, 1890)
Eulima distorta var. *gracilis* Forbes & Hanley, 1853

Diagnostic characters
Shell like that of *P. sinuosa* (above) but usually with straight spire; varices opisthocline; outer lip in side view slants smoothly backwards to surface of last whorl. Animal reddish.

Other characters
The whorls are extremely flat-sided and the sutures correspondingly slight; there are only occasional indistinct growth lines. The protoconch axis may be a little oblique to that of the rest of the shell, which is a little broader

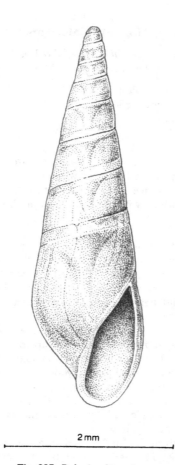

2 mm

Fig. 227. *Polygireulima sinuosa*.

than that of *P. sinuosa*. Whitish. Up to 6 mm high, 2 mm broad; last whorl occupies a little less than half of the shell height, the aperture between a quarter and a third; the breadth of the aperture is half its height, sometimes a little more.

This species occurs from the Mediterranean to Scandinavia. The animals have been met with, but rarely, in the northern parts of the North Sea, off south-western shores of England, and the Channel Islands, on soft and gravelly bottoms 20–120 m deep. Their way of life is unknown but they presumably parasitize some echinoderm.

Genus VITREOLINA Monterosato, 1884

Vitreolina philippii (Rayneval & Ponzi, 1854)
(Figs 228; 219, *10*; 220, *10*)

Eulima philippii Rayneval & Ponzi, 1854
Balcis devians (Monterosato, 1884)
Eulima distorta of authors

Diagnostic characters
Shell very glossy and almost glassily transparent; spire commonly bent, some-times more than once; whorls flat-sided without ornament, sutures insignifi-cant; some opisthocline varices present, the suture dipping markedly to their adapical end. Animal white, with red and yellow markings.

Other characters
The shell has 10–11 whorls, four belonging to the protoconch. The aperture is drop-shaped, narrow adapically; in side view the outer lip runs opistho-clinally to the surface of the last whorl just below which it shows a small anal sinus. Whitish or colourless. Up to 8 mm high, 2.5 mm broad; last whorl occupies nearly half shell height, aperture not quite a third, its breadth about half its height.

The head comprises a projecting ledge with the opening of an introvert on its underside and with divergent tentacles arising from its anterior edge. Each has a basal eye. The foot is narrow, its sole with a median groove; the opercular lobes are large and equal. Males have a recurved penis within the mantle cavity. The body is mainly white, but yellow spots lie along the tentacles; the eyes lie in white areas bordered by red; the front of the foot and the opercular lobes have red and yellow marks, its middle parts red streaks and its posterior end yellow ones.

V. philippii occurs from the Mediterranean to Norway. It is probably the commonest eulimid round the British Isles, but it is restricted to the western Channel, the Irish Sea and western Irish coasts, though empty shells have been found as far north as Shetland. Animals are occasionally found at L.W.S.T. but they are usually sublittoral, to depths of 200 m. They live on echinoderms and have been recorded (if correctly identified) on crinoids, holothurians, and ophiuroids, and so seem less restricted in their choice of host than most other species.

The veliger larva has been described (cited as *Balcis devians*) by Lebour (1935b) and Fretter & Pilkington (1970).

1 mm

Fig. 228. *Vitreolina philippii*.

Vitreolina collinsi (Sykes, 1903)
(Figs 229; 219, *11*; 229, *11*)

Eulima collinsi Sykes, 1903
Balcis collinsi (Sykes, 1903)

Diagnostic characters

Shell resembles that of *V. philippii* (p. 534) closely, but spire is usually less twisted, whorls more tumid, last whorl longer and broader, aperture longer, the outer lip in side view showing anal sinus, to which the adapical suture does not dip, and peripheral bulge.

Other characters

The shell is whitish. Up to 4 mm high, 1.3 mm broad; last whorl occupies a little more than half shell height, aperture rather more than a third, its breadth equal to about half its height.

The animal differs from *V. philippii* only in having no colour markings. The species is Mediterranean; shells have been found in the Channel Islands and from the west coast of Scotland. The animals appear to be rare and the details of their way of life are not known.

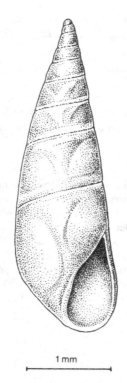

1 mm

Fig. 229. *Vitreolina collinsi*.

Vitreolina petitiana (Brusina, 1869)
(Figs 230; 219, *12*; 220, *13*)

Eulima petitiana Brusina, 1869
Balcis petitiana (Brusina, 1869)

Diagnostic characters
Shell a short and broad cone, its whorls slightly tumid; sutures indistinct; outer lip in side view a little opisthocline adapically, with peripheral bulge abapically.

Other characters
Though presenting the general appearance of other eulimids this species can hardly be confused with any except the next on account of its breadth and the shape of the aperture. The spire has 10–11 whorls and is sometimes bent a little at its apical end. Occasional varices occur. Whitish. Up to 5 mm high, 2 mm broad; last whorl occupies about half the shell height, aperture about a third; apertural breadth nearly equal to three quarters of its height.

V. petitiana is rare so that the range of the species is uncertain, though it is known to occur in the Mediterranean. British records seem to be restricted to Scilly. The animal and its habits are unknown.

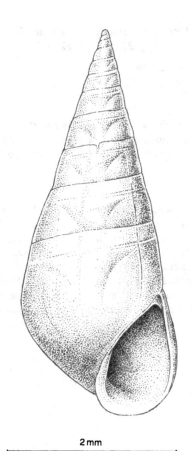

2 mm

Fig. 230. *Vitreolina petitiana*.

Vitreolina curva (Monterosato, 1874)
(Figs 231; 219, *13*; 220, *12*)

Eulima curva Monterosato, 1874
Balcis curva (Monterosato, 1874)

Diagnostic characters
Shell like that of *V. petitiana* (p. 538) in being short and broad, but more clearly twisted; whorls noticeably tumid, the last long and dilated. Outer lip in side view gently opisthocline adapically with shallow but distinct anal sinus.

Other characters
There are 9–10 whorls which, despite their tumid character, meet at inconspicuous sutures, and which show no ornament beyond a few growth lines and varices. White. Up to 3.5 mm high, 1.75 mm broad; last whorl constitutes about 60% of shell height, aperture about 40%; breadth of aperture about 60% of its height. The animal of this species and its habits are unknown.

V. curva occurs between the Mediterranean and the Channel Islands and Isles of Scilly. The animals, so far as is known, live on soft substrata to depths of 70 m, but they are so rare that this statement may well have to be modified.

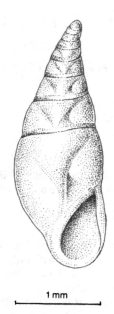

1 mm

Fig. 231. *Vitreolina curva*.

Family STILIFERIDAE H. & A. Adams, 1853

Genus PELSENEERIA Köhler & Vaney, 1908

Pelseneeria stylifera (Turton, 1825)
(Fig. 232)

Phasianella stylifera Turton, 1825
Stilifer turtoni Broderip, 1832
Stylifer turtoni of authors

Diagnostic characters

Found only on the test of regular sea urchins. Shell rather globose, the first 3–4 whorls forming a narrow pillar-like apex; base of shell covered by pedal lobes as the animal creeps.

Other characters

The shell is thin, glossy, semitransparent, has 6–7 swollen whorls, the last much the biggest. It is nearly smooth, with only some growth lines and weak spiral striae. The aperture is oval, broader basally, where the thin lip is a little flattened. Amber. Up to 5 mm high, 3 mm broad; last whorl occupies three quarters of shell height, aperture about half.

The head forms a short snout with the mouth on its underside. The bases of the tentacles touch and each has an eye behind it. Males have a large penis in the mantle cavity, armed with several spines. The foot has an anterior lobe to which the anterior pedal gland opens, a creeping sole marked with an anterior pit and a median groove, and lateral folds which may extend so as to cover the shell. The body is white.

P. stylifera is found amongst the spines and pedicellariae of regular urchins, mainly *Echinus esculentus* and *Psammechinus miliaris* but also (off south-west England and Ireland) *Paracentrotus lividus*, and (North Sea) *Strongylocentrotus drobachiensis*, feeding on the epidermis (Ankel, 1938). The animals are simultaneous hermaphrodites and lay eggs in triangular capsules attached to the test of the urchin (Lebour, 1932b). The veliger larvae which emerge are colourless and have a bilobed velum.

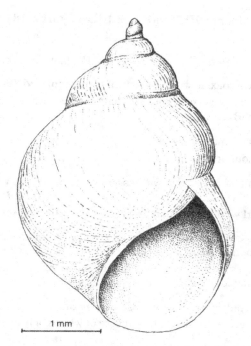

Fig. 232. *Pelseneeria stylifera*.

Family PYRAMIDELLIDAE Gray, 1840

Key to British genera of Pyramidellidae

1. Protoconch coaxial with teleoconch, apex intucked (Fig. 233A) **2**

   Protoconch axis perpendicular to that of teleoconch, apex exposed (Fig. 233B) **9**

2. Shell with prominent costae **3**

   Shell without prominent costae **6**

3. Upper half of last whorl with costae only, basal half with spiral ridges only .... *Partulida* (p. 562)

   Last whorl with at least part carrying both costae and spiral ridges **4**

4. Spiral ridges narrower than costae, confined to intercostal spaces below periphery .... *Chrysallida* (p. 546)

   Spiral ridges equal costae in breadth, cross costae and may be tuberculated there **5**

5. Shell turreted, spiral ridges spread over whorls .... *Ividella* (p. 560)

   Shell not turreted, spiral ridges at base of whorls only ... *Tragula* (p. 564)

6. Growth lines opisthocline **7**

   Growth lines prosocline **8**

7. Last whorl large; whorls (except *E. diaphana*) with strap-shaped spiral ridges on their basal parts; anal sinus on outer lip *Evalea* (p. 566)

   Last whorl not large; without spiral ridges; outer lip straight .... *Liostomia* (p. 574)

8. Ornament microscopic .... *Brachystomia* (p. 582)

   Ornament of readily visible spiral ridges and thickened growth lines .... *Jordaniella* (p. 578)

9. Costae present .... *Turbonilla* (p. 620)

   Costae absent **10**

10. Shell tall and slender, breadth 25–38% of its height **11**

    Shell breadth 40–50% of its height **12**

11. Shell very slender (breadth 25–30% of height); whorls markedly tumid .... *Ebala* (p. 618)

    Shell breadth more than 30% of height; whorls flat-sided or slightly convex .... *Eulimella* (p. 612)

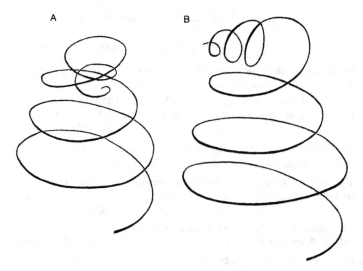

Fig. 233. Diagram of the two arrangements of heterostrophic shell found in pyramidellids. A, the protoconch is inverted and submerged in the apical whorls of the teleoconch (found in *Chrysallida, Ividella, Partulida, Tragula, Evalea, Liostomia, Jordaniella, Brachystomia*); B, the protoconch lies exposed across the apex of the teleoconch, the two axes at right angles to one another (found in *Odostomia, Noemiamea, Eulimella, Ebala, Turbonilla*).

12. Last whorl occupies 80–90% of shell height; ornament of distinct
    spiral ridges and grooves .... *Noemiamea* (p. 610)
  Last whorl occupies about 60% of shell height; ornament micro-
    scopic .... *Odostomia* (p. 592)

Genus CHRYSALLIDA Carpenter, 1857

Key to British species of Chrysallida

1. Whorls rather flat-sided; spiral ridges narrower than costae and confined to intercostal spaces **2**

   Whorls markedly swollen; spiral ridges as broad as costae and crossing them; 3–4 spiral ridges and 15 costae on last whorl .... *Chrysallida eximia* (p. 558)

2. More than five ridges on last whorl **3**

   1–3 spiral ridges on last whorl **4**

3. 5–10 spiral ridges on last whorl; base usually smooth; spire narrow, its breadth equal to 35–40% of shell height .... *Chrysallida indistincta* (p. 552)

   10–13 spiral ridges on last whorl; base crossed by costae; spire broad (45–50% of shell height) .... *Chrysallida decussata* (p. 556)

4. Base of last whorl crossed by costae **5**

   Base of last whorl smooth, not crossed by costae **6**

5. 2–3 spiral ridges on penult whorl; 19–22 broad costae on last whorl; no tooth .... *Chrysallida clathrata* (p. 554)

   1–2 spiral ridges on penult whorl; 26–27 costae on last whorl; slight tooth .... *Chrysallida suturalis* (p. 548)

6. Shell tall, last whorl less than half shell height, aperture less than one third .... *Chrysallida terebellum* (p. 550)

   Shell short, last whorl occupies more than half shell height, aperture more than one third .... *Chrysallida obtusa* (p. 546)

Chrysallida obtusa (Brown, 1827)
(Fig. 234)

Jaminia obtusa Brown, 1827
Parthenina obtusa (Brown, 1827)
Odostomia interstincta (Montagu, 1803)

Diagnostic characters
A moderately tall conical shell of 5–6 whorls; apex blunt with intucked larval shell; 20–25 costae on last whorl, ending at the periphery and not reaching base, mainly orthocline. Usually two spiral ridges near periphery of last whorl, one on each whorl in the spire. Umbilical groove usually ends in a chink. Tooth usually obvious on columella.

Other characters
The whorls dip sharply to the sutures but are otherwise only gently curved.

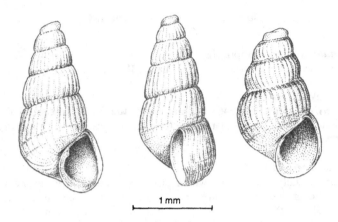

Fig. 234. *Chrysallida obtusa.*

The costae are flat-topped and about equal in breadth to the intervening hollows. Those on the last whorl tend to be flexuous and end with their summits level with the smooth base. The outer lip has a slight anal sinus and peripheral projection. White. Up to about 3 mm high, 1.5 mm broad; last whorl occupies about 60% of shell height, aperture 35–40%.

The head has two more or less triangular tentacles longitudinally grooved along their lateral margins, with their bases almost linked across the mid-line. The eyes are small and close together between the tentacle bases. A short and narrow mentum projects from under the tentacles with the proboscis opening at its base dorsally. The opening of a sac into which the penis is invaginated lies between the mentum and the dorsal surface of the foot. Foot short, with slight anterolateral points. Body translucent white.

C. obtusa has been recorded from most British and Irish coasts, though dead shells are more frequent than live animals, which live from low intertidal levels in rock pools to 90 m deep. They commonly associate with oysters but also with other bivalves (Cole & Hancock, 1955). The species occurs from the Mediterranean to northern Norway.

Chrysallida suturalis (Philippi, 1844)
(Fig. 235)

Delphinula suturalis Philippi, 1844
Odostomia interstincta var. *suturalis* Philippi, 1844

Diagnostic characters
Similar to *C. obtusa* (p. 546) but narrower; the 26–27 costae on the last whorl extend to the base of the shell and there are three spiral ridges on the last whorl.

Other characters
There are usually five whorls which dip to the sutures forming V-shaped nicks in the profile of the spire and are rather flat in between. The shell is more delicate than in *C. obtusa* and has a less obvious tooth. White. Up to 2.25 mm high, 0.8 mm broad; last whorl occupies about 55% of shell height, aperture about one third.

This species ranges from the Mediterranean to the British Isles but its distribution there is not fully known since it has often not been distinguished from *C. obtusa*.

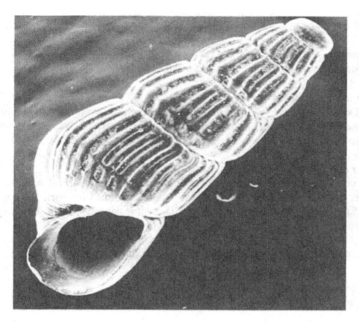

Fig. 235. *Chrysallida suturalis.* SEM × 45.

Chrysallida terebellum (Philippi, 1844)
(Fig. 236)

Chemnitzia terebellum Philippi, 1844
Odostomia interstincta var. *terebellum* Philippi, 1844

Diagnostic characters
Shell with tall, slender, cyrtoconoid spire; there are seven postlarval whorls. Costae mainly opisthocline, 21–26 on last whorl. Base of last whorl smooth; two spiral ridges on last whorl, one on each whorl of the spire.

Other characters
The costae tend to be extremely oblique on the more apical whorls, less so on basal ones and are distinctly flexuous on the last whorl. White. Up to 2.5 mm high, 0.9 mm broad; last whorl occupies about 45% of shell height, aperture 25–30%.

    C. terebellum occurs from the Mediterranean north to the Channel Islands; not recorded from mainland Britain or from Ireland.

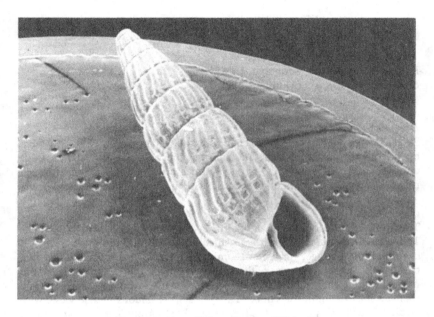

Fig. 236. *Chrysallida terebellum*. SEM × 32.

552

Chrysallida indistincta (Montagu, 1808)
(Fig. 237)

Turbo indistinctus Montagu, 1808
Parthenina indistincta (Montagu, 1808)
Chemnitzia indistincta (Montagu, 1808)
Odostomia indistincta (Montagu, 1808)

Diagnostic characters
Shell a tall, slender, blunt-tipped cone with intucked protoconch; 28–30 costae on last whorl; costae straight and opisthocline on upper whorls, flexuous on last whorl, sometimes fading, sometimes not, on the base; 5–10 spiral ridges on last whorl, confined to spaces between costae. Tooth rarely visible.

Other characters
The spire is cyrtoconoid but tends to have parallel sides in its basal half. There are 6–7 postlarval whorls of which the older ones are often flat-sided in profile, the younger ones more tumid; the apical sutures are shallow, the basal ones deeper. Whorls in the spire have two, sometimes three, spiral ridges on their abapical half. There is a small umbilical groove and, in some shells, a narrow umbilical chink. The outer lip is sinuous in side view, with an anal sinus and a peripheral bulge. White. Up to 4 mm high, 1.5 mm broad; last whorl occupies 45–50% of shell height, aperture 25–30%.

The animal is in most respects like *C. obtusa* (p. 546) but with shorter, broader tentacles, larger eyes, a longer mentum, a longer and narrower foot with better developed anterolateral processes. Body cream with opaque white flecks.

This species occurs between the Mediterranean and southern Norway and the Kattegat, though not in the southern parts of the North Sea. Animals have been collected off most parts of the British Isles, most commonly 7–100 m deep, but occasionally in rock pools near L.W.S.T. Their prey is not known.

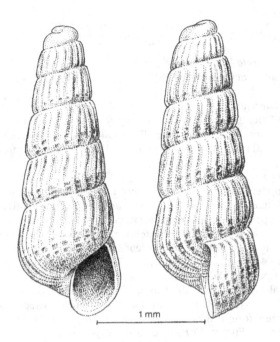

Fig. 237. *Chrysallida indistincta*.

Chrysallida clathrata (Jeffreys, 1848)
(Fig. 238)

Odostomia clathrata Jeffreys, 1848
Parthenina clathrata (Jeffreys, 1848)

Diagnostic characters
Shell tall and slender with crytoconoid spire; whorls moderately convex in profile; costae markedly prosocline, flattened, broader than the intervening spaces; 2–3 spiral ridges on last whorl; no tooth visible in aperture.

Other characters
The spire appears slightly turreted; there are 5–6 postlarval whorls. There are 19–22 costae on the last whorl and as they decrease little in number towards the apex they appear more crowded on apical whorls. The spiral ridges are narrow, confined to the intercostal spaces and lie rather far apart. The aperture is oval; a slight groove leads to a small but distinct umbilicus. White. Up to 4 mm high, 1.5 mm broad; last whorl occupies about half the shell height, aperture about 30%.

The animal has not been described. The species is spread between the Mediterranean and the British Isles, but it is doubtful whether any animals have ever been found alive in the latter locality. Dead shells have been recorded from the Firth of Forth, from off Lewis, and from Galway Bay.

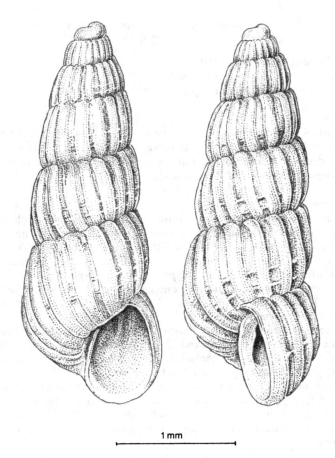

1 mm

Fig. 238. *Chrysallida clathrata*.

$$Chrysallida\ decussata\ (Montagu, 1803)$$
(Fig. 239)

Turbo decussatus Montagu, 1803
Odostomia decussata (Montagu, 1803)
Parthenina decussata (Montagu, 1803)

Diagnostic characters
Shell broadly conical with slightly turreted spire; 3–4 tumid postlarval whorls, last whorl large; 25–27 narrow, flexuous costae on last whorl with 10–13 spiral ridges extending from just above periphery to the base; tooth represented by slight thickening on columella.

Other characters
The first and sometimes the second postlarval whorls may be smooth, perhaps due to erosion of the costae. These are flat on top, about equal to the intervening spaces in breadth, prosocline and straight in the spire, but flexuous on the last whorl where they run to the base. There are 5–6 spiral ridges on the basal half of each whorl in the spire. Aperture oval, pointed adapically. There is a narrow umbilical groove, sometimes a small umbilical chink. White. Up to 3.75 mm high, 1.75 mm broad; last whorl occupies about two thirds of shell height, aperture about 40%.

The tentacles are short and triangular, their bases joined across the midline, their lateral margins grooved. The eyes are large and lie at the base of the median edges of the tentacles. The mentum is narrow, but expands and is slightly bifid distally. The foot is broad. White.

C. decussata has been recorded from the Mediterranean and off Atlantic coasts of Europe as far north as Shetland. The animals have been found alive on sandy and shelly bottoms 14–40 m deep, mainly off western coasts of England, Scotland and Ireland, but dead shells may be recovered elsewhere.

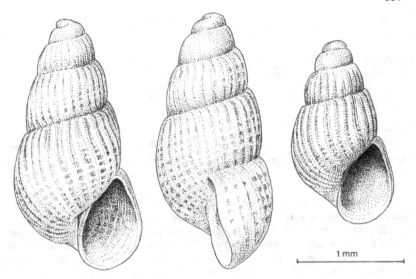

Fig. 239. *Chrysallida decussata.*

Chrysallida eximia (Jeffreys, 1849)
(Fig. 240)

Rissoa eximia Jeffreys, 1849
Odostomia eximia (Jeffreys, 1849)
Chemnitzia eximia (Jeffreys, 1849)
Pyrgulina eximia (Jeffreys, 1849)

Diagnostic characters
Shell small, with about three postlarval tumid whorls; last whorl with fifteen costae and 3–4 spiral ridges; protoconch with spiral ridges; no tooth visible.

Other characters
The costae and spiral ridges interact to give a square reticulation which may cover the whole surface of the whorls in the spire and the adapical half of the last whorl, the base of which is usually smooth. Where costae and ridges cross there are often slight tubercles. The aperture is usually rather broad and round but sometimes more elongate; a very narrow umbilical groove is present. White. About 1.5 mm high, 0.75 mm broad; last whorl occupies about 60% of total height; aperture about 40%.

The animal is undescribed so its allocation to the genus *Chrysallida* rests wholly on shell features. This is a northern species found between the British Isles and Norway, but possibly overlooked elsewhere because of its smallness. Shells have been found on soft and gravelly bottoms from 20 to more than 1000 m deep.

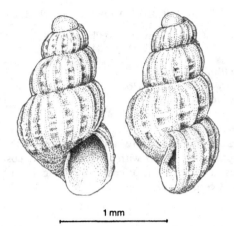

1 mm

Fig. 240. *Chrysallida eximia*.

Genus IVIDELLA Dall & Bartsch, 1909

Ividella excavata (Philippi, 1836)
(Fig. 241)

Rissoa excavata Philippi, 1836
Chrysallida excavata (Philippi, 1836)
Odostomia excavata (Philippi, 1836)

Diagnostic characters
Shell with markedly turreted spire and deep sutures, the five whorls with prominent sculpture; last whorl with 19–25 prosocline costae and five spiral ridges; tubercles at intersection of costae and ridges; umbilical groove and tooth both usually conspicuous.

Other characters
The profile of the spire is deeply notched by the sutures, below each of which is a marked subsutural shelf. In most shells the surface is marked by deep rectangular depressions bounded by the costae and spiral ridges. In the spire each whorl has two spiral ridges, one subsutural, the other peripheral. The aperture is squarish with the outer lip arising level with the third spiral ridge. White. Up to 3 mm high, 1.5 m broad; last whorl half the shell height or a little over, aperture about one third.

The animal is apparently not known. *I. excavata* is a southern species reaching from the Mediterranean as far north as south-western English and western Irish and Scottish coasts. But only dead shells appear to have been found in these localities.

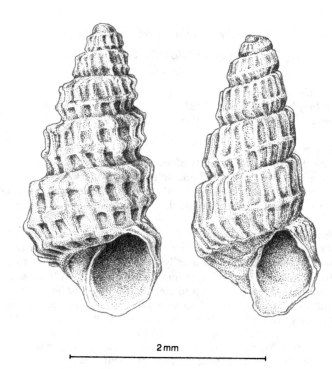

2 mm

Fig. 241. *Ividella excavata*.

Genus PARTULIDA Schaufuss, 1869

Partulida spiralis (Montagu, 1803)
(Fig. 242)

Turbo spiralis Montagu, 1803
Chrysallida spiralis (Montagu, 1803)
Odostomia spiralis (Montagu, 1803)

Diagnostic characters
Shell a short, broad cone; whorls of spire with numerous costae, last whorl with costae on the adapical half and spiral ridges on the basal half, the two sets not crossing; aperture a rounded lozenge shape, outer lip prosocline in side view, everted at base of columella; umbilicus marked and tooth visible.

Other characters
The apex is blunt, the spire slightly cyrtoconoid. There are four nearly flat-sided postlarval whorls with sutures lying nearly at right angles to the shell axis. There are about thirty costae on the last whorl, slightly prosocline and usually straight, together with up to eight spiral ridges of which that nearest the periphery is the broadest. Though costae and spiral ridges do not cross the latter may be slightly tuberculated in line with the costae. The outer lip arises level with the topmost ridge. The inner lip everts over a marked umbilical groove. White. Up to 3 mm high, 1.75 mm broad; last whorl occupies about two thirds of the total height, aperture about 45%.

The body has short tentacles, rather narrow, with a slight bulge at the base of the lateral marginal groove. An eye lies level with this bulge in line with the median edge of each tentacle. The mentum is narrow, its tip often expanded. The foot is rather broad, slightly concave anteriorly with weak anterolateral corners. White or cream with opaque white flecks, a white line along each tentacle and a white spot at its tip.

This species ranges from the western Mediterranean to Norway and Denmark. It has been found alive, moderately frequently, on all coasts of the British Isles, though only dead shells occur in the eastern Channel. The animals live from L.W.S.T. to depths of about 120 m, commonly in association with colonies of *Pomatoceros* and *Sabellaria*, on which they feed.

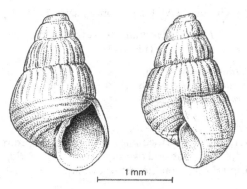

Fig. 242. *Partulida spiralis*.

Genus TRAGULA Monterosato, 1884

Tragula fenestrata (Jeffreys, 1848)
(Fig. 243)

Odostomia fenestrata Jeffreys, 1848
Chemnitzia fenestrata (Jeffreys, 1848)
Turbonilla fenestrata (Jeffreys, 1848)

Diagnostic characters
Shell tall, slender, slightly cyrtoconoid, with intucked protoconch; whorls moderately tumid, with costae and two suprasutural spiral ridges in the spire, three peripheral spiral ridges on the last whorl. Aperture small oval to rhomboidal.

Other characters
The apex is pointed. There are about eight whorls meeting at deep sutures which lie rather obliquely. Each whorl has a flat profile from the adapical suture to the spiral ridges, then it curves rapidly to the abapical suture. The costae are undulate, equal the interstices in breadth, and are usually slightly opisthocline; they end at the most basal ridge on the last whorl, though traces may occasionally extend to the base. There are tubercles at the intersections of costae and ridges. The initial part of the outer lip is straight and it is slightly flared at its base. The umbilicus is absent and usually so is the tooth. White. Up to 3.5 mm high, 1 mm broad; last whorl occupies about 40% of shell height, aperture about 25%.

The head bears long, narrow, diverging tentacles, each grooved laterally; their bases join across the mid-line and the eyes lie close together between the bases. The mentum is narrow, but deep dorsoventrally, tapering to its tip. A lobe on the right of the mantle skirt marks the end of the exhalant respiratory channel. The foot is broad anteriorly with projecting lateral points, lanceolate posteriorly. White or yellowish with many black points on the dorsal surface of the foot, the sides of the foot by the tentacle bases, and the floor of the mantle cavity. Tentacle tips white.

This is a southern species found between the Black Sea and the British Isles where it is limited to south-west England and western Ireland, off which the animals are found at depths of 10–25 m. Its host is not known.

1 mm

Fig. 243. *Tragula fenestrata*.

Genus EVALEA A. Adams, 1860

Key to British species of Evalea

1. Spiral ornament clear on basal parts of whorls **2**

   Spiral ornament absent or nearly so **3**

2. Shell delicate, spire a little turreted; six spiral bands on each whorl
   of the spire; umbilicus obvious .... *Evalea warreni* (p. 572)

   Shell solid, spire not turreted; 3–4 spiral bands on each whorl of
   the spire; umbilicus small .... *Evalea divisa* (p. 566)

3. Apex blunt with flat protoconch; spire cyrtoconoid; umbilicus small
   or absent .... *Evalea diaphana* (p. 568)

   Apex with oblique protoconch; spire coeloconoid; umbilicus
   conspicuous .... *Evalea obliqua* (p. 570)

Evalea divisa (J. Adams, 1797)
(Fig. 244)

Turbo divisus J. Adams, 1797
Menestho divisa (J. Adams, 1797)
Odostomia insculpta (Montagu, 1803)

Diagnostic characters
Shell a moderately tall cone with blunt apex and intucked protoconch; shallow
spiral grooves on basal half of each whorl (four on the penult) and opisthocline
growth lines; last whorl large. Aperture elongate, rather narrow; umbilicus
small; outer lip with anal sinus; tooth a low bulge. Foot bifid posteriorly.

Other characters
When fully grown the shell has four tumid postlarval whorls and the spire
is rather narrow. The spiral grooves (16–20 on the last whorl) are separated
by low, strap-shaped spiral ridges. The base of the aperture flares outwards
a little to form a rounded spout. White to cream. Up to nearly 4 mm high,
1.75 mm broad; last whorl occupies about two thirds of shell height, aperture
about 40%.

The animal has a short mentum deeply bifid distally, the two lobes diverging
markedly. The tentacles are short and broad, each with a ciliated lateral
groove, the bases joining across the mid-line where the eyes lie close together.
A projection from the mantle skirt at the right extremity of the mantle cavity
marks the end of an exhalant ciliary tract. Anteriorly the foot has a pronounced median bay and tentaculiform lateral points; posteriorly it is also
partially split in the same way. White, with numerous opaque white points
and a reddish patch on the head.

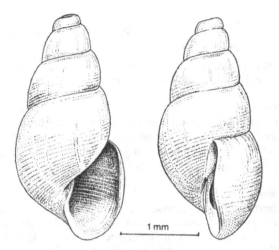

Fig. 244. *Evalea divisa.*

This species ranges from the Bay of Biscay to the north of Norway. It has been recorded from around the British Isles except for the southern North Sea, but most records are of dead shells and it is never common. The finds are made on sandy or gravelly mud from 18 to 200 m deep. Its host is not known.

Evalea diaphana (Jeffreys, 1848)
(Fig. 245)

Odostomia diaphana Jeffreys, 1848
Menestho diaphana (Jeffreys, 1848)
Odostomia perezi Dautzenberg & Fischer, 1925

Diagnostic characters
Shell somewhat ovoid with blunt apex; whorls with only delicate spiral lines plainer on base of last whorl; last whorl elongated; aperture a long oval with basal shelving spout; umbilical groove and umbilicus inconspicuous; tooth only a slight thickening.

Other characters
The spire is short and rather narrow. There are 3–4 gently swollen postlarval whorls marked with opisthocline growth lines. In side view the outer lip shows an anal sinus and, below it, a peripheral bulge. White. Up to 4 mm high, nearly 2 mm broad; last whorl occupies about three quarters of total height, aperture rather less than half.

The mentum is split deeply into right and left lobes which diverge markedly. The tentacles are rather long and curve outwards and a little backwards. The eyes are large and oval, placed close to one another between the tentacle bases. The foot is moderately broad, its anterior end embayed with small projecting lateral corners, and a little bifid posteriorly. White with opaque white spots.

E. diaphana is usually found living in association with *Phascolion strombi* in old *Turritella* or *Aporrhais* shells on soft ground 20–90 m deep, and presumably feeding on that animal (Gibbs, 1978). It is locally not uncommon. The species ranges from the British Isles south to the Mediterranean.

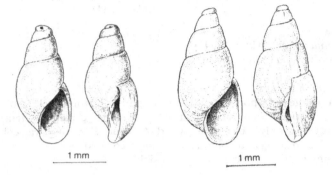

Fig. 245. *Evalea diaphana*.

Evalea obliqua (Alder, 1844)
(Fig. 246)

Odostomia obliqua Alder, 1844
Menestho obliqua (Alder, 1844)

Diagnostic characters
Shell with large last whorl and rather small spire with protoconch set obliquely
at apex; whorls swollen, meeting at oblique sutures; ornament slight, spiral
grooves obvious only at base of last whorl. Aperture elongated oval; columella
often with fold along it; umbilical groove often covered by everted lip, though
a small umbilicus may occur. No tooth visible.

Other characters
The shell has 4–5 tumid postlarval whorls meeting at rather deep sutures
which may lie at varying angles to the shell axis. The spire is often coeloconoid
in profile because of the great breadth of the last whorl. Most of the ornament
(opisthocline growth lines, spiral ridges and grooves) is microscopic. In young
shells the umbilical groove and umbilicus are clear, becoming covered only
in old ones. The outer lip shows a shallow anal sinus and slight peripheral
bulge. White or cream. Up to nearly 4 mm high, 2 mm broad; last whorl
occupies 75–80% of shell height, aperture 55–60%.

The animal resembles that of other *Evalea* species (p. 566–572) in having
a markedly bilobed mentum, but the posterior end of the foot is not bifid.
White with opaque white spots.

E. obliqua occurs from the Bay of Biscay north to southern Scandinavia.
In the British Isles it is limited to south-west England, western Irish and
Scottish coasts, where it has been dredged from gravelly or sandy mud 30–60 m
deep. It is rare and its host is not known.

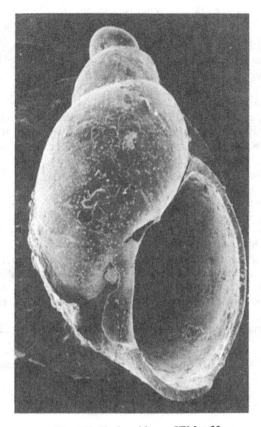

Fig. 246. *Evalea obliqua*. SEM × 30.

Evalea warreni (Thompson, 1845)
(Fig. 247)

Rissoa warreni Thompson, 1845
Menestho warreni (Thompson, 1845)
Odostomia warreni (Thompson, 1845)

Diagnostic characters

Shell delicate, a rather broad cone with blunt apex, spire slightly turreted, whorls distinctly tumid with spiral lines on their basal half; first postlarval whorl with its whole surface marked with spirals. Aperture moderately broad with basal spout; umbilical groove well marked and umbilicus rather large. No tooth visible.

Other characters

There are four postlarval whorls meeting at rather deep sutures. White. Up to 5.5 mm high and 2.75 mm broad; last whorl occupies two thirds of the total shell height, aperture about 45%.

The animal is white with opaque white spots. It has a bifid mentum, moderately long tentacles, the eyes between their bases; the foot is slightly concave anteriorly, rounded posteriorly.

E. warreni occurs off western British and Irish coasts on gravelly and sandy mud, 30–60 m deep. Shells are not uncommon in some places but living animals are rare. The species ranges southwards into the Bay of Biscay.

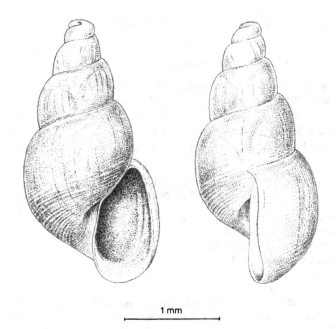

1 mm

Fig. 247. *Evalea warreni.*

Genus LIOSTOMIA G. O. Sars, 1878

Key to the British species of Liostomia

Shell with 3 postlarval whorls, the last about 75% of shell height, aperture about 50%, breadth about 60%; only found once ....
.... *Liostomia oblongula* (p. 576)
Shell with 4–5 postlarval whorls, the last about 60% of shell height, aperture about 40%, breadth about 40% .... *Liostomia clavula* (p. 574)

Liostomia clavula (Lovén, 1846)
(Fig. 248)

Turbonilla clavula Lovén, 1846
Menestho clavula (Lovén, 1846)
Odostomia clavula (Lovén, 1846)
Eulimella clavula (Lovén, 1846)

Diagnostic characters
Shell a moderately broad cone with blunt apex and intucked protoconch; whorls gently convex to flat in profile; ornament microscopic with opisthocline growth lines; a small subsutural shelf. Aperture oval; umbilical groove broad and deep and umbilicus conspicuous. Tooth absent.

Other characters
There are 4–5 postlarval whorls meeting at incised sutures set nearly at right angles to the shell axis. The last whorl may show a slight peripheral angulation. In side view the outer lip is nearly straight; it flares slightly at the base. White to cream. Up to 3.5 mm high, 1.5 mm broad; last whorl occupies about two thirds of the shell height, the aperture 40%.

The animal is said to have a narrow mentum, perhaps bilobed anteriorly, short tentacles with their lateral grooves well-marked and their bases joined across the mid-line. The eyes lie rather far apart. An accessory tentacle on the left side is mentioned in Lovén's original description. White with opaque white speckling.

L. clavula has been found from the Mediterranean to southern Scandinavia, sometimes in association with pennatulids, on soft bottoms 30–90 m deep. There are few local records – western Channel, west coasts of Ireland and Scotland.

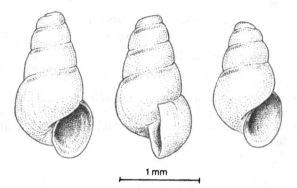

Fig. 248. *Liostomia clavula*.

Liostomia oblongula (Marshall, 1895)
(Fig. 249)

Odostomia oblongula Marshall, 1895

Diagnostic characters
Shell ovoid, spire cyrtoconoid, tip blunt, last whorl and aperture both large; ornament microscopic with opisthocline growth lines; whorls convex. Aperture elongated; umbilical groove narrow, umbilicus a chink. A small thickening on the columella.

Other characters
There are three postlarval whorls meeting at sutures which lie more obliquely than in *L. clavula* (p. 574). Though generally microscopic, spiral lines are sometimes more prominent on the base of the last whorl. The outer lip has a shallow anal sinus and a peripheral bulge, and is a little out-turned at its base. White to cream. Up to about 3 mm high, 1.75 mm broad; last whorl occupies about 70% of shell height, aperture about 50%.

The claim of this species to inclusion in the British fauna rests upon a single find of eight shells in The Minch by J. T. Marshall in 1895. Its true systematic position is doubtful.

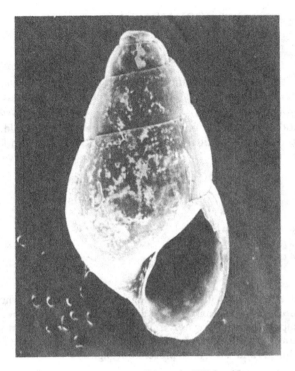

Fig. 249. *Liostomia oblongula*. SEM × 35.

Genus JORDANIELLA Chaster, 1898

Key to the British species of Jordaniella

Shell breadth equals about half its height; whorls rather flat, the last
with prominent spiral ridge(s) at periphery; tooth inconspicuous .....
.... *Jordaniella nivosa* (p. 578)
Shell breadth equals about one third of its height; whorls tumid, the
last with numerous slight spiral ridges and thickened growth lines;
tooth prominent .... *Jordaniella truncatula* (p. 580)

Jordaniella nivosa (Montagu, 1803)
(Fig. 250)

Turbo nivosus Montagu, 1803
Odostomia nivosa (Montagu, 1803)
Odostomia cylindrica Alder, 1844

Diagnostic characters
Shell small, conical, with blunt apex and intucked protoconch; whorls slightly
convex, sutures deeply incised; growth lines slightly prosocline, thickened
at their adapical ends; each whorl in the spire has one prominent spiral ridge
above suture, last whorl has 2–3 ridges at periphery. Aperture oval, flared
below; umbilicus and tooth both present but slight.

Other characters
The shell has four postlarval whorls, each with a slight flattened shelf below
the suture. The thickened ends of the growth lines mimic the appearance
of small costae. The umbilical groove is slight. The outer lip shows a short
anal sinus and a small peripheral bulge. White to cream. Up to 2.1 mm high,
1.2 mm broad; last whorl occupies about two thirds of shell height, aperture
about 40%.

The animal does not seem to have been described.

J. nivosa is a southern species extending north from the Mediterranean
to the British Isles where it has been found off most coasts, though most
finds are of dead shells. The animals live at and just below L.W.S.T.; their
host is not known, and they are rare.

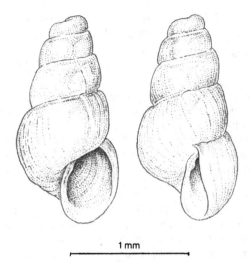

1 mm

Fig. 250. *Jordaniella nivosa*.

Jordaniella truncatula (Jeffreys, 1850)
(Fig. 251)

Odostomia truncatula Jeffreys, 1850

Diagnostic characters
Shell tall, slender, nearly columnar, with blunt, oblique apex; whorls rather swollen, sutures deep; ornament of exaggerated growth lines or weak costae, slightly opisthocline, and of spiral ridges, more prominent in basal half of each whorl. Aperture squarish to rectangular; outer lip with anal sinus and peripheral bulge, flaring basally; columella short; umbilical groove but no umbilicus. Tooth distinct.

Other characters
There are five postlarval whorls, often with a subsutural shelf and a flat peripheral region in profile. There are about twenty spiral ridges on the last whorl and about twelve on each in the spire save the most adapical. The growth lines (or incipient costae) tend to be more flexuous on the last whorl where they are briefly prosocline near the suture and again on the base. Cream to pale reddish brown. Up to 4.5 mm high, 1.75 mm broad; last whorl occupies about 60% of total height, aperture 35–40%. The appearance of the animal is not known.

This species seems to have a limited distribution: shells have been found in dredging to about 40 m depth only between the western parts of the Channel and the northern parts of the Bay of Biscay, though a single record from off north east England suggests that it may be more widespread. Its host is unknown and the animals must be rare. Its systematic position is uncertain.

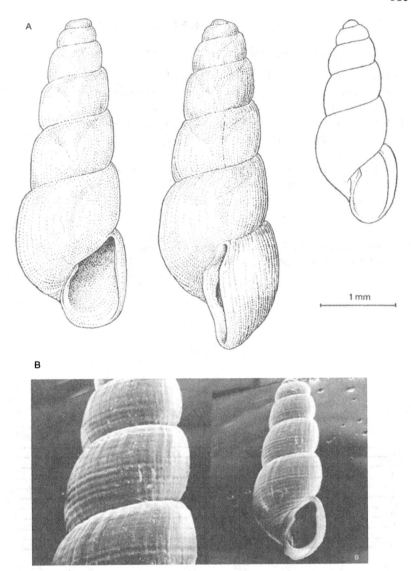

Fig. 251. *Jordaniella truncatula*. These drawings are good representations of the shell shape, but do not show the ornament; B, shell showing ornament. SEM × 13.

582

Genus BRACHYSTOMIA Monterosato, 1884

Key to the British species of Brachystomia

1. Shell breadth less than half its height; aperture equals 40% of height   **2**

   Shell breadth exceeds half the height; aperture exceeds 40% of height **3**

2. Whorls nearly flat-sided with inconspicuous subsutural shelf; growth lines very prosocline; no tooth visible (Fig. 252, *8*) ....
   .... *Brachystomia albella* (p. 588)

   Whorls tumid with marked subsutural shelf; growth lines moderately prosocline and curved; thickening on columella (Fig. 252, *11*) .....
   .... *Brachystomia rissoides* (p. 584)

3. Apex broad; whorls tumid; growth lines nearly orthocline; tooth and umbilicus clear (Fig. 252, *10*) .... *Brachystomia lukisi* (p. 590)

   Apex narrow; whorls flat-sided; growth lines very prosocline; tooth prominent but umbilicus usually absent (Fig. 252, *9*) ....
   .... *Brachystomia eulimoides* (p. 586)

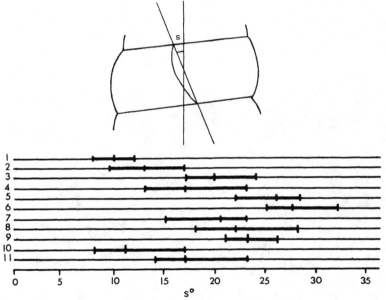

Fig. 252. Diagram to show the angle of inclination (*s*) of growth lines to shell axis in *Brachystomia* and *Odostomia* species. Above, how *s* is measured; below, each thick line gives the range and its central mark the average value of *s* for each species. *1, Odostomia acuta; 2, O. conoidea; 3, O. conspicua; 4, O. plicata; 5, O. turrita; 6, O. umbilicaris; 7, O. unidentata; 8, Brachystomia albella; 9, B. eulimoides; 10, B. lukisi; 11, B. rissoides.*

583

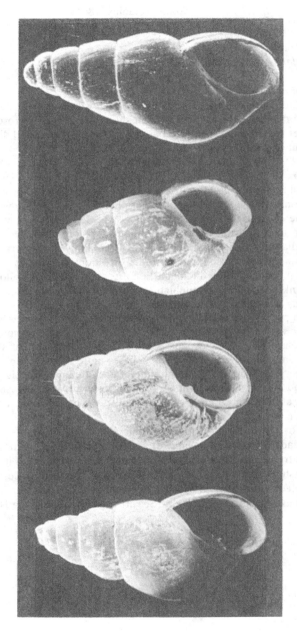

Fig. 253. *Brachystomia* species. Whole shells in apertural view, to facilitate identification of four very similar species. From left to right: *B. rissoides*, *B. eulimoides*, *B. lukisi*, *B. albella*.

Brachystomia rissoides (Hanley, 1844)
(Figs 254; 253; 252, *11*)

Odostomia rissoides Hanley, 1844
Odostomia scalaris Macgillivray, 1843

Diagnostic characters
Shell a moderately narrow cone, whorls tumid, each with a subsutural shelf; growth lines moderately prosocline, slightly flexuous, spiral lines microscopic. Aperture rather narrow, its lip everted basally; umbilical groove and distinct umbilicus usually present. Tooth only a slight thickening.

Other characters
The spire is tall and has a slightly turreted profile. There are five postlarval whorls, commonly a little flattened at the periphery. The sutures are rather deep and lie nearly at right angles to the shell axis. The umbilicus and the groove leading to it are inconspicuous in about 25% of shells. The tooth on the columella which appears only as a slight swelling in apertural view becomes prominent when examined from the direction of the outer lip. In side view the lip is gently curved and a little prosocline. White or yellowish, occasionally pinkish. Up to 3.25 mm high, 1.5 mm broad; last whorl occupies about two thirds of the shell height, aperture about 40%.

The mentum projects from below the tentacles as a narrow process which expands a little at the tip but is not bifid there; the proboscis opening lies dorsally at its base. The tentacles have a rather elongated triangular shape, each with a prominent lateral groove; their bases unite across the mid-line with a large eye placed opposite the mid-line of each tentacle. The foot has the anterior end almost straight, but is rounded posteriorly. Yellowish with brighter colour at the anterior end of the foot and at the base of each tentacle laterally. The tip of each tentacle is white.

This is one of the common littoral pyramidellids, usually found among the byssus threads of banks of *Mytilus* at about mid-tide level. It may also occur, but less frequently, with a series of other molluscs on which it feeds (Ankel & Christensen, 1963; Rasmussen, 1973). The species is widespread between the Mediterranean and southern Scandinavia, the Danish fjords, and the most western parts of the Baltic Sea. It occurs all round the British Isles.

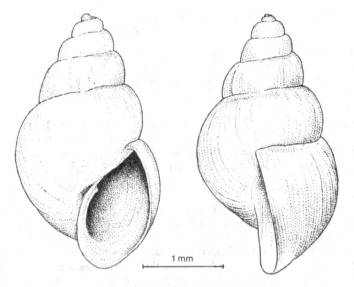

Fig. 254. *Brachystomia rissoides*.

Brachystomia eulimoides (Hanley, 1844)
(Figs 255; 253; 252, 9)

Odostomia eulimoides Hanley, 1844
Odostomia pallida (Montagu, 1803)
Odostomia ambigua (Maton & Rackett, 1807)

Diagnostic characters
Shell a rather broad cone with narrow apex and large last whorl; whorls nearly flat-sided in profile, growth lines straight and very prosocline. Aperture elongated, lip a little flared basally and prosocline in side view; umbilicus absent and groove slight. Tooth prominent.

Other characters
The shell is not as translucent as that of most other *Brachystomia* species. There are six postlarval whorls meeting at slightly channelled sutures. The growth lines are sometimes exaggerated to produce wrinkles, especially in the adapical half of a whorl; spiral lines are always microscopic. In younger shells the lip alongside the columella is thin and leaves a narrow umbilical groove exposed, but with age the lip thickens and the groove is usually hidden. White or cream, some shells with an obscure brownish spiral round the base of the last whorl; this colour may occur elsewhere. Up to 5.5 mm high, 3 mm broad; last whorl occupies about 70% of shell height, aperture 45–50%.

The tentacles are rather long, triangular, prominently grooved laterally, united to one another basally. The eyes are small, placed far apart in line with the sides of the mentum, which is narrow and expands to form a rounded tip. The anterior end of the foot is nearly straight with blunt lateral projecting corners, broadly rounded posteriorly. Head and foot yellowish; viscera showing through the shell pinkish.

B. eulimoides occurs between the Mediterranean and the Arctic. It is found, fairly commonly, off all parts of the British Isles except in the southern part of the North Sea. The animals are frequent on the ears of shells of *Pecten maximus* and *Chlamys opercularis*, but have also been found with oysters and *Turritella* (Ankel, 1959). Eggs are laid in gelatinous masses about 1 mm across attached to the shells, especially the ears, of the animals with which the adults live. Veliger larvae hatch, colourless, with a bilobed velum.

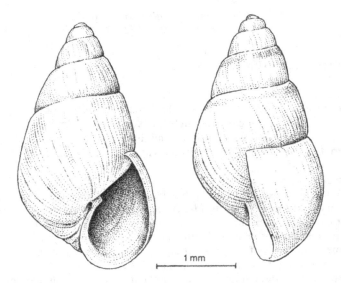

Fig. 255. *Brachystomia eulimoides*.

Brachystomia albella (Lovén, 1846)
(Figs 256; 253; 252, *8*)

Turbonilla albella Lovén, 1846
Odostomia albella (Lovén, 1846)

Diagnostic characters
Shell a rather narrow cone with rounded apex, the whorls moderately convex; growth lines very prosocline and slightly flexuous. Aperture short and narrow, often spout-like basally with narrow umbilical groove and chink-like umbilicus; tooth not visible in direct apertural view.

Other characters
There are five postlarval whorls showing characters to some extent intermediate between those of *B. rissoides* (p. 584) and *B. eulimoides* (p. 586): their convexity is less than that of *rissoides* but more marked than in *eulimoides*, the sutures are less oblique to the shell axis than in *eulimoides* but more oblique than those of *rissoides*, the growth lines are more prosocline than in *rissoides*, about as prosocline as those of *eulimoides* but more flexuous. The spiral lines, though microscopic, are a little coarser than in other species of *Brachystomia*. The aperture is relatively short. White or cream. Up to about 3.75 mm high, 1.8 mm broad; last whorl occupies about two thirds of shell height, aperture 40%.

The animal has short tentacles and its eyes lie close together: otherwise it is as in the other species. Yellowish with brighter patches on the head, the base of the tentacles, and the sides of the foot.

B. albella ranges from the Mediterranean to northern Norway. It has been found off most British and Irish coasts except those of the southern North Sea. It occurs from L.W.S.T. to depths of about 70 m. Its host is not known.

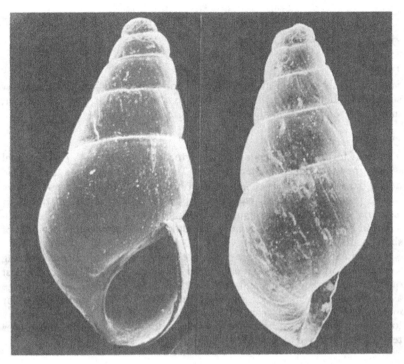

Fig. 256. *Brachystomia albella.* SEM × 20.

Brachystomia lukisi (Jeffreys, 1858)
(Figs 257; 253; 252, *10*)

Odostomia lukisi Jeffreys, 1858

Diagnostic characters
Shell relatively broad, with blunt apex; whorls somewhat tumid with narrow subsutural shelf; growth lines nearly orthocline, straight. Aperture short; umbilical groove and umbilicus well-marked. Tooth obvious.

Other characters
The shell forms a short cone with a cyrtoconoid spire. The 4–5 postlarval whorls meet at slightly incised sutures. The aperture may be a little flared at the base. The tooth is clearly visible in a strictly apertural view, though not so prominent as in *B. eulimoides* (p. 586). In side view the outer lip is nearly straight and very slightly prosocline. White. Up to a little less than 4 mm in height and about 2 mm in breadth; last whorl occupies 65–70% of shell height, aperture 40–45%.

The animals of this species are more or less identical in form and in colouring to those of other *Brachystomia* species. They are amongst the commonest of intertidal pyramidellids and may be found amid the tubes of fairly large assemblages of *Pomatoceros*, on which they feed; they also occur on similar masses of *Serpula* and *Spirorbis*. The species extends from the Bay of Biscay to the west coast of southern Norway (Höisaeter, 1968a). In the British Isles records are confined to west coast localities.

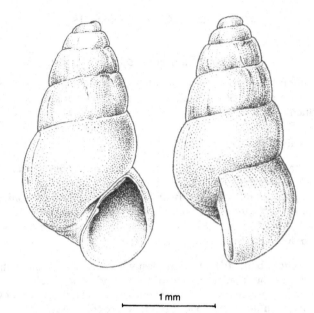

Fig. 257. *Brachystomia lukisi*.

Genus ODOSTOMIA Fleming, 1813

Key to British species of Odostomia

1. Shell a short broad cone, cyrtoconoid; breadth about 60% of height, last whorl about 66% of shell height; whorls rather swollen, with distinctly prosocline growth lines; umbilicus and tooth distinct (Fig. 252, *6*) .... *Odostomia umbilicaris* (p. 608)
   Shell not like this **2**

2. Shell breadth 45–50% of height, last whorl about 60% of height; whorls nearly flat-sided with (in most shells) enlarged peripheral spiral groove and only slightly prosocline growth lines; umbilicus a chink; tooth prominent; throat often ridged; mentum grooved dorsally; animal white (Fig. 252, *2*) .... *Odostomia conoidea* (p. 600)
   Shell not like this **3**

3. Shell breadth about 50% of height, last whorl about 60% of height, sometimes not broader than the penult; whorls moderately convex with very prosocline growth lines; umbilicus usually absent; tooth distinct; mentum not grooved; animal white (Fig. 252, *5*) ....
   .... *Odostomia turrita* (p. 598)
   Shell not like this **4**

4. Shell breadth about 40% of height, last whorl about 50% of height; whorls nearly flat-sided with slightly prosocline growth lines; last whorl rounded at periphery; umbilicus small or absent; tooth prominent; mentum not grooved; animal with yellow markings (Fig. 252, *4*) .... *Odostomia plicata* (p. 596)
   Shell not like this, last whorl usually angulated **5**

5. Shell pointed, often with brown or pink tinge, breadth about 50% of height, last whorl about 60% of height; whorls slightly swollen with straight, nearly orthocline growth lines; aperture oval, its base rounded without spout; umbilicus and tooth distinct; mentum grooved dorsally; animal with brown markings (Fig. 252, *1*) ....
   .... *Odostomia acuta* (p. 602)
   Shell and animal not like this **6**

(Key continued on p. 594)

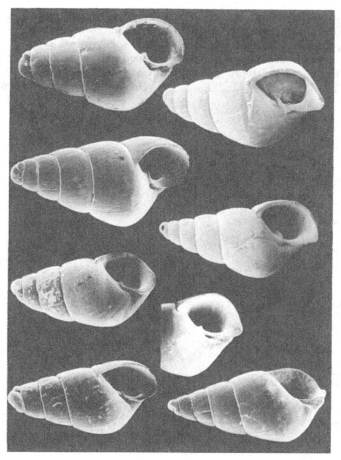

Fig. 258. *Odostomia* species. Whole shells in apertural view. From left to right, top row: *O. plicata, O. turrita, O. acuta, O. umbilicaris*; bottom row: *O. unidentata, O. unidentata* aperture, *O. conspicua, O. conoidea.*

6. Shell whitish, breadth about 50% of height, last whorl about 60% of height and usually keeled at its periphery; whorls nearly flat-sided with very prosocline growth lines; aperture lozenge-shaped with basal spout; umbilicus absent or a chink; tooth prominent; mentum shallowly grooved dorsally; animal white or grey (Fig. 252, 7) .... *Odostomia unidentata* (p. 604)

Shell brownish or pinkish, breadth about 50% of height, last whorl 55–65% of height, not keeled at its periphery; whorls nearly flat-sided, the basal one with distinct subsutural shelf; growth lines markedly prosocline; aperture lozenge-shaped with basal spout; umbilicus small and tooth prominent; throat sometimes ridged (Fig. 252, 3) .... *Odostomia conspicua* (p. 606)

Odostomia plicata (Montagu, 1803)
(Figs 259; 258; 252, *4*)

Turbo plicatus Montagu, 1803

Diagnostic characters
Shell a relatively narrow cone; whorls nearly flat-sided in profile with moderately prosocline growth lines. Aperture small, oval; umbilicus and tooth rather prominent. Without ridges within the throat.

Other characters
There are 5–6 postlarval whorls and the spire is slightly cyrtoconoid; the sutures are a little channelled; the aperture is both rather short and narrow, sometimes with a basal spout-like region. In side view the outer lip is prosocline and nearly straight. The umbilical groove is deep and leads to a small umbilicus. White or yellowish, sometimes with darker yellow along the sutures. Up to 5 mm high, 1.75 mm broad; last whorl occupies 50–55% of shell height, aperture 30–35%.

The animal has a short mentum somewhat dilated at its distal end and with the opening of the proboscis dorsally at its base. The tentacles are broad and have a pronounced lateral groove. The eyes lie between their bases. Anteriorly the foot is straight-edged, the lateral corners extended a little to form blunt lobes; posteriorly it is broadly rounded. White with much yellow speckling, especially on the tentacles and the sides of the head.

This is a common intertidal pyramidellid, usually found with *Pomatoceros*, on which it feeds (Ankel, 1959). It occurs all round the British Isles and the species ranges from the Black Sea to southern Scandinavia.

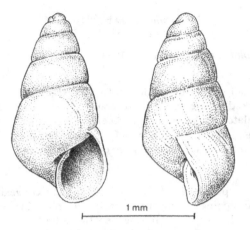

Fig. 259. *Odostomia plicata*.

Odostomia turrita Hanley, 1844
(Figs 260; 258; 252, 5)

Odostomia striolata Forbes & Hanley, 1853

Diagnostic characters

A moderately tall shell with cyrtoconoid spire; whorls only slightly convex in profile, marked with extremely prosocline growth lines; last whorl may be a little angulated at the periphery. Aperture rather small with spout-like flare at the base; umbilical groove well marked, but no umbilicus. Tooth moderately prominent, linked by ridge to edge of peristome.

Other characters

The shell has five postlarval whorls, variable in their convexity even in the same shell, and dipping to obvious sutures. The spiral lines are often more distinct than in related species. In side view the outer lip is markedly prosocline. White to cream. Up to about 3.5 mm high, 1.6–1.7 mm broad; last whorl occupies about 60% of shell height, aperture about 40%.

There is a rather long mentum with a slightly expanded distal end. The tentacles too are long, with small eyes lying between their bases. The pallial projection at the right extremity of the mantle cavity is marked. The foot broadens anteriorly, has a median bay but no prominent lateral corners; posteriorly it has an asymmetrically lobed end. White with scattered opaque white spots.

O. turrita extends from the Mediterranean to northern Norway; animals have been found off most parts of the British Isles apart from the southern parts of the North Sea and the eastern half of the Channel. They have usually been found on weed or on clay bottoms from L.W.S.T. downwards. Their normal host is not known. They have been recorded on the lobster *Homarus* (Sneli, 1972), but this is probably not their main host since they are commonly found where lobsters are unlikely to occur.

Fig. 260. *Odostomia turrita*. SEM × 100.

Odostomia conoidea (Brocchi, 1814)
(Figs 261; 258; 252, 2)

Turbo conoideus Brocchi, 1814

Diagnostic characters
Shell a moderately tall and moderately broad cone; spire slightly cyrtoconoid; whorls gently convex, suture slightly incised; growth lines only a little prosocline and usually curved; one prominent spiral groove near periphery of last whorl in most shells. Aperture usually with gently flaring base and often with spiral ridges on inside of outer lip. Umbilical groove narrow and deep, umbilicus a chink. Tooth prominent, arising at edge of inner lip.

Other characters
There are 5–6 whorls, those in the spire often with a pronounced spiral groove just adapical of the suture. The last whorl may be smoothly rounded at the periphery or show a distinct angulation there. The outer lip may have a basal spout-like out-turning; if it bears internal ridges these number 6–9, each ending in a small projection placed a short distance within the edge of the lip. Cream. Up to 3.75 mm high, 1.75 mm broad; last whorl occupies about 60% of shell height, aperture 35–40%.

The mentum is grooved dorsally and has a bifid tip, each lobe rather tentacle-like. The tentacles are flattened, triangular, each with a lateral groove and slightly swollen tip; their bases are joined across the mid-line and the large eyes lie close to one another in line with the inner edges of the tentacles. The foot is large, its anterior edge embayed in the mid-line and expanded at the lateral margins. White, with numerous opaque white markings, one at the tip of each tentacle.

O. conoidea occurs from the Mediterranean to Norway, but is absent from the North Sea. In the British Isles it is restricted to some southern and western areas. The animals may be found 10–150 m deep, usually in association with the starfish *Astropecten*, on which they live and feed.

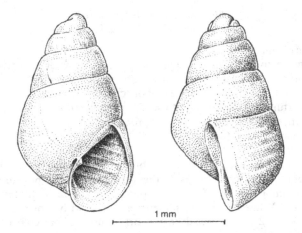

Fig. 261. *Odostomia conoidea*.

Odostomia acuta Jeffreys, 1848
(Figs 262; 258; 252, *1*)

Diagnostic characters

Shell a regular pointed cone, straight-sided in profile; whorls slightly swollen, each with a small subsutural shelf; growth lines nearly straight and only slightly prosocline; sutures channelled; last whorl commonly angulated at the periphery. Aperture a broad oval, not usually much flared at the base; umbilical groove deep, umbilicus distinct. Tooth prominent.

Other characters

There are 5–6 postlarval whorls. The outer lip is slightly prosocline and a little flexuous. The tooth arises at the very edge of the peristome at the level of the umbilicus. In a few shells spiral ridges occur in the throat under the outer lip. White or cream, sometimes tinged with brown or pink. Up to 5.5 mm high, 2.75 mm broad; last whorl occupies about 60% of the shell height, the aperture 35–40%.

The animal has rather stout, short tentacles with large eyes between their bases. The mentum is long, its lateral edges turning dorsally to form the sides of a groove which supports the proboscis when everted. There is a prominent tubular process on the extreme right of the mantle skirt. The front of the foot shows a medial bay and short lateral points; its posterior end is bluntly rounded. Pale cream with reddish brown marks on the mentum, the anterior end of the foot and its sides, as well as on the right projection of the mantle; tentacles pale.

O. acuta occurs in the Black Sea, the Mediterranean, and off European coasts as far north as the Lofoten Islands. Recorded occasionally from many parts of the British Isles, but not from the southern North Sea. The animals are not intertidal and live 20–80 m deep, perhaps with bryozoans.

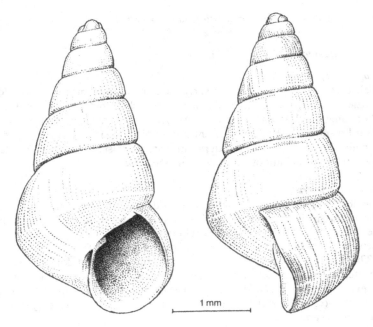

Fig. 262. *Odostomia acuta*.

Odostomia unidentata (Montagu, 1803)
(Figs 263; 258; 252, 7)

Turbo unidentatus Montagu, 1803

Diagnostic characters
Shell a regular pointed cone, nearly straight-sided in profile; whorls only a little swollen, growth lines distinctly prosocline and a little curved; last whorl angulated at the periphery. Aperture lozenge-shaped with angle in outer lip and spout at base of columella; umbilical groove narrow, only occasionally ending in a chink. Tooth prominent, not arising at edge of lip, placed closer to adapical limit of aperture than to its base.

Other characters
The shell has six postlarval whorls, each dipping to a well marked suture. The outer lip is strongly prosocline in side view. Ridges do not occur in the throat. White, sometimes grey. Up to 6 mm high, 3 mm broad; last whorl occupies about 60% of shell height, aperture 40%.

The mentum is rather short, with a blunt tip, and is not dorsally grooved. The tentacles are short and triangular, joined basally across the mid-line, an eye level with the median edge of each. The foot is small, has a nearly straight anterior edge and a blunt posterior end. White, with a grey or bluish tint in many animals.

This is a common and widespread intertidal pyramidellid, occurring near L.W.S.T. and extending to 100 m deep. It is most readily found on boulders with large numbers of *Pomatoceros* or other serpulid tubes amongst which it lurks, feeding on the worms when they emerge.

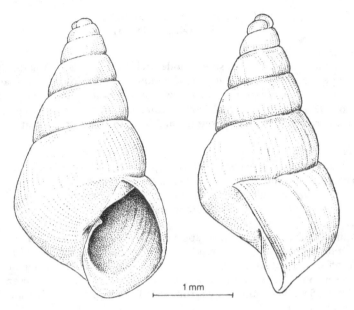

Fig. 263. *Odostomia unidentata*.

Odostomia conspicua Alder, 1850
(Figs 264; 258; 252, *3*)

Diagnostic characters

Shell a tall, pointed cone, solid, moderately broad; whorls slightly tumid, flatter peripherally, last whorl often roundly keeled; growth lines markedly prosocline. Aperture lozenge-shaped with basal spout; umbilical groove narrow and deep, umbilicus small. Tooth prominent, linked to edge of peristome by strong ridge, and placed about the middle of the apertural height.

Other characters

There are 6–7 postlarval whorls forming a more or less straight-sided spire and meeting at rather deep sutures below each of which is a narrow subsutural shelf. Some large shells show ridges in the throat. Cream, sometimes brownish. Up to 5 mm high, 2.5 mm broad; last whorl occupies 55–65% of shell height, aperture, about 40%.

Animals of this species have apparently never been described. They are rare, recent records from around the British Isles being confined to a few areas off west coasts. Elsewhere *O. conspicua* has been found from the Mediterranean north to southern Scandinavia but it is apparently absent from the North Sea. The animals live from 18–100 m deep on muddy sand. Their host is unknown.

Fig. 264. *Odostomia conspicua*. SEM × 23.

Odostomia umbilicaris (Malm, 1863)
(Figs 265, 258; 252, 6)

Turbonilla umbilicaris Malm, 1863

Diagnostic characters
Shell a short, broad cone with clearly cyrtoconoid profile; whorls tumid with slight subsutural shelf and deep sutures; last whorl large; growth lines markedly prosocline. Aperture a broad oval, flared at the base; umbilical groove and umbilicus marked. Tooth prominent, reaching the edge of the peristome.

Other characters
The shell has four whorls and a blunt tip. The tooth is placed towards the adapical end of the aperture; in side view the outer lip is markedly prosocline and often curved. White or cream. Up to 3 mm high, 1.75 mm broad; last whorl occupies nearly 70% of the shell height, aperture 40%.

The appearance of the animal seems to be unknown. In the neighbourhood of the British Isles shells are rare and there are no recent findings of live animals. All local records are from west coast localities. The species is northern in its distribution, with the British Isles at its southern limit. Shells have been found from 20 to 275 m deep; the animals may be associated with bivalves.

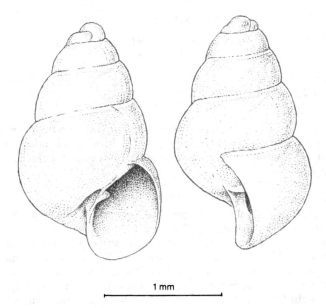

Fig. 265. *Odostomia umbilicaris*.

Genus NOEMIAMEA Hoyle, 1886

Noemiamea dolioliformis (Jeffreys, 1848)
(Fig. 266)

Odostomia dolioliformis Jeffreys, 1848
Menestho dolioliformis (Jeffreys, 1848)

Diagnostic characters
Shell ovoid, tending to globular, with very large last whorl and small spire; protoconch lies across apex of spire, exposed; ornament of many strap-shaped spiral ridges. Aperture a broad oval; umbilical groove and umbilicus distinct. Tooth obvious.

Other characters
There are 2–3 postlarval whorls which are tumid and have a slight subsutural shelf and peripheral flattening. The sutures are deep and a little incised. There are about twenty spiral ridges on the last whorl and about ten on the penult; in addition, prosocline growth lines are present, sometimes dividing the spiral ridges into squarish areas. The aperture is narrow adapically, broad at the base, where the lip is often out-turned. The outer lip is often rather straight towards the periphery and shows an anal sinus and peripheral bulge. Cream. Up to 2 mm high, 1.3 mm broad; last whorl occupies about 85% of shell height, aperture 55–60%.

The animal has short, rather broad tentacles, each with a shallow lateral groove. The eyes lie between their bases and the mentum stretches forwards from under the bases as a rounded structure tapering to its tip. The end of the pallial exhalant channel is visible at the right side of the mantle skirt. The foot has a narrow, nearly straight anterior end, the lateral corners hardly projecting; posteriorly it broadens and is rounded. Body milk-white.

N. dolioliformis is a southern species ranging from the Mediterranean to a northern limit in the British Isles. Shells have been found off most coasts but recent and live finds are limited to south-western coasts, where the animals may be found, but rarely, at L.W.S.T. and below. The host and details of the animals' reproduction are unknown.

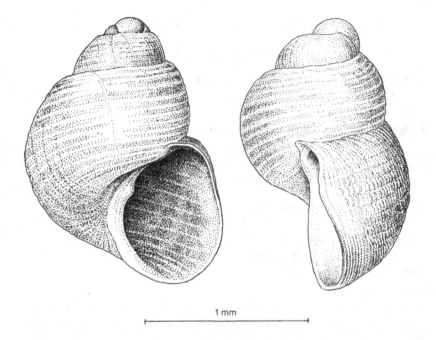

1 mm

Fig. 266. *Noemiamea dolioliformis*.

Genus EULIMELLA Gray, 1847

Key to British species of Eulimella

1. Whorls distinctly tumid, smooth to naked eye, semitransparent; shell
   without spiral lines .... *Eulimella ventricosa* (p. 616)
   Whorls flat or only slightly tumid **2**
2. Whorls flat-sided, shell a regular cone with flattish base; last whorl
   peripherally angulated .... *Eulimella scillae* (p. 612)
   Whorls slightly tumid; last whorl rounded; microscopic spiral lines
   present .... *Eulimella laevis* (p. 614)

Eulimella scillae (Scacchi, 1835)
(Fig. 267)

Melania scillae Scacchi, 1835
Odostomia scillae (Scacchi, 1835)
Eulimella macandrei (Forbes, 1844)

Diagnostic characters
Shell a tall regular cone with flattened base and rounded apex on which
the protoconch lies transversely; whorls nearly flat-sided in profile, last whorl
small with angulated periphery; whorls smooth to naked eye. Aperture small,
squarish; umbilical groove short, shallow; umbilicus and tooth absent, perhaps
a slight thickening.

Other characters
There are 9–10 postlarval whorls of which the oldest 2–3 are more tumid
than the rest; the axis of the protoconch is tilted, with the apex usually downward. If growth lines are visible (they need magnification and reflected light
to be seen and are easily eroded) they are slightly opisthocline or orthocline
in the main part of each whorl though prosocline at the adapical suture.
The outer lip in apertural view is nearly straight, as are the base of the
aperture and the columella. White. Up to about 9 mm high, 5.5 mm broad;
last whorl occupies 40–45% of shell height, aperture 25–30%.

The tentacles are triangular, not very long, with a heavily ciliated groove
on the lateral margin. A large eye lies at the base of the median edge of
each, the two eyes close together. The mentum extends from under the
tentacular bases, dorsally grooved and bifid distally, with the opening of
the proboscis between the lobes. A ciliated lobe lies on the mantle margin
at the extreme right. The anterior end of the foot is axe-shaped with projecting
lateral points, the posterior end narrowly pointed. White, with many opaque
white points.

E. scillae ranges from the Mediterranean, Madeira and the Canary Islands
to arctic Norway. There are records from most parts of the British Isles

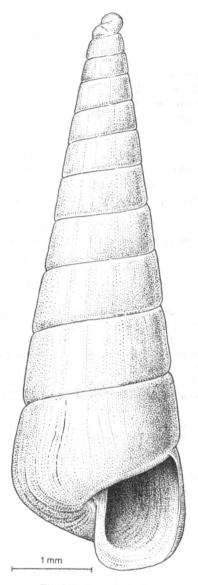

Fig. 267. *Eulimella scillae*.

though not the eastern Channel or southern North Sea; most refer to dead shells. Five living specimens have recently been taken off Galway, dredged on sand 51–81 m deep; other catches have been from sand 20–400 m deep. The host is not known.

Eulimella laevis (Brown, 1827)
(Fig. 268)

Pyramis laevis Brown, 1827
Odostomia laevis (Brown, 1827)
Odostomia acicula (Philippi, 1836)
Eulimella acicula (Philippi, 1836)

Diagnostic characters
Shell a tall, slender cone with blunt tip; whorls slightly convex in profile, often flatter at periphery, sutures rather deep; last whorl rounded rather than angulated; whorls smooth to naked eye but marked with microscopic spiral and growth lines. Aperture oval, no umbilicus. Tooth slight.

Other characters
There are 7–8 postlarval whorls, each in some shells with a minute subsutural shelf. The base of the aperture is rounded and may be a little out-turned and there may be a shallow umbilical groove. White. Up to about 3 mm high, 1 mm broad; last whorl occupies 40% of shell height, aperture about a quarter.

The head has two moderately long triangular tentacles, each grooved later-ally, and each with an eye at the base of its median edge. The mentum is grooved dorsally and is bifid at its tip where the proboscis opens. The exhalant projection at the right extremity of the mantle skirt is obvious. White in general, mentum grey, each tentacle with a central white line along it.

This is a southern species with a range from the Black Sea to southern Scandinavia, excluding the southern half of the North Sea. There are records from most parts of the British Isles but it is more common off Scottish coasts than elsewhere. The animals live on muddy sand 20–400 m deep. Their prey is not known.

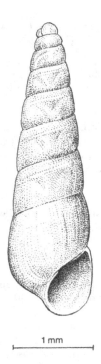

1 mm

Fig. 268. *Eulimella laevis*.

Eulimella ventricosa (Forbes, 1843)
(Fig. 269)

Parthenia ventricosa Forbes, 1843
Odostomia acicula var. *ventricosa* Jeffreys, 1869
Eulimella gracilis Jeffreys, 1847

Diagnostic characters
Shell a tall, slender cone; whorls distinctly tumid, rarely with a subsutural
shelf, smooth to the naked eye and nearly transparent; protoconch with a
flattened coil. Aperture small, rhomboidal, somewhat flared basally; umbilical
groove shallow, no umbilicus. No tooth usually visible.

Other characters
There are commonly 7–8 postlarval whorls of which the oldest 2–3 are more
tumid than younger ones. The protoconch axis is at right angles to that of
the rest of the shell and as its second whorl is large it often stands as an
erect point to the shell. Any ornament requires reflected light and some
magnification to be seen, and spiral lines are not normally present; growth
lines are more obvious, slightly prosocline in direction. White. Up to 4.5 mm
high, 1.5 mm broad; last whorl occupies 40–45% of shell height, aperture
25%.

The tentacles are short and triangular and are held so as to point laterally;
small eyes lie between their bases. The mentum is long, narrow, dorsally
grooved, bifid at the tip where the proboscis opens. A tubular projection
lies at the right margin of the mantle skirt. The foot is long, broad anteriorly
and pointed posteriorly. White with scattered opaque white marks.

E. ventricosa spreads from the Mediterranean to north Norway, though
absent from the North Sea. There are only a few records from the British
Isles, all recent living ones from near the Shetlands; all southern records
are of dead shells. The animals live on muddy sand 20–400 m deep. Their
food is unknown.

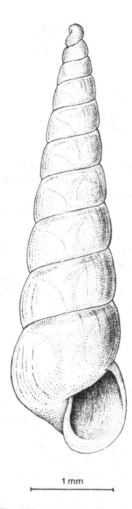

1 mm

Fig. 269. *Eulimella ventricosa*.

Genus EBALA Gray, 1847

Ebala nitidissima (Montagu, 1803)
(Fig. 270)

Turbo nitidissimus Montagu, 1803
Odostomia nitidissima (Montagu, 1803)
Eulimella nitidissima (Montagu, 1803)
Aclis nitidissima (Montagu, 1803)

Diagnostic characters
Shell small, tall and extremely slender, with upstanding apex; whorls swollen, sutures deep and oblique; fine spiral lines, especially on basal half of each whorl. Aperture small, oval; umbilical groove slight; no umbilicus and no tooth visible.

Other characters
This is a very delicate shell of 7–8 postlarval whorls. The protoconch lies at the apex of the spire, its axis at right angles to that of the teleoconch, forming a nipple-like projection; it has two whorls lying almost in the same plane. The ornament is fine but clear, the growth lines prosocline. White; brownish if it contains an animal. Up to 2.5 mm high, 0.7 mm broad; last whorl occupies 40–45% of shell height, aperture 20–25%.

The head is narrow and bears two triangular tentacles; below these lies a short mentum with laterally projecting ciliated lobes at its tip between which the proboscis opens. An eye lies at the base of each tentacle. Ciliated lobes project at the right end of the mantle cavity. The foot is short, truncated anteriorly and rounded posteriorly. Grey, darker on the mentum, the sides of the foot and on the mantle skirt.

This species ranges from the western Mediterranean to southern Norway, Isefjord (Denmark), and into the western end of the Baltic. There are records from all round the British Isles but the animals have rarely been found alive, and only once recently. They may occasionally occur amongst algae at L.W.S.T. but are commoner on muddy, sandy, or shelly bottoms 5–50 m deep.

Spawning occurs late spring–early summer when eggs were observed (Rasmussen, 1944) laid in an oval jelly mass attached to an aquarium wall. Veliger larvae hatched with a sinistral shell with a knobbly surface and a bilobed velum.

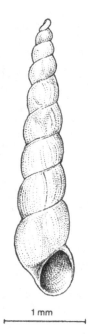

1 mm

Fig. 270. *Ebala nitidissima*.

Genus TURBONILLA Risso, 1826

Key to species of Turbonilla

1. Shell without macroscopic spiral ridges, and usually without spiral
   colour bands 2

   Shell with macroscopic spiral ridges between costae and usually with
   spiral colour bands 4

2. Basal half of shell usually with nearly parallel sides apical half conical; costae flexuous on basal whorls, opisthocline on apical ones;
   sometimes an umbilicus .... *Turbonilla pusilla* (p. 626)

   Shell more or less regularly conical; costae straight or flexuous; umbilicus absent 3

3. Shell of 10–13 whorls, apex upstanding; 18–22 costae on last whorl,
   flexuous .... *Turbonilla lactea* (p. 622)

   Shell of 10 whorls, apex rounded; 21–26 costae on last whorl, usually
   straight .... *Turbonilla acuta* (p. 624)

4. Shell tall and very slender, with sharp apex and 11–12 whorls; 20–26
   costae on last whorl; peripheral brown spiral on last whorl ....
   .... *Turbonilla crenata* (p. 628)
   .... and *Turbonilla fulvocincta* (p. 630)

   Shell relatively broader; not more than 7 whorls 5

5. Whorls moderately tumid with subsutural shelf, sutures lying just
   below periphery of upper whorl; 16–22 costae on last whorl;
   usually 2 brown spirals on last whorl; columella with slight bulge
   .... *Turbonilla jeffreysi* (p. 632)

   Whorls distinctly tumid and lacking subsutural shelf, sutures placed
   well below periphery of upper whorl; 20–22 costae on last whorl;
   usually 3 brown spirals on last whorl; columella without bulge
   .... *Turbonilla rufescens* (p. 634)

Turbonilla lactea (Linné, 1758)
(Fig. 271)

Turbo lacteus Linné, 1758
Turbonilla elegantissima (Montagu, 1803)
Chemnitzia elegantissima (Montagu, 1803)

Diagnostic characters

Shell a tall slender, rather regular cone with narrow apex; protoconch upstanding; whorls slightly tumid, meeting at well marked sutures, ornamented with numerous costae which are broader than the intervening spaces and are flexuous on the basal whorls, opisthocline on adapical ones. Aperture small, squarish, with rather straight outer lip and rounded, slightly flared base; umbilical groove small, umbilicus and tooth absent. Without colour bands.

Other characters

There are up to thirteen postlarval whorls which show no obvious subsutural shelf. The basal half of the shell is regularly conical, not columnar. On the last whorl there are 18–22 costae, the number decreasing slowly in older whorls. Microscopic spiral lines lie in the spaces between costae. On the last whorl the costae end abruptly, their summits level with the smooth base of the whorl. White. Up to 8.5 mm high, 2.5 mm broad; last whorl occupies about one third of the shell height, aperture about one fifth.

The head carries two triangular tentacles, diverging and flattened, each with a blunt tip and lateral groove; the eyes lie between their bases. The mentum extends forwards from under their bases, is flattened, grooved dorsally and bifid distally with the proboscis opening between the lobes. The tip of an exhalant gutter projects from the mantle skirt on the right. The penis is armed with rows of minute cuticular teeth. Foot short, truncated anteriorly with short projecting lateral points, pointed posteriorly. White.

T. lactea occurs from the Mediterranean to north Norway but is absent from the North Sea. The animals live from L.W.S.T. to about 80 m deep. Intertidally they occur under stones in silty places, in sand and silt between reefs; sublittoral habitats are soft, muddy, sandy, and stony bottoms. The animals avoid exposed shores. They feed on the polychaetes *Cirratulus*, *Audouinia* and *Amphitrite* (Fretter, 1951c), and are not uncommon on southern and western coasts of the British Isles.

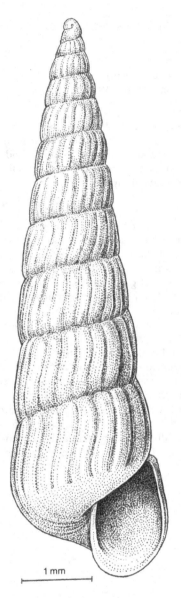

Fig. 271. *Turbonilla lactea*.

Turbonilla acuta (Donovan, 1804)
(Fig. 272)

Turbo acutus Donovan, 1804
Turbonilla gracilis (Philippi, 1836)
Turbonilla delicata Monterosato, 1884

Diagnostic characters
Like *T. lactea*, but whorls nearly flat in profile with a short subsutural shelf, costae opisthocline, straight on last whorl, their adapical ends often a little raised. Aperture rectangular with curved outer lip and distinctly flared, rather narrow base; umbilical groove small, no umbilicus; tooth appears as thickening on columella. Without colour bands.

Other characters
There are ten whorls, flatter in the basal half of the shell than adapically, with about 25 costae on the last whorl. The raised ends of the costae (not invariably present) give the sutures a wavy appearance. On the last whorl the base is smooth, as in *T. lactea* (p. 622), but the intercostal grooves in *acuta* tend to merge with it gradually instead of ending abruptly as they do in *lactea*. White. Up to 4 mm high, 1.25 mm broad; last whorl occupies about 40% of shell height, aperture 25%.

The animal has the same appearance as *T. lactea*. Its general colour is white, with a greyish tint, marked along the lateral margin of each tentacle, the sides of the foot and those of the mentum.

This species has been found in the Black Sea, the Mediterranean and the Atlantic as far north as the south-western parts of the British Isles where the animals occur on soft bottoms. Its food is unknown.

It should be added that there is doubt as to whether the animal known as *T. acuta* is distinct from *T. lactea*, as shells can be found that seem to be intermediate. The differences rest primarily on whorl shape, costal number and orientation, the last two, however, being features which are notoriously variable in this species.

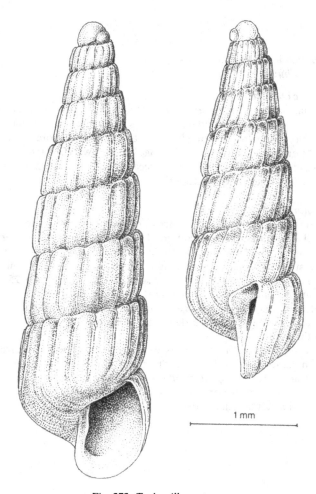

Fig. 272. *Turbonilla acuta*.

Turbonilla pusilla (Philippi, 1844)
(Fig. 273)

Odostomia pusilla Philippi, 1844
Turbonilla innovata Monterosato, 1884

Diagnostic characters

Shell small, a tall cone tapering to a narrow apex, its basal half columnar, with nearly parallel sides; adapical whorls moderately tumid, basal ones flat-sided in profile; costae opisthocline in adapical whorls, more flexuous in basal ones. Aperture rectangular or lozenge-shaped with straight outer lip and flare base; umbilical groove slight, occasionally a small umbilicus present. Tooth a slight thickening. Without colour bands.

Other characters

As a consequence of the differences between the apical and basal parts of the shell it is markedly cyrtoconoid. There are about nine postlarval whorls the last with 20–24 costae, the number decreasing to about seventeen on the oldest, very variable in direction. White. Up to 4.25 mm high, 1.2 mm broad; last whorl occupies 35–40% of shell height, aperture 20–25%.

The animal is like that of other *Turbonilla* species, but has pointed rather than blunt tentacles and a longer foot with a narrow pointed posterior end. The flesh is white; there is a dark line along each tentacle and along each side of the mentum.

T. pusilla is a southern species extending from the Mediterranean to a northern limit off the south-western parts of the British Isles. It lives on soft bottoms 5–50 m deep. Its food is not known.

Whether this species and *T. innovata* Monterosato are properly synony-mized must await detailed scrutiny.

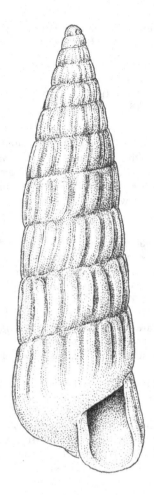

Fig. 273. *Turbonilla pusilla.*

Turbonilla crenata (Brown, 1827)
(Fig. 274)

Pyramis crenatus Brown, 1827
Turbonilla formosa (Jeffreys, 1848)
Turbonilla rufa (Philippi, 1836) (part) ?
Chemnitzia rufa Philippi, 1836

Diagnostic characters
Shell a tall and very slender cone, sharply pointed; whorls slightly convex
in profile, ornamented with numerous costae and spiral ridges in the intercostal spaces. Aperture small, nearly square, broad basally and a little flared.
Umbilicus absent. Tooth a slight thickening. A spiral orange brown band
round the periphery of each whorl.

Other characters
There are 11–12 postlarval whorls with moderately deep sutures. The costae
are straight and usually orthocline or nearly so. There are 20–26 on the last
whorl, reducing to about twenty on adapical whorls. They are narrower than
the intervening spaces and on the last whorl may either end at the periphery
or extend across the base. The spiral ridges are low, sometimes double, variable in width, separated by narrow grooves, and confined to the spaces
between costae; there are 14–18 on the last whorl, 7–9 on each whorl of
the spire. Cream, with peripheral brown band on each whorl. Up to 9 mm
high, 2.5 mm broad; last whorl occupies about 30–38% of shell height,
aperture 15–25%.

The body of the animal exhibits the same external features as the other
Turbonilla species. It is a translucent white with numerous opaque white
or yellow spots.

T. crenata occurs in the Mediterranean and has been found off European
shores from there to Norway. It is absent from the North Sea and so is
confined to western British and Irish coasts. The animals live on fine sandy
bottoms 15–350 m deep. Their food is not known.

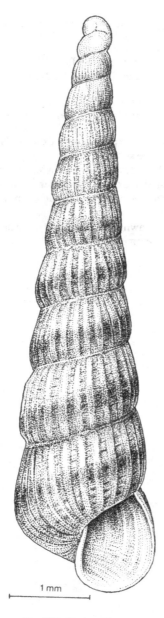

1 mm

Fig. 274. *Turbonilla crenata*.

Turbonilla fulvocincta (Thompson, 1840)

Turritella fulvocincta Thompson, 1840
Chemnitzia fulvocincta (Thompson, 1840)
Turbonilla rufa var. *fulvocincta* Thompson, 1840

Diagnostic characters
This differs from *T. crenata* (of which it may be only a form) in the following points:
(1) the number of costae tends to be greater on the last whorl than in shells of *T. crenata* of the same size;
(2) the sutures cut the shell axis at a lower angle, i.e. they are more oblique;
(3) the spiral ridges usually exhibit a regular pattern – a wide subsutural one, then some narrower ones, 1–2 broad peripheral ones, with narrow ridges on the base;
(4) the shell is generally tawny with a narrow subsutural brown spiral and a broad peripheral one on each whorl.

Turbonilla jeffreysi Forbes & Hanley, 1850
(Fig. 275)

Chemnitzia scalaris Philippi, 1836
Turbonilla scalaris (Philippi, 1836)
Odostomia scalaris (Philippi, 1836)

Diagnostic characters

Shell a moderately tall, rather broad cone; whorls with subsutural shelf, sutures just below periphery of each whorl; ornament of narrow, mainly orthocline costae and spiral ridges in the intercostal spaces. Aperture square to rhomboidal; no true umbilical groove or umbilicus. Tooth absent though usually a slight bulge on columella. Cream with two, occasionally three brown spiral bands on the last whorl.

Other characters

The spire is noticeably cyrtoconoid and has a turreted profile. There are seven postlarval whorls, those of the protoconch often lost, meeting at sutures made sinuous by the uprising ends of the costae. The costae are much narrower than the intervening spaces; there are 16–22 on the last whorl where they fade below the periphery, 17–19 on the penult whorl. The last whorl has 20–25 spiral ridges, the penult 10–12. One coloured spiral band lies at the periphery, another is subperipheral and these two are regularly present; a third is subsutural but this may be absent. Up to about 6 mm high, 2 mm broad; last whorl occupies about half the shell height (usually slightly less), aperture about 30% or a little less.

The tentacles are triangular, grooved along the lateral margin, rather longer and narrower than in most species. The mentum is long and narrow, dorsally grooved and bifid distally where the proboscis opens. The eyes lie between the tentacle bases. There is a lobed projection on the mantle skirt on the right. The foot is short, truncated anteriorly, with a median embayment and extended lateral points, pointed posteriorly. White with numerous white spots and a general reddish tinge; the foot has white streaks alongside the operculum and there is yellow on the mantle edge.

T. jeffreysi ranges from the Mediterranean to southern Scandinavia but there are few records from the North Sea, none from the eastern Channel and Irish Sea. The animals occur 5–10 m deep, usually with the hydroids *Halecium* and *Antennularia*, on which they feed (Fretter & Graham, 1949).

Fig. 275. *Turbonilla jeffreysi*.

Turbonilla rufescens (Forbes, 1846)
(Fig. 276)

Chemnitzia rufescens Forbes, 1846
Turbonilla scalaris var. *rufescens* Forbes, 1846

Diagnostic characters

Shell a moderately tall, rather narrow cone; whorls distinctly tumid, spire not turreted in profile; sutures far below periphery of each whorl; ornament of narrow costae with spiral ridges in the interstices. Aperture oval, without marked umbilical groove, umbilicus or visible tooth. Tawny with three orange-brown spiral bands on last whorl.

Other characters

The shell is similar to that of *T. jeffreysi* (p. 632) in general appearance but is narrower and thinner, and its seven whorls differ in details of shape. The costae are commonly flexuous, narrower and more close-set than in *jeffreysi*, and number 20–22 on the last whorl, 19–20 on the penult. The spiral bands of colour are of regular occurrence. Up to 6 mm high, 2 mm broad; last whorl occupies just less than half the shell height, aperture about 30%.

The tentacles are long, narrow, triangular, and are grooved laterally. The mentum is long and narrow, grooved dorsally and bifid distally. The eyes are close together. The foot is short, broad anteriorly and slightly concave with anterolateral points, narrow and pointed posteriorly. Flesh a pale reddish brown.

The geographical range of this species is inaccurately known since it has not always been distinguished from *T. jeffreysi*. There are few British or Irish records and these are limited to Scilly, Donegal, and the Scottish west and north coasts, where it occurs in similar habitats to *T. jeffreysi*. Details of its food and reproductive habits are not known.

1 mm

Fig. 276. *Turbonilla rufescens*.

Acknowledgements

I have been fortunate in being able to illustrate the text with the extremely beautiful drawings of shells by the Danish artist Poul Winther. Their history is this: in the 1960's Prof. Gunnar Thorson proposed to write an account of the marine prosobranchs found in Denmark, a plan later enlarged to include British and Irish species not found there, and asked Winther to make the drawings to illustrate the book. Unfortunately the project was never completed, Winther dying in 1966 and Thorson in 1971, leaving a nearly complete series of drawings, many notes, but no accompanying text. The late Dr Henning Lemche suggested to Dr V. Fretter and I that we might find some way to make use of the drawings and we have indeed done so in a series of descriptive publications in the *Journal of Molluscan Studies*. I am very grateful to the Malacological Society of London, who own the copyright, for permission to use Winther's drawings: no matter how one tries it is often not possible to find easily quantifiable features for the identification of shells, and a figure is much better than words for the description of the complexity of shape and the minutiae of structure that a prosobranch shell may exhibit. Some extra details not shown in the original figures have been added, and where there were no figures drawn by Winther, or where these referred explicitly to shells such as would be found in Danish rather than in British and Irish waters, I have used SEM photographs; for some of these I am indebted to the staff of the Electron Microscope Unit, Royal Holloway and Bedford New College (University of London). I would also like to express my gratitude to Dr D. M. Kermack, whose editorial skills have improved the text and eliminated many mistakes.

636

Glossary

abapertural Away from the aperture.

abapical Away from the apex; the youngest part of the shell.

abaxial Further away from the axis of shell coiling.

accessory boring organ A gland used in boring the shell of prey, housed in a pit on the sole of the foot in muricids, on the ventral lip of the mouth in naticids.

acrembolic proboscis A proboscis retracted by muscles attached to its tip so as to form an introvert with the mouth at its innermost part.

adapertural Close to the aperture.

adapical Towards the apex; the oldest part of the shell.

adaxial Nearer or towards the axis of shell coiling.

anal sinus A bay or slit on the outer lip near the adapical end, lying over the exhalant current from the mantle cavity.

anterior pedal gland Opens between propodium and mesopodium, secreting the mucus on which a snail crawls.

aperture The opening at the youngest part of the shell through which the body is extruded and withdrawn.

apex The first formed part of the shell, usually pointed.

aspidobranch A ctenidium with double row of lamellae on the axis.

axial In or parallel to the axis of shell coiling.

bipectinate = Aspidobranch, ctenidium with double row of lamellae on the axis.

cancellated Shell surface divided into chambers by the intersection of costae and spiral ridges.

cephalic lappet A fold of skin lying at or near the base of a cephalic tentacle.

cephalic tentacle A tentacle on the head.

coeloconoid Used to describe a spire with concave profile.

columella The central pillar of a shell formed by meeting of the adaxial walls of the whorls; may be solid or hollow.

columellar lip That part of the inner lip lying alongside the columella.

columellar muscle A muscle running from the columella of the shell to the head and foot, withdrawing them on contraction.

convolute A shell with the spire totally concealed by overgrowth of the last whorl.

costa (*pl.* **costae**) A ridge across a whorl.

costella A little costa.

ctenidium A gill composed of an axis bearing either a double or single row of lamellae, located in the mantle cavity.

cyrtoconoid Used to describe a spire with a convex profile.

dextral Used to describe the direction of shell coiling: clockwise when the shell is viewed from above the apex.

637

echinospira A modified veliger larva characterized by having a double shell; found only in a few higher prosobranchs.

ectoparasitic A parasite found on the surface of its host, though often feeding internally by means of a proboscis.

egg capsule A protective secretion enclosing one or more eggs with a supply of food.

epipodial tentacle A tentacle on the epipodium.

epipodium A fold or series of tentacles or both along the sides of the foot, often with special sense organs.

exhalant Describing a channel for the outward flow of water.

eye stalk A tentaculiform process carrying an eye placed on the outer side of a cephalic tentacle; fused with that in higher prosobranchs.

false suture Part of the internal partition separating whorls seen in transparent shells and imitating a true suture.

food egg An egg which cannot, or fails to develop and is used as food by those which do within an egg capsule.

foot The muscular and glandular ventral part of the body on which a prosobranch moves.

genital opening The external opening of the genital duct and, sometimes, of associated glands.

growth line A transverse mark on the surface of a whorl indicating a previous position of the outer lip.

head-foot The body of a prosobranch excluding the visceral mass, extrusible from the shell and concerned with feeding, sensation, and locomotion.

height The total distance, measured as a straight line, between the apical and basal extremities of a shell.

hermaphrodite An animal which either consecutively or simultaneously produces spermatoza and ova; not usually self-fertile.

heterostrophic A condition in which the protoconch appears to coil in a different direction from the teleoconch.

hypobranchial gland A secretory field between the ctenidium and the rectum on the roof of the mantle cavity.

inhalant Describing a channel for the inward flow of water.

inner lip The adaxial edge of the aperture, extending from the base of the columella to the origin of the outer lip, composed of the columellar lip and the parietal lip.

introvert The tube between the apparent mouth and the true mouth formed by the retraction of an acrembolic proboscis.

labial varix A varix lying along or close to the outer lip.

last whorl The youngest and usually the largest whorl of a shell.

mantle The outer wall of the visceral mass; it secretes the shell.

mantle cavity The cavity under the mantle skirt housing the pallial complex; pallial cavity is a synonym.

mantle skirt A fold from the edge of the mantle projecting over the mantle cavity and forming its roof.

mentum A transverse fold on the propodium lying on the anterodorsal surface of the foot under the front of the head.

mesopodium The part of the foot forming the sole on which a prosobranch creeps.

metapodial tentacle   A tentacle on the metapodium.

metapodium   A posterodorsal lobe of the foot secreting and carrying the operculum.

monopectinate   = Pectinibranch, a ctenidium with one row of lamellae.

nacreous   Made of mother-of-pearl.

neck lobe   A fold on each side of the head near the mouth of the mantle cavity.

nuchal cavity   That part of the mantle cavity overlying the head in limpets.

operculum   A horny plate, sometimes strengthened with calcareous material, carried on the metapodium, closing the shell aperture when the animal retracts.

opisthocline   Inclined so that the abapical end of a shell feature (e.g. costa) is in advance of its adapical end (i.e. nearer the base of the shell spiral).

ornament   The sculptured pattern on the surface of a shell.

orthocline   Parallel to the shell axis.

osphradium   A chemosensory organ associated with a ctenidium.

outer lip   The abaxial edge of the aperture, extending from its origin on the last whorl to the base of the columella.

ovipositor   A structure used in depositing spawn; in littorinids a modified area of body wall behind and below the right tentacle; in other groups a cavity, sometimes with internal papilla, opening to the sole of the foot.

ovoviviparity   A reproductive strategy in which fertilized eggs are retained within the female duct until juveniles hatch, so omitting a free larval stage.

pallial complex   The structures contained within the mantle cavity: osphradium, ctenidium, hypobranchial gland, kidney openings, rectum with anus, and genital duct and pore.

pallial gill   A respiratory lamella on the inner surface of the edge of the mantle skirt.

pallial tentacle   A tentacle on the edge of the mantle skirt.

parietal lip   that part of the inner lip lying on the surface of the last whorl.

pectinibranch   A ctenidium with only a single row of lamellae.

pedal gland   A gland on the foot, usually secreting mucus for locomotion.

penis   A male copulatory organ, usually formed from the foot, but in some species a modified cephalic tentacle.

periostracum   A horny layer overlying the calcareous matter of the shell.

periphery   The part of a shell, or of a whorl, lying furthest from the central axis of coiling.

peristome   The whole margin of the aperture.

planorboid (discoidal)   With whorls all in one plane like shells of the freshwater snail *Planorbis*.

pleurembolic proboscis   A proboscis retracted by muscles attached to its side walls so as to form a partial introvert in which lies the most anterior part of the proboscis.

proboscis   An elongation of the snout anterior to the cephalic tentacles, usually retractile into the body and extended only when feeding.

propodium   The most anterior part of the foot, separated from the sole by a groove into which the anterior pedal gland discharges.

prosocline   Inclined so that the adapical end of a shell feature (e.g. a costa) lies further forward (i.e. nearer the base of the shell spiral) than its abapical end.

protoconch   The apical whorls of a shell produced during the embryonic and larval stages of the life history.

scaphoconch The outer of the two shells of an echinospira larva, lost at meta-morphosis.

seminal groove An open groove found in some species running from the male pore in the mantle cavity to the tip of the penis, conducting sperm during copulation.

sinistral Used to describe the direction of shell coiling: anticlockwise when viewed from above the shell apex.

siphon Any extension of the edge of the mantle skirt associated with incurrent or excurrent respiratory streams; usually refers to an extension on the left anterior part of the mantle skirt forming an inhalant channel, often supported by an extension of the outer lip of the shell, the siphonal canal.

siphonal canal A grooved, occasionally tubular extension of the aperture at the base of the columella in which the inhalant siphon lies.

siphonal fasciole A spiral ridge near the columella marking previous positions of the end of a siphonal canal.

siphonal notch A bay at the base of the columella in which the inhalant siphon lies.

slit band A spiral band of growth lines marking previous positions of a slit on the outer lip.

spermatozeugma (*pl.* **spermatozeugmata**) A structure formed from a greatly enlarged apyrene sperm to which many eupyrene sperm attach which transports them to the female; found only in some secondarily aphallic prosobranchs.

spire The whorls of the shell, excluding the last, considered together.

stria (*pl.* **striae**) A very narrow spiral groove.

subsutural ramp or **shelf** A more or less horizontal area on a whorl alongside its adapical suture.

suture The line of contact of adjacent whorls.

teleoconch The whole shell less the protoconch.

throat That part of the inner surface of the shell seen through the aperture.

tooth Used to indicate papilliform or ridge-like thickenings within the outer lip; or the end of a fold on the columella which projects like a tooth into the throat.

trochophore The early free-swimming larval stage of a prosobranch, found only in primitive forms; in others a stage which the embryo passes through before hatching.

tubercle A moderately prominent rounded elevation on the shell surface usually where costae and spiral ridges cross.

tumid Ventricose, swollen, with a convex profile.

turreted Descriptive of a shell with a stepped profile.

umbilical groove A groove alongside the columellar lip leading to the umbilicus, or to its site if closed.

umbilicus A space around the shell axis formed when the adaxial walls of whorls do not meet, or the external opening to this space near the base of the shell.

varix (*pl.* **varices**) A ridge across a whorl more prominent than a costa; varices lie further apart than costae.

veliger A later larva than the trochophore, with more powerful locomotor organs.

velum A ciliated extension of the head region of a veliger larva acting simultaneously as locomotor and feeding organ; usually 2-lobed.

visceral hump or **mass** That part of the body containing the viscera; it never leaves the shelter of the shell.

whorl One complete turn of the spiral of the shell.

References

Anderson, A. 1971. Intertidal activity, breeding and the floating habit of *Hydrobia ulvae* in the Ythan estuary. *J. mar. biol. Ass. U.K.*, **51**, 423–437.

Ankel, F. and Christensen, A. M. 1963. Non-specificity in host selection by *Odostomia scalaris*. *Vidensk. Meddr. dansk naturh. Foren.*, **125**, 321–325.

Ankel, W. E. 1926. Spermiozeugmenbildung durch atypische (apyrene) und typische Spermien bei *Scala* und *Janthina*. *Verh. dt. zool. Ges. 1926, Zool. Anz.* Suppl. **2**, 193–202.

Ankel, W. E. 1929. Über die Bildung der Eikapsel bei *Nassa*-Arten. *Verh. dt. zool. Ges. 1929, Zool. Anz.* Suppl. **4**, 219–230.

Ankel, W. E. 1935. Das Gelege von *Lamellaria perspicua*. *Z. Morph. Ökol. Tiere*, **30**, 635–647.

Ankel, W. E. 1936. Prosobranchia. In *Die Tierwelt der Nord- und Ostsee* (G. Grimpe and E. Wagler eds), IXb1. Leipzig, Akademische Verlagsgesellschaft.

Ankel, W. E. 1937. Der feinere Bau des Kokons der Purpurschnecke *Nucella lapillus* (L.) und seine Bedeutung für das Laichleben. *Verh. dt. zool. Ges. 1937, Zool. Anz.* Suppl. **10**, 77–86.

Ankel, W. E. 1938. Erwerb und Aufnahme der Nahrung bei den Gastropoden. *Verh. dt. zool. Ges. 1938, Zool. Anz.* Suppl. **11**, 223–295.

Ankel, W. E. 1959. Beobachtungen an Pyramidelliden des Gullmar-Fjordes. *Zool. Anz.*, **162**, 1–21.

Arnold, D. C. 1957. The reactions of the limpet, *Patella vulgata* L., to waters of different salinities. *J. mar. biol. Ass. U.K.*, **36**, 121–128.

Arnold, D. C. 1972. Salinity tolerances of some common prosobranchs. *J. mar. biol. Ass. U.K.*, **52**, 475–486.

Barnes, H. and Bagenal, T. B. 1952. The habits and habitats of *Aporrhais pes-pelicani* (L.). *Proc. malac. Soc. Lond.*, **29**, 101–105.

Barnes, R. S. K. 1981a. An experimental study of the patttern and significance of the climbing behaviour of *Hydrobia ulvae*. *J. mar. biol. Ass. U.K.*, **61**, 285–299.

Barnes R. S. K. 1981b. Factors affecting climbing in the coastal gastropod *Hydrobia ulvae*. *J. mar. biol. Ass. U.K.*, **61**, 301–306.

Barnes R. S. K. 1987. Coastal lagoons of East Anglia. *J. coast. Res.*, **3**, 417–427.

Bayer, F. M. 1963. Observations on pelagic mollusks associated with the siphonophores *Velella* and *Physalia*. *Bull. mar. Sci. Gulf Caribb.*, **13**, 454–466.

Beer, S. A., Korolova, Y. M. and Lifshits, A. V. 1969. Age of *Bithynia leachi* determined by shell rings. *Zool. Zh.*, **48**, 1401–1404. [In Russian.]

Berry, R. J. and Crothers, J. H. 1968. Stabilizing selection in the dog-whelk (*Nucella lapillus*). *J. Zool. Lond.*, **155**, 5–17.

Bishop, M. J. 1976. *Hydrobia neglecta* in the British Isles. *J. moll. Stud.*, **42**, 319–326.

Blaber, S. J. M. 1970. The occurrence of a penis-like outgrowth behind the right tentacle in spent females of *Nucella lapillus* (L.). *Proc. malac. Soc. Lond.*, **39**, 231–233.

Bondesen, P. 1940. Preliminary investigations into the development of *Neritina*

642

fluviatilis L. in brackish and fresh water. *Vidensk. Medd. dansk naturh. Foren.*, **104**, 283–318.

Bouchet, P. 1977. *Le développement larvaire des Gastéropodes de grande profondeur. Considérations sur la spéciation.* Thèse (3me cycle), Université Pierre et Marie Curie, Paris.

Bouchet, P., Danrigal, F. and Huyghens, C. 1979. *Living Seashells. Molluscs of the English Channel and Atlantic Coasts.* Poole, Dorset, Blandford Press.

Bouchet, P. and Guillemot, H. 1978. The *Triphora perversa*-complex in Western Europe. *J. moll. Stud.*, **44**, 344–356.

Bouchet, P. and Warén, A. 1980. Revision of the North-east Atlantic bathyal and abyssal Turridae (Mollusca, Gastropoda). *J. moll. Stud.* Suppl. **8**, 1–119.

Boutan, L. 1885. Recherches sur l'anatomie et le développement de la Fissurelle. *Arch. Zool. exp. gén.*(2), **3** bis, 1–173.

Bowman, R. S. 1981. The morphology of *Patella* spp. juveniles in Britain, and some phylogenetic inferences. *J. mar. biol. Ass. U.K.*, **61**, 647–666.

Bowman, R. S. and Lewis, J. R. 1977. Annual fluctuations in the recruitment of *Patella vulgata* L. *J. mar. biol. Ass. U.K.*, **57**, 793–815.

Boycott, A. E. 1936. *Neritina fluviatilis* in Orkney. *J. Conch. Lond.*, **20**, 199–200.

Boyle, P. R. 1981. *Molluscs and Man.* The Institute of Biology's Studies in Biology **134**, 1–60. London, Arnold.

Brehaut, R. N. 1973. The occurrence of *Charonia lampas* (L.) at Guernsey. *J. Conch. Lond.*, **28**, 41–42.

Brown, B. and Rittschof, D. 1984. Effects of flow and concentration of attractant on newly hatched oyster drills, *Urosalpinx cinerea* (Say). *Mar. Behav. Physiol.*, **11**, 75–93.

Bryan, G. W. 1969. The effects of oil-spill removers ('detergents') on the gastropod *Nucella lapillus* on a rocky shore and in the laboratory. *J. mar. biol. Ass. U.K.*, **49**, 1067–1092.

Bryan, G. W., Gibbs, P. E., Hummerstone, L. G. and Burt, G. R. 1986. The decline of the gastropod *Nucella lapillus* around south-west England: evidence for the effect of tributyltin from antifouling paints. *J. mar. biol. Ass. U.K.*, **66**, 611–640.

Cabioch, L., Grainger, J. N. R., Keegan, B. F. and Könnecker, G. 1978. *Balcis alba* a temporary ectoparasite on *Neopentadactyla mixta* Östergren. In *Physiology and Behaviour of Marine Organisms* (D. S. McLusky and A. J. Berry eds), 237–241. Oxford, Pergamon Press.

Carriker, M. R. 1981. Shell penetration and feeding by naticacean and muricacean predatory gastropods: a synthesis. *Malacologia*, **20**, 403–422.

Chatfield, J. E. 1972. Studies on variation and life history in the prosobranch *Hydrobia ulvae* (Pennant). *J. Conch. Lond.*, **27**, 463–473.

Cherrill, A. J. and James, R. 1985. The distribution and habitat preferences of four species of Hydrobiidae in East Anglia. *J. Conch. Lond.*, **32**, 123–133.

Chipperfield, P. N. J. 1951. The breeding of *Crepidula fornicata* (L.) in the river Blackwater, Essex. *J. mar. biol. Ass. U.K.*, **30**, 49–71.

Choquet, M. 1967. Gamétogenèse in vitro au cours du cycle sexuel chez *Patella vulgata* L. en phase mâle. *C.R. Acad. Sci. Paris*, **265D**, 333–335.

Choquet, M. 1970. Etude cytologique de la gonade de *Patella vulgata* L. au cours du changement de sexe naturel. *C.R. Acad. Sci. Paris*, **271D**, 1287–1290.

Clark, R. B. 1968. Biological causes and effects of paralytic shell-fish poisoning. *Lancet*, **1968**, **2**, 770–772.

Cleland, D. M. 1954. A study of the habits of *Valvata piscinalis* (Müller), and the structure and function of the alimentary canal and reproductive system. *Proc. malac. Soc. Lond.*, **30**, 167–203.

Coe, W. R. 1953. Influences of association, isolation and nutrition on the sexuality of snails of the genus *Crepidula. J. exp. Zool.*, **122**, 5–19.

Cole, H. A. and Hancock, D. A. 1955. *Odostomia* as a pest of oysters and mussels. *J. mar biol. Ass. U.K.*, **34**, 25–31.

Collyer, D. M. 1961. Differences revealed by paper partition chromatography between the gastropod *Nassarius reticulatus* (L.) and specimens believed to be *N. nitida* (Jeffreys). *J. mar. biol. Ass. U.K.*, **41**, 683–693.

Cook, A., Bamford, O. S., Freeman, J. D. B. and Teideman, D. J. 1969. A study of the homing habit of the limpet. *Anim. Behav.*, **17**, 330–339.

Cook, P. M. 1949. A ciliary feeding mechanism in *Viviparus viviparus* (L.). *Proc. malac. Soc. Lond.*, **27**, 265–271.

Coombs, V.-A. 1973. A quantitative system of age analysis for the dog-whelk, *Nucella lapillus. J. Zool., Lond.*, **171**, 437–448.

Cousin, C. G. 1975. Etude de la croissance d'un gastéropode prosobranche gonochorique: *Littorina littorea* L. *Cah. Biol. mar.*, **16**, 483–494.

Cowell, E. B. and Crothers, J. H. 1970. On the occurrence of multiple rows of 'teeth' in the shell of the dog-whelk *Nucella lapillus. J. mar. biol. Ass. U.K.*, **50**, 1101–1111.

Creek, G. A. 1951. The reproductive system and embryology of the snail *Pomatias elegans* (Müller). *Proc. zool. Soc. Lond.*, **121**, 599–640.

Creek, G. A. 1953. The morphology of *Acme fusca* (Montagu) with special reference to the genital system. *Proc. malac. Soc. Lond.*, **29**, 228–240.

Crisp, D. J. 1964. The effects of the severe winter of 1962–63 on marine life in Britain. *J. Anim. Ecol.*, **33**, 165–210.

Crisp, D. J. and Southward, A. J. 1958. The distribution of intertidal organisms along the coast of the English Channel. *J. mar. biol. Ass. U.K.*, **37**, 157–208.

Crisp, M. 1971. Structure and abundance of receptors of the unspecialized external epithelium of *Nassarius reticulatus* [Gastropoda, Prosobranchia]. *J. mar. biol. Ass. U.K.*, **51**, 865–890.

Crisp, M. 1972. Photoreceptive function of an epithelial receptor of *Nassarius reticulatus* [Gastropoda, Prosobranchia]. *J. mar. biol. Ass. U.K.*, **54**, 437–442.

Crofts, D. R. 1929. *Haliotis. LMBC Memoir* **29**, 1–74.

Crofts, D. R. 1937. The development of *Haliotis tuberculata*, with special reference to the organogenesis during torsion. *Phil. Trans. R. Soc. Lond.* B **208**, 219–268.

Crowley, T. E. 1961. F. C. Lukis and the Triton. *J. Conch. Lond.*, **25**, 17–20.

Daguzan, J. 1976a. Contribution à l'écologie des Littorinidae (Mollusques Gastéropoides Prosobranches). I. *Littorina neritoides* (L.) et *L. saxatilis* (Olivi). *Cah. Biol. mar.*, **17**, 213–236.

Daguzan, J. 1976b. Contribution à l'écolgie des Littorinidae (Mollusques Gastéropodes Prosobranches). II. *Littorina littorea* (L.) et *L. littoralis* (L.). *Cah. Biol. mar.*, **17**, 275–293.

Dall, W. H. 1918. Notes on *Chrysodomus* and other mollusks from the North Pacific Ocean. *Proc. U.S. Nat. Mus.*, **54**, 207–234.

Dautzenberg, P. and Fischer, H. 1914. Etude sur le *Littorina obtusata* et ses variations. *J. Conchyliol.*, **62**, 87–128.

Dembski, W. J. 1968. Histeochemische Untersuchungen über Funktion und Verleib eu- und oligo-pyrener Spermien von *Viviparus contectus* (Millet, 1813). *Z. Morph. Ökol. Tiere*, **89**, 151–179.

Desai, B. N. 1966. The biology of *Monodonta lineata* (da Costa). *Proc. malac. Soc. Lond.*, **37**, 1–17.

Diehl, M. 1956. Die Raubschnecke *Velutina velutina* das Feind und Bruteinmieter der Ascidie *Styela coriacea*. *Kieler Meeresforsch.*, **12**, 180–185.

Dunkin, S. de B. and Hughes, R. N. 1984. Behavioural components of prey-selection by dogwhelks, *Nucella lapillus* (L.), feeding on barnacles, *Semibalanus balanoides* (L.), in the laboratory. *J. exp. mar. Biol. Ecol.*, **79**, 91–103.

Dussart, G. J. B. 1977. The ecology of *Potamopyrgus jenkinsi* (Smith) in North West England with a note on *Marstoniopsis scholtzi* (Schmidt). *J. moll. Stud.*, **43**, 208–216.

Ebling, F. J., Kitching, J. A., Purchon, R. D. and Bassindale, R. 1948. The ecology of the Lough Ine rapids with special reference to water currents. 2. The fauna of the *Saccorhiza* canopy. *J. Anim. Ecol.*, **17**, 223–244.

Feare, C. J. 1970. The reproductive cycle of the dog whelk (*Nucella lapillus*). *Proc. malac. Soc. Lond.*, **39**, 125–137.

Fenchel, T. 1975. Character displacement and coexistence in mud snails (Hydrobiidae). *Oecologia, Berlin*, **20**, 19–32.

Fenchel, T., Kofoed, L. H. and Lappalainen, A. 1976. Particle selection of two deposit feeders: the amphipod *Corophium volutator* and the prosobranch *Hydrobia ulvae*. *Marine Biology*, **30**, 119–128.

Féral, C. and Gall, S. le. 1982. Induction expérimentale par un pollutant marin (le tributylétain), de l'activité neuroendocrine contrôlant la morphogenèse du pénis chez les femelles d'*Ocenebra erinacea* (Mollusque, Prosobranche gonochorique). *C.R. Acad. Sci.*, **295**, 627–630.

Fioroni, P. 1966. Zur Morphologie und Embryogenese des Darmtraktus und der transitorischen Organe bei Prosobranchiern (Mollusca, Gastropoda). *Rev. suisse Zool.*, **73**, 652–876.

Fischer-Piette, E. and Gaillard, J. M. 1956. Sur l'écologie comparée de *Gibbula umbilicalis* da Costa et *Gibbula pennanti* Phil. *J. Conchyliol.*, **96**, 115–118.

Fish, J. D. 1972. The breeding cycle and growth of open coast and estuarine populations of *Littorina littorea*. *J. mar. biol. Ass. U.K.*, **52**, 1011–1019.

Fish, J. D. and Fish, S. 1977. The veliger larva of *Hydrobia ulvae* with observations on the veliger of *Littorina littorea* (Mollusca: Prosobranchia). *J. Zool. Lond.*, **182**, 495–503.

Fleming, C. 1971. Case of poisoning from red whelk. *Brit. Med. J.*, **3**, 520–521.

Forbes, E. and Hanley, S. 1849–53. *A History of British Mollusca, and their Shells*. 4 vols.: **2**, 1849; **3**, 1850; **4**, 1852–3. London, Van Voorst.

Forster, G. R. 1962. Observations on the ormer population of Guernsey. *J. mar. biol. Ass. U.K.*, **42**, 493–498.

Fraenkel, G. 1927. Biologische Beobachtungen an *Ianthina*. *Z. Morph. Ökol. Tiere*, **7**, 597–608.

Franc, A. 1940. Recherches sur le développement d'*Ocinebra aciculata*, Lamarck (Mollusque Gastéropode). *Bull. biol.*, **74**, 327–345.

Franc. A. 1948. Note sur deux Homalogyridés: *H. Fischeriana* et *H. atomus* (Gastéropodes Prosobranches) et sur leur développement. *Bull. Soc. Hist. nat. Afr. N.*, **39**, 142–145.

Franc, A. 1952a. Notes éthologiques et anatomiques sur *Tritonalia (Ocinebrina) aciculata* (Lk.) (Mollusque Prosobranche). *Bull. Lab. marit. Dinard*, **36**, 31–34.

Franc, A. 1952b. Notes éthologiques et anatomiques sur *Philbertia purpurea* (Montagu) (Moll. Ctenobr.). *Bull. Mus. Hist. nat. Paris*, **24**, 302–305.

Fretter, V. 1941. The genital ducts of some British stenoglossan prosobranchs. *J. mar. biol. Ass. U.K.*, **25**, 173–211.

Fretter, V. 1946. The genital ducts of *Theodoxus*, *Lamellaria* and *Trivia*, and a discussion on their evolution in the prosobranchs. *J. mar. biol. Ass. U.K.*, **26**, 312–351.

Fretter, V. 1948. The structure and life history of some minute prosobranchs of rock pools: *Skeneopsis planorbis* (Fabricius), *Omalogyra atomus* (Philippi), *Rissoella diaphana* (Alder) and *Rissoella opalina* (Jeffreys). *J. mar. biol. Ass. U.K.*, **27**, 597–632.

Fretter, V. 1951a. Some observations on the British cypraeids. *Proc. malac. Soc. Lond.*, **29**, 14–20.

Fretter, V. 1951b. Observations on the life history and functional morphology of *Cerithiopsis tubercularis* (Montagu) and *Triphora perversa* (L.). *J. mar. biol. Ass. U.K.*, **29**, 567–586.

Fretter, V. 1951c. *Turbonilla elegantissima* (Montagu), a parasitic opisthobranch. *J. mar. biol. Ass. U.K.*, **30**, 37–47.

Fretter, V. 1955. Some observations on *Tricolia pullus* (L.) and *Margarites helicinus* (Fabricius). *Proc. malac. Soc. Lond.*, **31**, 159–162.

Fretter, V. 1956. The anatomy of the prosobranch *Circulus striatus* (Philippi) and a review of its systematic position. *Proc. zool. Soc. Lond.*, **126**, 369–381.

Fretter, V. and Graham, A. 1949. The structure and mode of life of the Pyramidellidae, parasitic opisthobranchs. *J. mar. biol. Ass. U.K.*, **28**, 493–532.

Fretter, V. and Graham, A. 1962. *British Prosobranch Molluscs*. London, Ray Society.

Fretter, V. and Graham, A. 1977. The prosobranch molluscs of Britain and Denmark. Part 2 – Trochacea. *J. moll. Stud.* Suppl. **3**, 39–100.

Fretter, V. and Graham, A. 1978. The prosobranch molluscs of Britain and Denmark. Part 3 – Neritacea, Viviparacea, Valvatacea, terrestrial and freshwater Littorinacea and Rissoacea. *J. moll. Stud.* Suppl. **5**, 101–152.

Fretter, V. and Graham, A. 1982. The prosobranch molluscs of Britain and Denmark. Part 7 – 'Heterogastropoda' (Cerithiopsacea. Triforacea, Epitoniacea, Eulimacea). *J. moll. Stud.* Suppl. **11**, 363–434.

Fretter, V., Graham, A. and Andrews, E. B. 1986. The prosobranch molluscs of Britain and Denmark. Part 9 – Pyramidellacea. *J. moll. Stud.* Suppl. **16**, 557–649.

Fretter, V. and Manly, R. 1977a. The settlement and early benthic life of *Littorina neritoides* (L.) at Wembury, S. Devon. *J. moll. Stud.*, **43**, 255–262.

Fretter, V. and Manly, R. 1977b. Algal associations of *Tricolia pullus* (L.), *Lacuna vincta* (Montagu) and *Cerithiopsis tubercularis* (Montagu) with special reference to the settlement of their larvae. *J. mar. biol. Ass. U.K.*, **57**, 999–1017.

Fretter, V. and Patil, A. M. 1961. Observations on some British rissoaceans and a record of *Setia inflata* Monterosato, new to British waters. *Proc. malac. Soc. Lond.*, **34**, 212–223.

Fretter, V. and Pilkington, M. C. 1970. Prosobranchia. Veliger larvae of Taenioglossa and Stenoglossa. *Conseil international pour l'exploration de la mer. Fiches d'identification. Zooplankton*, 129–132.

Fretter, V. and Shale, D. 1973. Seasonal changes in population density and vertical distribution of prosobranch veligers in offshore plankton at Plymouth. *J. mar. biol. Ass. U.K.*, **53**, 471–492.

Frömming, E. 1956. *Biologie der mitteleuropäischen Süsswasserschnecken*. Berlin, Duncker and Humblot.

Funke, W. 1964. Untersuchungen zur Heimfindverhalten und zur Ortstreue von *Patella* L. (Gastropoda, Prosobranchia). *Verh. dt. zool. Ges. 1964, Zool. Anz.* Suppl. **29**, 411–418.

Funke, W. 1968. Heimfindvermögen und Ortstreue bei *Patella* L. (Gastropoda, Prosobranchia). *Oecologia*, **2**, 19–142.

Garwood, P. R. and Kendall, M. A. 1985. The reproductive cycles of *Monodonta lineata* and *Gibbula umbilicalis* on the coast of mid-Wales. *J. mar. biol. Ass. U.K.*, **65**, 993–1008.

Gersch, M. 1936. Der Genitalapparat und die Sexualbiologie der Nordseetrochiden. *Z. Morph. Ökol. Tiere*, **31**, 106–150.

Gibbs, P. E. 1978. *Menestho diaphana* (Gastropoda) and *Montacuta phascolionis* (Lamellibranchia) in association with the sipunculan *Phascolion strombi* in British waters. *J. mar. biol. Ass. U.K.*, **58**, 683–685.

Gibbs, P. E. and Bryan, G. W. 1986. Reproductive failure in populations of the dog-whelk, *Nucella lapillus*, caused by imposex induced by tributyltin from antifouling paints. *J. mar. biol. Ass. U.K.*, **66**, 767–777.

Giglioli, M. E. C. 1955. The egg masses of the Naticidae (Gastropoda). *J. Fish. Res. Bd. Canada*, **12**, 287–327.

Glynne-Williams, J. and Hobart, J. 1952. Studies on the crevice fauna of a selected shore in Anglesey. *Proc. zool Soc. Lond.*, **122**, 797–824.

Goodwin, B. J. 1975. *Studies on the biology of* Littorina obtusata *and* L. mariae *(Mollusca: Gastropoda)*. Ph.D. thesis, University of Wales.

Goodwin, B. J. 1978. The growth and breeding cycle of *Littorina obtusata* (Gastropoda: Prosobranchia) from Cardigan Bay. *J. moll. Stud.*, **44**, 231–242.

Goodwin, B. J. 1979. The egg mass of *Littorina obtusata* and *Lacuna pallidula* (Gastropoda: Prosobranchia). *J. moll. Stud.*, **45**, 1–11.

Goodwin, B. J. and Fish, J. D. 1977. Inter- and intraspecific variation in *Littorina obtusata* and *L. mariae* (Gastropoda: Prosobranchiata). *J. moll. Stud.*, **43**, 241–254.

Götze, E. 1938. Bau und Leben von *Caecum glabrum* (Montagu). *Zool. Jb. (Syst.)*, **71**, 55–122.

Gould, H. N. 1952. Studies on sex in the hermaphrodite mollusk *Crepidula plana*. IV. Internal and external factors influencing growth and sex development. *J. exp. Zool.*, **119**, 93–160.

Graham, A. 1938. On a ciliary process of food-collecting in the gastropod *Turritella communis* Risso. *Proc. zool. Soc. Lond.* A **108**, 456–463.

Graham, A. 1939. On the structure of the alimentary canal of style-bearing prosobranchs. *Proc. zool. Soc. Lond.* B **109**, 75–112.

Graham, A. 1982. *Tornus subcarinatus* (Prosobranchia, Rissoacea), anatomy and relationships. *J. moll. Stud.*, **48**, 144–147.

Graham, A. and Fretter, V. 1947. The life history of *Patina pellucida* (L.). *J. mar. biol. Ass. U.K.*, **26**, 590–601.

Grahame, J. 1970. Shedding of the penis in *Littorina littorea*. *Nature, Lond.*, **221**, 976.

Grahame, J. 1975. Spawning in *Littorina littorea* (L.) (Gastropoda: Prosobranchiata). *J. exp. mar. Biol. Ecol.*, **18**, 185–196.

Hadlock, R. P. 1980. Alarm response of the intertidal snail *Littorina littorea* (L.) to predation by the crab *Carcinus maenas* (L.). *Biol. Bull.*, **159**, 269–279.

Hancock, D. A. 1959. The biology and control of the American whelk tingle *Urosalpinx cinerea* (Say). *Fish. Invest. Lond.* (2), **22**, No. 10, 1–66.

Hancock, D. A. 1960. The ecology of the molluscan enemies of the edible mollusc. *Proc. malac. Soc. Lond.*, **34**, 123–143.

Hannaford Ellis, C. J. H. 1984. Ontogenetic change of shell colour patterns in *Littorina neglecta* (Bean, 1844). *J. Conch., Lond.*, **31**, 343–347.

Hannaford Ellis, C. J. H. 1985. The breeding migration of *Littorina arcana* Hannaford Ellis, 1978 (Prosobranchia: Littorinidae). *Zool. J. Linnean Soc.*, **84**, 91–96.

Harris, G. J. 1985. *Pseudamnicola confusa* rediscovered in the Thames estuary. *J. Conch., Lond.*, **32**, 147.

Hartnoll, R. G. and Wright, J. R. 1977. Foraging movements and homing in the limpet *Patella vulgata* L. *Anim. Behav.*, **25**, 806–810.

Hawthorne, J. B. 1965. The eastern limit of distribution of *Monodonta lineata* (da Costa) in the English Channel. *J. Conch., Lond.*, **25**, 348–352.

Hayashi, I. 1980a. The reproductive biology of the ormer, *Haliotis tuberculata*. *J. mar. biol. Ass. U.K.*, **60**, 415–430.

Hayashi, I. 1980b. Structure and growth of the shore population of the ormer, *Haliotis tuberculata*. *J. mar. biol. Ass. U.K.*, **60**, 431–437.

Heller, J. 1975. The taxonomy of some British *Littorina* species, with notes on their reproduction (Mollusca: Prosobranchia). *Zool. J. Linn. Soc.*, **56**, 131–151.

Henschel, J. 1932. Untersuchungen über den chemischen Sinn von *Nassa reticulata*. *Wiss. Meeresunters. Abt. Kiel*, **21**, 131–159.

Hoagland, K. E. 1978. Protandry and the evolution of environmentally-mediated sex change: a study of the Mollusca. *Malacologia*, **17**, 365–391.

Höisaeter, T. 1968a. *Skenea nitens, Ammonicera rota, Odostomia lukisi* and *Eulimella nitidissima* new to the Norwegian fauna. *Sarsia*, **31**, 25–33.

Höisaeter, T. 1968b. Taxonomic notes on the North European species of 'Cyclostrema' sensu Jeffreys, 1863 (Prosobranchia: Diotocardia). *Sarsia*, **33**, 43–58.

Houbrick, R. S. and Fretter, V. 1969. Some aspects of the functional anatomy and biology of *Cymatium* and *Bursa*. *Proc. malac. Soc. Lond.*, **38**, 415–430.

Hughes, R. N. 1980. Population dynamics, growth and reproductive rates of *Littorina nigrolineata* Gray from a moderately sheltered locality in North Wales. *J. exp. mar. Biol. Ecol.*, **44**, 211–228.

Hughes, R. N. and Drewett, D. 1985. A comparison of the foraging behaviour of dogwhelks, *Nucella lapillus* (L.), feeding on barnacles or mussels on the shore. *J. moll. Stud.*, **51**, 73–77.

Hughes, R. N. and Dunkin, S. de B. 1984a. Behavioural components of prey-selection by dogwhelks, *Nucella lapillus*, feeding on mussels, *Mytilus edulis* (L.), in the laboratory. *J. exp. mar. Biol. Ecol.*, **77**, 45–68.

Hughes, R. N. and Dunkin, S. de B. 1984b. Effect of dietary history on selection of prey and foraging behaviour among patches of prey, by the dogwhelk, *Nucella lapillus* (L.). *J. exp. mar. Biol. Ecol.*, **79**, 159–172.

Hughes, R. N. and Elner, R. W. 1979. Tactics of a predator, *Carcinus maenas*, and morphological responses of the prey, *Nucella lapillus*. *J. Anim. Ecol.*, **48**, 65–78.

Hughes, R. N. and Hughes, H. P. I. 1981. Morphological and behavioural aspects of feeding in the Cassidae (Tonnacea, Mesogastropoda). *Malacologia*, **20**, 385–402.

Hughes, R. N. and Roberts, D. W. 1980. Growth and reproductive rates of *Littorina neritoides* (L.) in North Wales. *J. mar. biol. Ass. U.K.*, **60**, 591–599.

Jackson, J. W. and Taylor, F. 1904. Observations on the habits and reproduction of *Paludestrina taylori*. *J. Conch., Lond.*, **11**, 9–11.

Jeffreys, J. G. 1862–9. *British Conchology*, vols 1–5. **1** (1862), **2** (1863), **3** (1865), **4** (1867), **5** (1869). London, Van Voorst.

Jensen, K. T. and Siegismund, H. R. 1980. The importance of diatoms and bacteria in the diet of *Hydrobia*-species. *Ophelia*, Suppl. **1**, 193–199.

Jones, H. D. 1984. Shell cleaning behaviour of *Calliostoma zizyphinum. J. moll. Stud.*, **50**, 245–247.

Jones, H. D. and Trueman, E. R. 1970. Locomotion of the limpet, *Patella vulgata* L. *J. exp. Biol.*, **52**, 201–216.

Jones, N. S. 1949. Biological note on *Capulus ungaricus. Ann. Rep. mar. biol. Sta. Port Erin, Isle of Man*, 29–30.

Kantor, Y. I. 1985. Feeding and some features of functional morphology of the molluscs of the subfamily Volutopsiinae (Gastropoda, Pectinibranchia). *Zool. Zh.*, **64**, 1640–1647. [In Russian.]

Kantor, Y. I. 1986. Egg capsules and intracapsular development of young in the subfamily Volutopsiinae (Gastropoda: Prosobranchia: Buccinidae). *Zh. obshch. Biol.*, **47**, 411–416. [In Russian.]

Kerney, M. P. 1976. *Atlas of the Non-Marine Mollusca of the British Isles*. London, Conchological Society of Great Britain and Ireland.

Kristensen, E. 1959. The coastal waters of the Netherlands as an environment of molluscan life. *Basteria*, **23**, Suppl. 18–46.

Largen, M. J. 1971. Genetic and environmental influences upon the expression of shell sculpture in the dog-whelk (*Nucella lapillus*). *Proc. malac. Soc. Lond.*, **39**, 393–398.

Laursen, D. 1953. The genus *Ianthina. Dana-Report*, **38**, 1–40.

Lebour, M. V. 1931a. The larval stages of *Nassarius reticulatus* and *Nassarius incrassatus. J. mar. biol. Ass. U.K.*, **17**, 797–818.

Lebour, M. V. 1931b. The larval stages of *Trivia europaea. J. mar. biol. Ass. U.K.*, **17**, 819–832.

Lebour, M. V. 1932a. The larval stages of *Simnia patula. J. mar. biol. Ass. U.K.*, **18**, 107–115.

Lebour, M. V. 1932b. The eggs and early larval stages of the commensal gastropods, *Stilifer stylifer* and *Odostomia eulimoides. J. mar. biol. Ass. U.K.*, **18**, 117–122.

Lebour, M. V. 1933a. The larval stages of *Erato voluta* (Montagu). *J. mar. biol. Ass. U.K.*, **18**, 485–490.

Lebour, M. V. 1933b. The life-histories of *Cerithiopsis tubercularis* (Montagu), *C. barleei* Jeffreys and *Triphora perversa* (L.). *J. mar. biol. Ass. U.K.*, **18**, 491–498.

Lebour, M. V. 1933c. The eggs and larvae of *Turritella communis* Lamarck and *Aporrhais pes-pelicani* (L.). *J. mar. biol. Ass. U.K.*, **18**, 499–506.

Lebour, M. V. 1933d. The eggs and larvae of *Philbertia gracilis* (Montagu). *J. mar. biol. Ass. U.K.*, **18**, 507–510.

Lebour, M. V. 1934. The eggs and larvae of some British Turridae. *J. mar. biol. Ass. U.K.*, **19**, 541–554.

Lebour, M. V. 1935a. The echinospira larvae of Plymouth. *Proc. zool. Soc. Lond.*, 163–174.

Lebour, M. V. 1935b. The larval stages of *Balcis alba* and *B. devians. J. mar. biol. Ass. U.K.*, **20**, 65–70.

Lebour, M. V. 1935c. The breeding of *Littorina neritoides. J. mar. biol. Ass. U.K.*, **20**, 373–378.

Lebour, M. V. 1936. Notes on the eggs and larvae of some Plymouth prosobranchs. *J. mar. biol. Ass. U.K.*, **20**, 547–565.

Lebour, M. V. 1937. The eggs and larvae of the British prosobranchs with special reference to those living in the plankton. *J. mar. biol. Ass. U.K.*, **22**, 105–166.

Lilly, M. M. 1953. The mode of life and the structure and functioning of the reproductive ducts of *Bithynia tentaculata* (L.). *Proc. malac. Soc. Lond.*, **30**, 87–110.

Little, C. and Stirling, P. 1985. Patterns of foraging activity in the limpet *Patella vulgata* L. – a preliminary study. *J. exp. mar. Biol. Ecol.*, **89**, 283–296.

Lopez, G. R. and Kofoed, L. H. 1980. Epipsammic browsing and deposit-feeding in mud snails (Hydrobiidae). *J. mar. Res.*, **38**, 585–599.

Lysaght, A. M. 1941). The biology and trematode parasites of the gastropod *Littorina neritoides* (L.) on the Plymouth Breakwater. *J. mar. biol. Ass. U.K.*, **25**, 41–67.

MacGinitie, N. 1959. Marine Mollusca of Point Barrow, Alaska. *Proc. U.S. Nat. Mus.*, **109**, 59–208.

McMillan, N. F. 1939. The British species of *Lamellaria*. *J. Conch. Lond.*, **21**, 170–173.

McMillan, N. F. 1944. The distribution of *Monodonta (Trochus) lineata* (da Costa) in Britain. *NWest Nat.*, **19**, 290–292.

McMillan, N. F. 1968. *British Shells*. London, F. Warne.

McMillan, N. F. 1981. A contribution to the marine census – *Littorina mariae* Sacchi and Rastelli. *Conch. Newsletter*, **79**, 353–354.

Marshall, J. T. 1911. Additions to 'British Conchology'. *J. Conch. Lond.*, **13**, 192–209.

Massy, A. L. 1930. Mollusca (Pelecypoda, Scaphopoda, Gastropoda, Opisthobranchia) of the Irish Atlantic Slope, 50–1, 500 fathoms. *Proc. R. Irish Acad.*, B **39**, 231–342.

Meier-Brook, C. and Kim, C. H. 1977. Notes on ciliary feeding in two Korean *Bithynia* species. *Malacologia*, **16**, 159–163.

Mistakidis, M. N. 1951. Quantitative studies of the bottom fauna of Essex oyster grounds. *Fish. Invest. Lond.* (2) **17**, No. 6, 1–47.

Mistakidis, M. N. and Hancock, D. A. 1955. Reappearance of *Ocenebra erinacea* (L.) off the east coast of England. *Nature, Lond.*, **175**, 734.

Moore, H. B. 1936. The biology of *Purpura lapillus*. I. Shell variation in relation to environment. *J. mar. biol. Ass. U.K.*, **21**, 61–89.

Moore, H. B. 1938. The biology of *Purpura lapillus*. Part II. Growth. *J. mar. biol. Ass. U.K.*, **23**, 57–66.

Morgan, P. R. 1972. *Nucella lapillus* (L.) as a predator of edible cockles. *J. exp. mar. Biol. Ecol.*, **8**, 45–52.

Morton, J. E. 1954. The crevice faunas of the upper intertidal zone at Wembury. *J. mar. biol. Ass. U.K.*, **33**, 187–224.

Morton, J. E. 1964. Locomotion. In *Physiology of Mollusca* (K. M. Wilbur and C. M. Yonge eds), **1**, 383–423.

Moyse, J. and Nelson-Smith, A. 1963. Zonation of animals and plants on rocky shores around Dale, Pembrokeshire. *Field Studies*, **1**, 5, 1–31.

Muus, B. J. 1963. Some Danish Hybrobiidae with the description of a new species, *Hydrobia neglecta*. *Proc. malac. Soc. London*, **35**, 131–138.

Nekrassow, A. D. 1929. Vergleichende Morphologie der Laiche von Süsswassergastropoden. *Z. Morph. Ökol. Tiere*, **13**, 1–35.

Nelson-Smith, A. 1967. Marine biology at Milford Haven: the distribution of littoral animals and plants. *Field Studies*, **2**, 435–77.

Neumann, D. 1959. Morphologische und experimentelle Untersuchungen über die Variabilität der Farbmuster auf der Schale von *Theodoxuz fluviatilis* (L.). *Z. Morph. Ökol. Tiere*, **48**, 349–411.

Newell, R. C. 1962. Behavioural aspects of the ecology of *Peringia* (= *Hydrobia*) *ulvae* (Pennant) (Gastropoda, Prosobranchia). *Proc. zool. Soc. London*, **142**, 85–106.

Nicol, E. A. T. 1935. The ecology of a salt-marsh. *J. mar. biol. Ass. U.K.*, **20**, 203–262.

650

Nicol, E. A. T. 1936. The brackish-water lochs of North Uist. *Proc. R. Soc. Edinburgh*, **56**, 169–195.

Nicol, E. A. T. 1938. The brackish-water lochs of Orkney. *Proc. R. Soc. Edinburgh*, **58**, 181–191.

Nielsen, C. 1975. Observations on *Buccinum undatum* L. attacking bivalves and on prey responses, with a short review on attack methods of other prosobranchs. *Ophelia*, **13**, 87–108.

Nordsieck, F. 1972. *Die europäischen Meeresschnecken*. Stuttgart, Gustav Fischer Verlag.

Ockelmann, K. W. and Nielsen, C. 1981. On the biology of the prosobranch *Lacuna parva* in the Öresund. *Ophelia*, **20**, 1–16.

O'Riordan, C. E. 1972. Two species of gastropod new to the Irish fauna. *J. Conch. Lond.*, **27**, 371–372.

O'Riordan, C. E. 1984. Some interesting fishes and other marine fauna from the Porcupine Bank. *Ir. Nat. J.*, **221**, 321–323.

O'Riordan, C. E. 1985. Some observations on the occurrence of *Galeodea rugosa* off the Irish coast. *Conch. Newsletter*, **95**, 307–308.

Orton, J. H. 1909. On the occurrence of protandric hermaphroditism in the mollusc *Crepidula fornicata*. *Proc. R. Soc. London*, B **81**, 468–484.

Orton, J. H. 1912. The mode of feeding in *Crepidula*, with an account of the current-producing mechanism in the mantle cavity, and some remarks on the mode of feeding in gastropods and lamellibranchs. *J. mar. biol. Ass. U.K.*, **9**, 444–478.

Orton, J. H. 1928. Observations on *Patella vulgata*. Part I. Sex-phenomena, breeding and shell growth. *J. mar. biol. Ass. U.K.*, **15**, 851–862.

Orton, J. H. 1929. Observations on *Patella vulgata*. Part II. Habitat and habits. *J. mar. biol. Ass. U.K.*, **16**, 277–288.

Orton, J. H. 1930. On the oyster drills in Essex estuaries. *Essex Nat.*, **22**, 298–306.

Orton, J. H., Southward, A. J. and Dodd, J. M. 1956. Studies on the biology of limpets. Part II. The breeding of *Patella vulgata* in Britain. *J. mar. biol. Ass. U.K.*, **35**, 149–176.

Orton, J. H. and Winckworth, R. 1928. The occurrence of the American oyster pest *Urosalpinx cinerea* (Say) on English oyster beds. *Nature, Lond.*, **122**, 241.

Pain, S. 1986. Are British shellfish safe to eat? *New Scientist*, No. 1523, 29–33.

Pain, T. 1977. The genus *Neptunea* Roeding, 1798 in Western Europe (Prosobranchia – Buccinacea). *Conchiglia*, **9**, No. 101, 9–14.

Pain, T. 1978. The genus *Colus* Roeding, 1798 in Western Europe (Prosobranchia – Buccinoidea). *Conchiglia*, **10** (114–115), 3–7.

Pain, T. 1979. The genus *Buccinum* Linné, 1758 in Western Europe (Prosobranchia – Buccinoidea). Part 1: from the British seas to the Mediterranean. *Conchiglia*, **11**, Nos 126–127, 15–18.

Pearce, J. B. 1966. On *Lora trevelliana* (Turton) (Gastropoda: Turridae). *Ophelia*, **3**, 81–91.

Pearce, J. B. and Thorson, G. 1967. The feeding and reproductive biology of the red whelk, *Neptunea antiqua* (L.) (Gastropoda: Prosobranchia). *Ophelia*, **4**, 277–314.

Pelseneer, P. 1935. Essai d'éthologie zoologique d'après l'étude des Mollusques. *Acad. R. Belg. Cl. Sci. Fondation Agathon de Potter*, **1**, 1–662.

Perron, F. 1975. Carnivorous *Calliostoma* (Prosobranchia: Trochidae) from the northeastern Pacific. *Veliger*, **18**, 52–54.

Perron, F. 1978. The habitat and feeding behaviour of the wentletrap *Epitonium greenlandicum*. *Malacologia*, **17**, 63–72.

Perron, F. and Turner, R. D. 1978. The feeding behaviour and diet of *Calliostoma occidentale*, a coelenterate associated prosobranch gastropod. *J. moll. Stud.*, **44**, 100–103.

Petpiroon, S. and Morgan, E. 1983. Observations on the tidal activity rhythm of the periwinkle *Littorina nigrolineata* (Gray). *Mar. Behav. Physiol.*, **9**, 171–192.

Pilkingtom, M. C. 1971. The veliger stage of *Hydrobia ulvae* (Pennant). *Proc. malac. Soc. Lond.*, **39**, 281–287.

Ponder, W. F. 1985. A review of the genera of the Rissoidae (Mollusca: Mesogastropoda: Rissoacea). *Rec. Aust. Mus.* Suppl. **4**, 1–221.

Portmann, A. 1925. Der Einfluss der Nähreier auf die Larvenentwicklung von *Buccinum* und *Purpura*. *Z. Morph. Ökol. Tiere*, **3**, 526–541.

Pratt, D. M. 1976. Intraspecific signalling of hunting success or failure in *Urosalpinx cinerea* (Say). *J. exp. mar. Biol. Ecol.*, **21**, 7–9.

Quick, H. E. 1924. Length of life of *Paludestrina ulvae*. *J. Conch. Lond.*, **17**, 169.

Raffaelli, D. G. 1976. *The determinants of zonation patterns of* Littorina neritoides *and the* Littorina saxatilis *species-complex*. Ph.D. thesis, University of Wales.

Raffaelli, D. G. 1978. Factors affecting the population structure of *Littorina neglecta* Bean. *J. moll. Stud.*, **44**, 223–230.

Raffaelli, D. G. 1982. Recent ecological research on some European species of *Littorina*. *J. moll. Stud.*, **48**, 342–354.

Rasmussen, E. 1944. Faunistic and biological notes on marine invertebrates. I. *Vidensk. Medd. dansk naturh. Foren.*, **107**, 207–233.

Rasmussen, E. 1951. Faunistic and biological notes on marine invertebrates. II. The eggs and larvae of some Danish marine gastropods. *Vidensk. Medd. dansk naturh. Foren.*, **113**, 201–249.

Rasmussen, E. 1973. Systematics and ecology of the Isefjord marine fauna (Denmark). *Ophelia*, **11**, 1–495.

Rehfeldt, N. 1968. Reproductive and morphological variations in the prosobranch 'Rissoa membranacea'. *Ophelia*, **5**, 157–173.

Rendall, R. 1936. *Simnia patula* (Pennant) in Orkney. *J. Conch. Lond.*, **20**, 283–285.

Richter, G. and Thorson, G. 1975. Pelagische Prosobranchier-Larven des Golfes von Neapel. *Ophelia*, **13**, 109–185.

Rittschof, D., Williams, L. G., Brown, B. and Carriker, M. R. 1983. Chemical attraction of newly hatched oyster drills. *Biol. Bull.*, **164**, 493–505.

Robert, A. 1902. Recherches sur le développement des troques. *Arch. Zool. exp. gén.* (3) **10**, 269–538.

Robertson, A. I. and Mann, K. H. 1982. Population dynamics and life history adaptations of *Littorina neglecta* Bean in an eelgrass meadow (*Zostera marina* L.) in Nova Scotia. *J. exp. mar. Biol. Ecol.*, **63**, 151–171.

Russell Hunter, W. D. 1961. Life cycles of four freshwater snails in limited populations in Loch Lomond with a discussion of infraspecific variation. *Proc. zool. Soc. Lond.*, **137**, 135–171.

Russell-Hunter, W. D. and Russell-Hunter, M. 1968. Pedal expansion in the naticid snails. I. Introduction and weighing experiments. *Biol. Bull.*, **135**, 548–562.

Sacchi, C. F. 1975. *Littorina nigrolineata* (Gray) (Gastropoda: Prosobranchia). *Cah. Biol. mar.*, **16**, 111–120.

Sacchi, C. F., Testard, P. and Voltalina, D. 1977. Recherches sur le spectre trophique comparé de *Littorina saxatilis* (Olivi) et de *L. nigrolineata* (Gray) (Gastropoda, Prosobranchia) sur la grève de Roscoff. *Cah. Biol. mar.*, **18**, 499–505.

Salvini-Plawen, L. von. 1972. Cnidaria as food sources for marine invertebrates, *Cah. Biol. mar.*, **13**, 385–400.

Sander, K. 1950. Beobachtungen zur Fortpflanzung von *Assiminea grayana* Leach. *Arch. Molluskenk.*, **79**, 147–149.

Sander, K. 1952. Beobachtungen zur Fortpflanzung von *Assiminea grayana* Leach. *Arch. Molluskenk.*, **81**, 133–134.

Sander, K. and Sibrecht, L. 1967. Das Schlupfen der Veliger-Larve von *Assiminea grayana* Leach (Gastropoda Prosobranchia). *Z. Morph. Ökol. Tiere*, **60**, 141–52.

Schäfer, H. 1952. Ein Beitrag zur Ernährungsbiologie von *Bithynia tentaculata* (Gastropoda Prosobranchia). *Zool. Anz.* **148**, 299–303.

Schäfer, H. 1953a. Beobachtungen zur Ökologie von *Bithynia tentaculata*. *Arch. Molluskenk.*, **82**, 67–70.

Schäfer, H. 1953b. Beiträge zur Ernährungsbiologie einheimischer Süsswasserprosobranchier. *Z. Morph. Ökol. Tiere*, **41**, 247–264.

Schäfer, W. 1955. Über die Bildung der Laichballen der Wellhorn-Schnecke. *Natur und Volk*, **85**, 92–97.

Schiemenz, P. 1884. Über die Wasseraufnahme bei Lamellibranchiaten und Gastropoden (einschliesslich der Pteropoden). *Mitt. zool. Sta. Neapel*, **5**, 509–543.

Schiemenz, P. 1887. Über die Wasseraufnahme bei Lamellibranchiaten und Gastropoden (einschliesslich der Pteropoden). Zweiter Theil. *Mitt. zool. Sta. Neapel*, **7**, 423–472.

Seaward, D. R. (ed.) 1982. *Sea Area Atlas of the Marine Molluscs of Britain and Ireland*. Shrewsbury, Conchological Society of Great Britain and Ireland and the Nature Conservancy Council.

Sharman, M. 1956. Note on *Capulus ungaricus* (L.). *J. mar. biol. Ass. U.K.*, **35**, 445–450.

Shimek, R. L. and Kohn, A. J. 1981. Functional morphology and evolution of the toxoglossan radula. *Malacologia*, **20**, 423–438.

Smith, B. S. 1971. Sexuality in the American mud snail, *Nassarius obsoletus* Say. *Proc. malac. Soc. Lond.*, **39**, 377–378.

Smith, B. S. 1981. Tributyltin compounds induce male characteristics on female mud snails *Nassarius obsoletus* = *Ilyanassa obsoleta*. *J. appl. Toxicol.*, **1**, 141–144.

Smith, D. A. S. 1973. The population biology of *Lacuna pallidula* (da Costa) and *Lacuna vincta* (Montagu) in north-east England. *J. mar. biol. Ass. U.K.*, **53**, 493–520.

Smith, D. A. S. 1976. Disruptive selection and morph-ratio clines in the polymorphic snail *Littorina obtusata* (L.) (Gastropoda: Prosobranchia). *J. moll. Stud.*, **42**, 114–135.

Smith, E. H. 1967. Two new species of British turrids. *Veliger*, **10**, 1–4.

Smith, J. E. 1981. The natural history and taxonomy of shell variation in the periwinkles *Littorina saxatilis* and *Littorina rudis*. *J. mar. biol. Ass. U.K.*, **61**, 215–241.

Smith, J. E. and Newell, G. E. 1955. The dynamics of the zonation of the common periwinkle (*Littorina littorea* (L.)) on a stony beach. *J. Anim. Ecol.*, **24**, 35–56.

Smith, S. M. 1970. *Rissoa violacea* Desmarest, *Rissoa lilacina* Récluz, *Rissoa rufilabrum* Alder and *Rissoa porifera* Lovén and their distribution. *J. Conch. Lond.*, **27**, 235–48.

Sneli, J.-A. 1972. *Odostomia turrita* found on *Homarus gammarus*. *Nautilus*, **86**, 23–24.

Southgate, T. 1982. The biology of *Barleeia unifasciata* (Gastropoda: Prosobranchia) in red algal tufts in S.W. Ireland. *J. mar. biol. Ass. U.K.*, **62**, 461–468.

Starmühlner, F. 1952. Zur Anatomie, Histologie und Biologie einheimischer Prosobranchier. *Öst. zool. Z.*, **3**, 546–590.

Stephenson, T. A. 1924. Notes on *Haliotis tuberculata* L. *J. mar. biol. Ass. U.K.*, **13**, 480–495.

Sylvest, O. 1949. Copulation observed in *Viviparus fasciatus* (O. F. Müller). *Hydrobiologia*, **1**, 309–311.

Tattersall, W. M. 1920. Notes on the breeding habits and life history of the periwinkle. *Sci. Invest. Fish. Bd. Ire.*, **1**, 1–11.

Taylor, J. D. 1978. The diet of *Buccinum undatum* and *Neptunea antiqua* (Gastropoda: Buccinidae). *J. Conch. Lond.*, **29**, 309–318.

Thiriot-Quiévreux, C. 1969. Caractéristiques morphologiques des véligères planctoniques de gastéropodes de la région de Banyuls-sur-Mer. *Vie et Milieu*, B **20**, 333–366.

Thiriot-Quiévreux, C. 1973. Heteropoda. *Oceanog. mar. Biol., ann. Rev.*, **11**, 237–261.

Thiriot-Quiévreux, C. 1975. Observations sur les larves et les adultes des Carinariidae (Mollusca: Heteropoda) de l'Océan Atlantique Nord. *Mar. Biol.*, **32**, 379–388.

Thiriot-Quiévreux, C. 1976. Description de la larve d'*Aporrhais serresianus* (Michaud) (Mollusca Prosobranchia) dans le plancton méditerranéen. *Vie et Milieu*, A **26**, 299–304.

Thiriot-Quiévreux, C. and Babio, C. R. 1975. Etude des protoconques de quelques prosobranches de la région de Roscoff. *Cah. Biol. mar.*, **16**, 135–148.

Thorson, G. 1935. Studies on the egg-capsules and development of Arctic marine prosobranchs. *Medd. Grønland*, **100** (5), 1–71.

Thorson, G. 1940. Notes on the egg-capsules of some North-Atlantic prosobranchs of the genus *Troschelia*, *Chrysodomus*, *Volutopsis*, *Sipho* and *Trophon*. *Vidensk. Medd. dansk naturh. Foren.*, **104**, 251–266.

Thorson, G. 1944. The zoology of East Greenland. Marine Gastropoda Prosobranchiata. *Medd. Grønland*, **121** (13), 1–181.

Thorson, G. 1946. Reproduction and larval development of Danish marine bottom invertebrates. *Medd. Komm. Havundersøg. Kbh.*, Ser. Plankton, **4** (1), 1–523.

Thorson, G. 1965. A neotenous dwarf-form of *Capulus ungaricus* (L.) (Gastropoda, Prosobranchia) commensalistic on *Turritella communis* Risso. *Ophelia*, **2**, 175–210.

Tsikhon-Likanina, E. A. 1961. On the filtration method of feeding in *Bithynia tentaculata* (L.) and *Valvata piscinalis* (Müller) (Gastropoda, Prosobranchiata). *Byull. Inst. Biol. Vodokhran.*, **10**, 28–30. [In Russian.]

Turk, S. M. 1976. *Charonia lampas* (L.) (Gastropoda: Cymatiidae) living off the Cornish coast. *J. Conch. Lond.*, **29**, 29–30.

Underwood, A. J. 1972. Observations on the reproductive cycles of *Monodonta lineata*, *Gibbula umbilicalis* and *G. cineraria*. *Mar. Biol.*, **17**, 333–340.

Vader, W. I. M. 1964. A preliminary investigation into the reactions of the infauna of the tidal flats to tidal fluctuations in water levels. *Netherlands J. Sea Res.*, **2**, 189–222.

Vahl, O. 1971. Growth and density of *Patina pellucida* (L.) (Gastropoda: Prosobranchia) on *Laminaria hyperborea* (Gunnerus) from western Norway. *Ophelia*, **9**, 31–50.

Vestergaard, K. 1935. Über den Laich und die Larven von *Scalaria communis* (Lam.), *Nassarius pygmaeus* (Lam.) und *Bela turricula* (Mont.). *Zool. Anz.*, **109**, 217–222.

Wallace, C. 1986. On the distribution of the sexes of *Potamopyrgus jenkinsi* (Smith). *J. moll. Stud.*, **51**, 290–296.

Walne, P. R. 1956. The biology and distribution of the slipper limpet *Crepidula fornicata* in Essex rivers with notes on the distribution of the larger epi-benthic invertebrates. *Fish. Invest. Lond.* (2), **20**, No. 6, 1–50.

Warén, A. 1974. Revision of the Arctic-Atlantic Rissoidae (Gastropoda Prosobranchia). *Zool. Scripta*, **3**, 121–135.

Warén, A. 1983a. A generic revision of the family Eulimidae (Gastropoda, Prosobranchia). *J. moll. Stud.* Suppl. **13**, 1–96.

Warén, A. 1983b. An anatomical description of *Eulima bilineata* Alder with remarks on and a revision of *Pyramidelloides* Nevill (Mollusca, Prosobranchia, Eulimidae). *Zool. Scripta*, **12**, 273–294.

Watson, D. C. and Norton, T. A. 1985. Dietary preferences of the common periwinkle *Littorina littorea* (L.). *J. exp. mar. Biol. Ecol.*, **88**, 193–211.

Weber, H. 1925. Über arhythmische Fortbewegung bei einige Prosobranchiern. Ein Beitrag zur Bewegungsphysiologie der Gastropoden. *Z. vergl. Physiol.*, **2**, 109–121.

Werner, B. 1952. Ausbildungsstufen der Filtrationsmechanismen bei filtrierenden Prosobranchiern. *Verh. dt. zool. Ges. 1952, Zool. Anz.* Suppl. **17**, 529–546.

Werner, B. 1953. Über den Nahrungserwerb der Calyptraeidae (Gastropoda Prosobranchia). Morphologie, Histologie und Funktion der am Nahrungserwerb beiteiligten Organe. *Helgoländer wiss. Meeresunters.*, **4**, 260–315.

Werner, B. 1959. Das Prinzip des endlosen Schleimfilters beim Nahrungserwerb wirbelloser Meerestiere. *Int. Revue ges. Hydrobiol. Hydrogr.*, **44**, 181–216.

Wigham, G. D. 1975a. The biology and ecology of *Rissoa parva* (da Costa) (Gastropoda: Prosobranchia). *J. mar. biol. Ass. U.K.*, **55**, 45–67.

Wigham, G. D. 1975b. Environmental influences upon the expression of shell form in *Rissoa parva* (da Costa) (Gastropoda: Prosobranchia). *J. mar. biol. Ass. U.K.*, **55**, 425–438.

Williams, E. E. 1964. The growth and distribution of *Littorina littorea* (L.) on a rocky shore in Wales. *J. Anim. Ecol.*, **33**, 413–432.

Williams, E. E. 1965. The growth and distribution of *Monodonta lineata* (da Costa) on a rocky shore in Wales. *Field Studies*, **2**, 189–198.

Williamson, P. and Kendall, M. A. 1981. Population age structure and growth of the trochid *Monodonta lineata* determined from shell rings. *J. mar. biol. Ass. U.K.*, 1011–1026.

Wilson, D. P. and Wilson, M. A. 1956. A contribution to the biology of *Ianthina janthina* (L.). *J. mar. biol. Ass. U.K.*, **35**, 291–305.

Winckworth, R. 1932. The British marine Mollusca. *J. Conch. Lond.*, **19**, 211–252.

Wit, W. F. de. 1965. Some observations on the reproduction of *Bithynia leachi* (Sheppard). *Basteria*, **29**, 72–75.

Woodward, M. F. 1899. On the anatomy of *Adeorbis subcarinatus* Montagu. *Proc. malac. Soc. Lond.*, **3**, 140–146.

Wright, J. R. and Hartnoll, R. G. 1981. An energy budget for a population of the limpet *Patella vulgata*. *J. mar. biol. Ass. U.K.*, **61**, 627–646.

Wyatt, H. V. 1960. Protandry and self-fertilization in the Calyptraeidae. *Nature, Lond.*, **187**, 520–521.

Yonge, C. M. 1937. The biology of *Aporrhais pes-pelecani* (L.) and *A. serresiana* (Mich.). *J. mar. biol. Ass. U.K.*, **21**, 687–704.

Yonge, C. M. 1938. Evolution of ciliary feeding in the Prosobranchia, with an account of ciliary feeding in *Capulus ungaricus*. *J. mar. biol. Ass. U.K.*, **22**, 453–68.

Yonge, C. M. 1946. On the habits of *Turritella communis* Risso. *J. mar. biol. Ass. U.K.*, **26**, 377–380.

Yonge, C. M. 1947. The pallial organs in the aspidobranch Gastropoda and their evolution throughout the Mollusca. *Phil. Trans. R. Soc. Lond.*, B **232**, 443–518.

Ziegelmeier, E. 1954. Beobachtungen über den Nahrungserwerb bei der Naticide *Lunatia nitida* Donovan (Gastropoda Prosobranchia). *Helgoländer wiss. Meeresunters.*, **5**, 1–33.

Ziegelmeier, E. 1961. Zur Fortpflanzungsbiologie der Naticiden (Gastropoda Prosobranchia). *Helgoländer wiss. Meeresunters.*, **8**, 94–118.

Systematic Index

Italic type indicates valid names as used here;
those in roman type are synonyms or superseded names

656

Printed in the United States
By Bookmasters